19

CRM SERIES

Centro
di Ricerca
Matematica
Ennio De Giorgi

Prof. Matteo Focardi
Dipartimento di Matematica "U. Dini"
Viale Morgagni 67/a
50134 Firenze, Italia

Prof. Giovanni Leoni
Department of Mathematical Sciences
Carnegie Mellon University
Pittsburgh, PA, USA

Prof. Massimiliano Morini
Dipartimento di Matematica e Informatica
Università degli Studi di Parma
Parco area delle Scienze, 53/A
43124 Parma, Italia

Free Discontinuity Problems

edited by
Nicola Fusco and Aldo Pratelli

EDIZIONI
DELLA
NORMALE

ISBN 978-88-7642-592-9
e-ISBN 978-88-7642-593-6

Contents

Introduction

The present volume collects the lectures notes of the courses given in July 2014 in the ERC "School on Free Discontinuity problems" that we organized at the Centro De Giorgi of the Scuola Normale Superiore at Pisa. The aim of the school was to present the main analytical and geometric ideas developed in the study of the so called free discontinuity problems by discussing three important examples: the Mumford-Shah model for image segmentation, a variational model for the epitaxial growth of thin films, and the sharp interface limit of an energy functional proposed by Ohta-Kawasaki to model pattern formation in dyblock copolymers. The common feature of these variational problems is the competition between volume and surface energies. The latter are concentrated on $(N-1)$-dimensional sets which are not given a priori and that indeed are the main unknown of the problem. They can be either rectifiable sets, as in the case of Mumford-Shah problem, or boundaries, as in the two other models.

The lectures were given by Matteo Focardi, Giovanni Leoni and Massimiliano Morini. They kindly agreed to write down in an extended form the content of their courses with the aim of both reviewing the main results of the theory and presenting the latest developments.

The volume starts with the contribution of Matteo Focardi "Fine regularity results for Mumford-Shah minimizers: porosity, higher integrability and the Mumford-Shah conjecture". He begins by presenting the weak formulation of the Mumford-Shah problem and the existence result of De Giorgi, Carriero and Leaci, which is showed by proving the equivalence between the weak and the strong formulation. In turn, this is an almost immediate consequence of a density lower bound for the jump set of the minimizer. Here, two different proofs of this density estimate are given, one due to De Lellis and Focardi and another one which follows from an almost monotonicity formula of Bucur and Luckhaus. Then, after recalling the partial regularity results proved in the '90s, the rest of the lecture focuses on the Mumford-Shah conjecture. The starting point is a result by Ambrosio, Fusco and Hutchinson which provides a quantitative link between the higher integrability of the gradient of a minimizer and the dimension of the singular set of the jump set. This result has been recently refined by De Lellis and Focardi who actually proved the equivalence between a weaker version of the Mumford-Shah conjecture and a precise degree of integrability of the gradient of the minimizers. A detailed proof of this equivalence is provided here. The last part of the lecture contains two higher integrability results. The first one, in two dimensions, is due to De Lellis and Focardi and it is based on Caccioppoli partitions. The second one, which applies to any dimension, was proved by De Philippis and Figalli using the porosity property of the jump set.

The lecture by Giovanni Leoni "Variational models for epitaxial growth" deals with a model for the epitaxial growth of thin films proposed by Spencer and Tersoff. They assume a planar symmetry of the three-dimensional configuration of the film. This leads to a two-dimensional model where the free surface of the film is represented as the graph of a periodic function over a reference interval and the substrate occupies an infinite strip. The total energy is given by the sum of two terms. The first one measures the length of the film weighed with a positive coefficient with a step discontinuity at the interface between the film and the substrate. The other one takes care of the elastic energy needed to deform the material. In the model the elastic properties of the film and the substrate are described by the same elasticity tensor. However, the interesting feature is the presence of a mismatch strain at the interface of the two materials. This mismatch is responsible of the so called islands formation, *i.e.*, non flat minimal configurations. The lecture starts with a detailed presentation of Korn's inequality in C^1 and Lipschitz domains which has its own interest besides the applications to thin films. Also of independent interest is the subsequent section which presents the classical results of Grisvard on the regularity of solutions of Lamé systems in polygonal domains. The second part of the lecture is devoted to the existence and regularity of minimizers. Here Leoni presents in an unified and comprehensive way various results scattered in the literature of the last fifteen years. This presentation can be very useful for young mathematicians interested in entering in a subject where mathematical research is still very active.

Massimiliano Morini presents a contribution on "Local and global minimality results for an isoperimetric problem with long-range interaction". The energy functional is given by the sum of the perimeter of a set E inside a fixed container Ω and the Dirichlet integral of the solution, under Neumann or periodic boundary conditions, of the Poisson equation with a right hand side depending on the characteristic function of E. When minimizing this energy under a volume constraint, one observes a wide class of configurations ranging from one or several almost spherical droplets to the union of cylinders or lamellae or even the more complicate patterns known as gyroids. The interesting feature of all these minimizers is that they are very close to periodic sets of constant mean curvature. Indeed various explicit constructions of stable critical configurations of the above type have been given in the last years. The lecture of Morini presents a second variation criterion for local minimality that he recently obtained in collaboration with Acerbi and Fusco. First, he derives the first and second variation formulae for the total energy established by Choksi and Sternberg. Then he proves that a strictly stable critical point, *i.e.*, a critical point with positive second variation, is a local minimizer under a volume constraint. To this aim he shows a quantitative estimate of the energy gap between the minimizer E and a competitor F in terms of the L^1 distance between the two sets. The second part of the lecture explains how this result can be applied to prove that certain lamellar and almost spherical configurations are indeed local and even global minimizers. Here one can find a unified approach to the recent results proved by Cicalese and Spadaro and by Morini and Sternberg.

We would like to conclude by thanking Matteo Focardi, Giovanni Leoni and Massimiliano Morini for preparing these very interesting lectures notes. We believe that they will be as successful as the beautiful courses that they gave two years ago at the school in Pisa.

Nicola Fusco
Aldo Pratelli

Fine regularity results for Mumford-Shah minimizers: porosity, higher integrability and the Mumford-Shah conjecture

Matteo Focardi

Abstract. We review some classical results and more recent insights about the regularity theory for local minimizers of the Mumford and Shah energy and their connections with the Mumford and Shah conjecture. We discuss in details the links among the latter, the porosity of the jump set and the higher integrability of the approximate gradient. In particular, higher integrability turns out to be related with an explicit estimate on the Hausdorff dimension of the singular set and an energetic characterization of the conjecture itself.

1 Introduction

The Mumford and Shah model is a prominent example of variational problem in image segmentation (see [69]). It is an algorithm able to detect the contours of the objects in a black and white digitized image. Representing the latter by a greyscale function $g \in L^\infty(\Omega, [0,1])$, a smoothed version of the original image is then obtained by minimizing the functional

$$(v, K) \to \mathscr{F}(v, K, \Omega) + \gamma \int_{\Omega \setminus K} |v - g|^2 dx, \tag{1.1}$$

with

$$\mathscr{F}(v, K, \Omega) := \int_{\Omega \setminus K} |\nabla v|^2 \, dx + \beta \, \mathcal{H}^1(K), \tag{1.2}$$

where $\Omega \subseteq \mathbb{R}^2$ is an open set, K is a relatively closed subset of Ω with finite $\mathcal{H}^1$ measure, $v \in C^1(\Omega \setminus K)$, β and γ are nonnegative parameters to be tuned suitably according to the applications. In our discussion we can set $\beta = 1$ without loss of generality.

The role of the squared L^2 distance in (1.1) is that of a fidelity term in order that the output of the process is close in an average sense to the original input image g. The set K represents the set of contours of the objects in the image, the length of which is kept controlled by the penalization of its $\mathcal{H}^1$ measure to avoid over segmentation, while the Dirichlet energy of v favors sharp contours rather than zones where a thin layer of gray is used to pass smoothly from white to black or vice versa.

We stress the attention upon the fact that the set K is not assigned a priori and it is not a boundary in general. Therefore, this problem is not a free boundary problem, and new ideas and techniques had to be developed to solve it. Since its appearance in the late 80's to today the research on the Mumford and Shah problem, and on related fields, has been very active and different approaches have been developed. In this notes we shall focus mainly on that proposed by De Giorgi and Ambrosio. This is only due to a matter of taste of the Author and it is also dictated by understandable reasons of space. Even more, it is not possible to be exhaustive in our (short) presentation, therefore we refer to the books by Ambrosio, Fusco and Pallara [7] and David [26] for the proofs of many results we shall only quote, for a more detailed account of the several contributions in literature, for the many connections with other fields and for complete lists of references (see also the recent survey [53] that covers several parts of the regularity theory that are not presented here).

Going back to the Mumford and Shah minimization problem and trying to follow the path of the Direct Method of the Calculus of Variations, it is clear that a weak formulation calls for a function space allowing for discontinuities of co-dimension 1 in which an existence theory can be established. Therefore, by taking into account the structure of the energy, De Giorgi and Ambrosio were led to consider the space SBV of *Special functions of Bounded Variation*, *i.e.* the subspace of BV functions with singular part of the distributional derivative concentrated on a 1-dimensional set called in what follows the *jump set* (throughout the paper we will use standard notations and results concerning the spaces BV and SBV, following the book [7]).

The purpose of the present set of notes is basically to resume and collect several of the regularity properties known at present for Mumford and Shah minimizers. More precisely, Section 2 is devoted to recalling basic facts about the functional setting of the problem and its weak formulation. The celebrated De Giorgi, Carriero and Leaci [33] regularity result implying the equivalence between the strong and weak formulations, is discussed in details. In Subsection 2.3 we provide a recent proof by De Lellis and Focardi valid in the 2d case that gives an explicit constant in the density lower bound, and in Subsection 2.4 we discuss the almost monotonicity formula by Bucur and Luckhaus. Next, we state the Mumford and Shah conjecture. The understanding of such a claim is the goal at which researchers involved in this problem are striving for. In this perspective well-established and more recent fine regularity results on the jump set of minimizers are discussed in Section 3. Furthermore, we highlight two different paths that might lead to the solution in positive of the Mumford and Shah conjecture: the complete characterization of blow ups in Subsection 2.6 and a sharp higher integrability of the (approximate) gradient in Theorem 3.11 together with the uniqueness of blow up limits. In particular, we discuss in details the latter by following the ideas introduced by Ambrosio, Fusco and Hutchinson [4] linking higher integrability of the gradient of a minimizer with the size of the *singular set* of the minimizer itself, *i.e.* the subset of points of the jump set having no neighborhood in which the jump set itself is a regular curve. An explicit estimate shows that the bigger the integrability exponent of the gradient is, the lower the Hausdorff dimension of the singular set is (*cf.* Theorem 3.10). Pushing forward this approach, an energetic characterization of a slightly weaker form of the Mumford and Shah conjecture can be found beyond the scale of L^p spaces (*cf.* Theorem 3.11). In particular, the quoted estimate on the Hausdorff dimension of the full singular set reduces to the higher integrability property of the gradient and a corresponding estimate on a special subset of singular points: those for which the scaled Dirichlet energy is infinitesimal. The latter topic is dealt with in full details in Section 4 in the setting of Caccioppoli partitions as done by De Lellis and Focardi in [35]. The analysis of Section 4 allowed the same Authors to prove the higher integrability property in 2-dimensions as explained in Section 5. A different path leading to higher integrability in any dimension is to exploit the porosity of the jump set. This approach, due to De Philippis and Figalli [37], is the object of Section 7. Some preliminaries on porous sets are discussed in Section 6.

To conclude this introduction it is worth mentioning that the Mumford and Shah energy and the theory developed in order to study it, have been employed in many other fields. The applications to Fracture Mechanics, both in a static setting and for quasi-static irreversible crack-growth for brittle materials according to Griffith are important instances of that (see in particular [12], [7, Section 4.6.6] and [15,21,60]). It is also valuable to recall that several contributions in literature are devoted to the asymptotic analysis or the variational approximation of free discontinuity energies by means of De Giorgi's Γ-*convergence* theory. We refer to the books by Braides [13–15] for the analysis of several interesting problems arising from models in different fields (for a quick introduction to Γ-convergence see [40], for a more detailed account consult the treatise [20]).

The occasion to write this set of notes stems from the course "*Fine regularity results for Mumford-Shah minimizers: higher integrability of the gradient and estimates on the Hausdorff dimension of the singular set*" taught by the Author in July 2014 at Centro De Giorgi in Pisa within the activities of the "School on Free Discontinuity problems", ERC Research Period on Calculus of Variations and Analysis in Metric Spaces. The material collected here covers entirely the six lectures of the course, additional topics and some more recent insights are also included for the sake of completeness and clarity. It is a pleasure to acknowledge the hospitality of Centro De Giorgi and to gratefully thank N. Fusco and A. Pratelli, the organizers of the school, for their kind invitation. Let me also thank all the people in the audience for their attention, patience, comments and questions. In particular, the kind help of R. Cristoferi and E. Radici who read a preliminary version of these notes is acknowledged. Nevertheless, the Author is the solely responsible for all the inaccuracies contained in them.

2 Existence theory and first regularity results

In this section we shall overview the first basic issues of the problem. More generally we discuss the n-dimensional case, though we shall often make specific comments related to the 2-dimensional setting of the original problem (and sometimes to the 3d case as well). We shall freely use the notation for BV functions and Caccioppoli sets adopted in the book by Ambrosio, Fusco and Pallara [7]. We shall always refer to it also for the many results that we shall apply or even only quote without giving a precise citation.

2.1 Functional setting of the problem

A function $v \in L^1(\Omega)$ belongs to $BV(\Omega)$ if and only if Dv is a (vector-valued) Radon measure on the non empty open subset Ω of $\mathbb{R}^n$. The distributional derivative of v can be decomposed according to

$$Dv = \nabla v\, \mathcal{L}^n \llcorner \Omega + (v^+ - v^-)\nu_v\, \mathcal{H}^{n-1} \llcorner S_v + D^c v,$$

where

(i) ∇v is the density of the absolutely continuous part of Dv with respect to $\mathcal{L}^n \llcorner \Omega$ (and the *approximate gradient* of v in the sense of Geometric Measure Theory as well);
(ii) S_v is the set of *approximate discontinuities* of v, an $\mathcal{H}^{n-1}$-rectifiable set (so that $\mathcal{L}^n(S_v) = 0$) endowed with approximate normal ν_v for $\mathcal{H}^{n-1}$ a.e. on S_v;
(iii) $v^\pm$ are the *approximate one-sided traces* left by v $\mathcal{H}^{n-1}$ a.e. on S_v;
(iv) $D^c v$ is the rest in the Radon-Nikodym decomposition of the singular part of Dv after the absolutely continuous part with respect to

$\mathcal{H}^{n-1} \llcorner S_v$ has been identified. Thus, it is a singular measure both with respect to $\mathcal{L}^n \llcorner \Omega$ and to $\mathcal{H}^{n-1} \llcorner S_v$ (for more details see [7, Proposition 3.92]).

By taking into account the structure of the energy in (1.1), only volume and surface contributions are penalized, so that it is natural to introduce the following subspace of BV.

Definition 2.1 ([32], Section 4.1 [7]). $v \in BV(\Omega)$ is a *Special function of Bounded Variation*, in short $v \in SBV(\Omega)$, if $D^c v = 0$, *i.e.* $Dv = \nabla v\, \mathcal{L}^n \llcorner \Omega + (v^+ - v^-)\nu_v\, \mathcal{H}^{n-1} \llcorner S_v$.

No Cantor staircase type behavior is allowed for these functions. Simple examples are collected in the ensuing list:

(i) if $n = 1$ and $\Omega = (\alpha, \beta)$, $SBV\big((\alpha, \beta)\big)$ is easily described in view of the well known decomposition of BV functions of one variable. Indeed, any function in $SBV\big((\alpha, \beta)\big)$ is the sum of a $W^{1,1}\big((\alpha, \beta)\big)$ function with one of pure jump, *i.e.* $\sum_{i\in\mathbb{N}} a_i \chi_{(\alpha_i, \alpha_{i+1})}$, with $\alpha = \alpha_0$, $\alpha_i < \alpha_{i+1} < \beta$, $(a_i)_{i\in I} \in \ell^\infty$;

(ii) $W^{1,1}(\Omega) \subset SBV(\Omega)$. Clearly, $Dv = \nabla v\, \mathcal{L}^n \llcorner \Omega$. In this case ∇v coincides with the usual distributional gradient;

(iii) let $(E_i)_{i\in I}$, $I \subseteq \mathbb{N}$, be a *Caccioppoli partition* of Ω, *i.e.* $\mathcal{L}^n\big(\Omega \setminus \cup_i E_i\big) = 0$ and $\mathcal{L}^n(E_i \cap E_j) = 0$ if $i \neq j$, with the E_i's sets of finite perimeter such that

$$\sum_{i\in I} \mathrm{Per}(E_i) < \infty.$$

Then, $v = \sum_{i\in I} a_i\, \chi_{E_i} \in SBV(\Omega)$ if $(a_i)_{i\in I} \in \ell^\infty$. In this case, if $J_{\mathscr{E}} := \cup_i \partial^* E_i$ denotes the *set of interfaces* of $\mathscr{E}$, with $\partial^* E_i$ the *essential boundary* of E_i, then $\mathcal{H}^{n-1}(S_v \setminus J_{\mathscr{E}}) = 0$ and

$$Dv = (v^+ - v^-)\nu_v\, \mathcal{H}^{n-1} \llcorner J_{\mathscr{E}}.$$

Functions of this type have zero approximate gradient, they are called *piecewise constant* and form a subspace denoted by $SBV_0(\Omega)$ (*cf.* [7, Theorem 4.23]);

(iv) the function $v(\rho, \theta) := \sqrt{\rho} \cdot \sin(\theta/2)$ for $\theta \in (-\pi, \pi)$ and $\rho > 0$ is in $SBV(B_r)$ for all $r > 0$. In particular, $v \in SBV(B_r) \setminus \big(W^{1,1}(B_r) \oplus SBV_0(B_r)\big)$.

A general receipt to construct interesting examples of SBV functions can be obtained as follows (see [7, Proposition 4.4]).

Proposition 2.2. *If $K \subset \Omega$ is a closed set such that $\mathcal{H}^{n-1}(K) < +\infty$ and $v \in W^{1,1} \cap L^\infty(\Omega \setminus K)$, then $v \in SBV(\Omega)$ and*

$$\mathcal{H}^{n-1}(S_v \setminus K) = 0. \tag{2.1}$$

Clearly, property (2.1) above is not valid for a generic member of SBV, but it does for a significant class of functions: local minimizers of the energy under consideration (see below for the definition), actually satisfying even a stronger property (*cf.* Proposition 2.9).

2.2 Tonelli's Direct Method and Weak formulation

The difficulty in applying the Direct Method is related to the surface term for which it is hard to find a topology ensuring at the same time lower semicontinuity and pre-compactness for minimizing sequences. Using the Hausdorff local topology requires a very delicate study of the latter ones to rule out typical counterexamples as shown by Maddalena and Solimini in [56]. Here, we shall follow instead the original approach by De Giorgi and Ambrosio [32].

Keeping in mind the example in Proposition 2.2, the weak formulation of the problem under study is obtained naively by taking $K = S_v$. Loosely speaking in this approach the set of contours K is identified by the (Borel) set S_v of (approximate) discontinuities of the function v that is not fixed a priori. This is the reason for the terminology *free discontinuity* problem coined by De Giorgi. The (weak counterpart of the) Mumford and Shah energy $\mathscr{F}$ in (1.2) of a function v in $SBV(\Omega)$ on an open subset $A \subseteq \Omega$ then reads as

$$\mathscr{F}(v, A) = \mathrm{MS}(v, A) + \gamma \int_A |v - g|^2 dx, \tag{2.2}$$

where

$$\mathrm{MS}(v, A) := \int_A |\nabla v|^2 dx + \mathcal{H}^{n-1}(S_v \cap A). \tag{2.3}$$

For the sake of simplicity in case $A = \Omega$ we drop the dependence on the set of integration.

In passing, we note that, the class $\{v \in BV(\Omega) : Dv = D^c v\}$ of Cantor type functions is dense in BV w.r.to the L^1 topology, thus it is easy to infer that

$$\inf_{BV(\Omega)} \mathscr{F} = 0,$$

so that the restriction to SBV is needed in order not to trivialize the problem.

Ambrosio's SBV closure and compactness theorem (see [7, Theorems 4.7 and 4.8]) ensures the existence of a minimizer of $\mathscr{F}$ on SBV.

Theorem 2.3 (Ambrosio [2]). *Let $(v_j)_j \subset SBV(\Omega)$ be such that*

$$\sup_j \big(\mathrm{MS}(v_j) + \|v_j\|_{L^\infty(\Omega)}\big) < \infty,$$

then there exists a subsequence $(v_{j_k})_k$ and a function $v \in SBV(\Omega)$ such that $v_{j_k} \to v$ $L^p(\Omega)$, for all $p \in [1, \infty)$.

Moreover, we have the separated lower semicontinuity estimates

$$\int_\Omega |\nabla v|^2 dx \leq \liminf_k \int_\Omega |\nabla v_{j_k}|^2 dx \tag{2.4}$$

and

$$\mathcal{H}^{n-1}(S_v) \leq \liminf_k \mathcal{H}^{n-1}(S_{v_{j_k}}). \tag{2.5}$$

Ambrosio's theorem is the natural counterpart of Rellich-Kondrakov theorem in Sobolev spaces. Indeed, for Sobolev functions, it reduces essentially to that statement provided that an L^p rather than an L^∞ bound is assumed. More generally, Ambrosio's theorem holds true in the bigger space $GSBV$. In particular, (2.4) and (2.5) display a separate lower semicontinuity property for the two terms of the energy in a way that the two terms cannot combine to create neither a contribution for the other nor a Cantor type one.

By means of the chain rule formula for BV functions one can prove that the functional under consideration is decreasing under truncation, *i.e.* for all $k \in \mathbb{N}$

$$\mathscr{F}(\tau_k(v)) \leq \mathscr{F}(v) \quad \forall v \in SBV(\Omega),$$

if $\tau_k(v) := (v \wedge k) \vee (-k)$.

Therefore, being $g \in L^\infty(\Omega)$, we can always restrict ourselves to minimize it over the ball in $L^\infty(\Omega)$ of radius $\|g\|_{L^\infty(\Omega)}$. In conclusion, Theorem 2.3 always provides the existence of a (global) minimizer for the weak formulation of the problem.

Once the existence has been checked, necessary conditions satisfied by minimizers are deduced. Supposing $g \in C^1(\Omega)$, by means of internal variations, *i.e.* constructing competitors to test the minimality of u by composition with diffeomorphisms of Ω arbitrarily close to the identity of the type $\mathrm{Id} + \varepsilon\,\phi$, the Euler-Lagrange equation takes the form

$$\begin{aligned}\int_{\Omega\setminus S_u}\Big((|\nabla u|^2+\gamma(u-g)^2)\mathrm{div}\phi-2\langle\nabla u,\nabla u\cdot\nabla\phi\rangle-2\gamma(u-g)\langle\nabla g,\phi\rangle\Big)dx \\ +\int_{S_u}\mathrm{div}^{S_u}\phi\, d\mathcal{H}^{n-1}=0\end{aligned} \tag{2.6}$$

for all $\phi \in C_c^1(\Omega, \mathbb{R}^n)$, $\mathrm{div}^{S_u}\phi$ denoting the tangential divergence of the field ϕ on S_u (*cf.* [7, Theorem 7.35]).

Instead, by using outer variations, *i.e.* range perturbations of the type $u + \varepsilon(v - u)$ for $v \in SBV(\Omega)$ such that $\mathrm{spt}(u - v) \Subset \Omega$ and $S_v \subseteq S_u$, we find

$$\int_\Omega \big(\langle \nabla u, \nabla(v-u)\rangle + \gamma\,(u-g)\,(v-u)\big)\,dx = 0. \tag{2.7}$$

2.3 Back to the strong formulation: the density lower bound

Existence of minimizers for the strong formulation of the problem is obtained via a regularity property enjoyed by (the jump set of) the minimizers of the weak counterpart. The results obtained in this framework will be instrumental also to establish way much finer regularity properties in the ensuing sections.

We start off analyzing the scaling of the energy in order to understand the local behavior of minimizers. This operation has to be done with some care since the volume and length terms in MS scale differently under affine change of variables of the domain. Let $v \in SBV(B_\rho(x))$, set

$$v_{x,\rho}(y) := \rho^{-1/2} v(x + \rho\, y), \tag{2.8}$$

then $v_{x,\rho} \in SBV(B_1)$, with

$$\mathrm{MS}(v_{x,\rho}, B_1) = \rho^{1-n}\mathrm{MS}(v, B_\rho(x))$$

and

$$\int_{B_1} |v_{x,\rho} - g_{x,\rho}|^2 dz = \rho^{-1-n} \int_{B_\rho(x)} |v - g|^2 dy.$$

Thus,

$$\rho^{1-n}\Big(\mathrm{MS}(v, B_\rho(x)) + \int_{B_\rho(x)} |v-g|^2 dz\Big)$$
$$= \mathrm{MS}(v_{x,\rho}, B_1) + \rho^2 \int_{B_1} |v_{x,\rho} - g_{x,\rho}|^2 dy.$$

By taking into account that $g \in L^\infty$ and that along the minimization process we are actually interested only in functions satisfying the bound $\|v\|_{L^\infty(\Omega)} \le \|g\|_{L^\infty(\Omega)}$, we get

$$\rho^2 \int_{B_1} |v_{x,\rho} - g_{x,\rho}|^2 dy \le 2\rho\, \|g\|^2_{L^\infty(\Omega)} = O(\rho) \qquad \rho \downarrow 0.$$

This calculation shows that, at the first order, the leading term in the energy $\mathscr{F}$ computed on $B_\rho(x)$ is that related to the MS functional, the other being a contribution of higher order that can be neglected in a preliminary analysis.

Motivated by this, we introduce a notion of minimality involving only the leading part of the energy. This corresponds to setting $\gamma = 0$ in the definition of $\mathscr{F}$ (*cf.* (2.2)).

Definition 2.4. A function $u \in SBV(\Omega)$ with $\mathrm{MS}(u) < \infty$[1] is a local minimizer of MS if

$$\mathrm{MS}(u) \le \mathrm{MS}(v) \quad \text{whenever } \{v \neq u\} \Subset \Omega.$$

In what follows, u will always denote a local minimizer of MS unless otherwise stated, and the class of all local minimizers shall be denoted by $\mathcal{M}(\Omega)$. Actually, we shall often refer to local minimizers simply as minimizers if no confusion can arise. In particular, regularity properties for minimizers of the whole energy can be obtained by perturbing the theory developed for local minimizers (see for instance Corollary 2.13 and Theorem 2.16 below).

Harmonic functions with small oscillation are minimizers as a simple consequence of (2.7).

Proposition 2.5 (Chambolle, see Proposition 6.8 [7]). *If u is harmonic in Ω', then $u \in \mathcal{M}(\Omega)$, for all $\Omega \Subset \Omega'$, provided*

$$\big(\sup_{\Omega} u - \inf_{\Omega} u\big)\|\nabla u\|_{L^\infty(\Omega)} \le 1. \tag{2.9}$$

Proof. Let $A \Subset \Omega$. By Theorem 2.3 it is easy to show the existence of a minimizer $w \in SBV(\Omega)$ of the Dirichlet problem $\min\big\{\mathrm{MS}(v) : v \in SBV(\Omega),\ v = u \text{ on } \Omega \setminus A\big\}$. Moreover, by truncation $\inf_\Omega u \le w \le \sup_\Omega u$ $\mathcal{L}^n$ a.e. on Ω.

By the arbitrariness of A, the local minimality of u follows provided we show that $\mathrm{MS}(u, \Omega) \le \mathrm{MS}(w, \Omega)$. To this aim, we use the Euler-Lagrange condition (2.7) with $\gamma = 0$, namely

$$\int_\Omega \langle \nabla w, \nabla(u - w)\rangle\, dx = 0 \iff \int_\Omega |\nabla w|^2\, dx = \int_\Omega \langle \nabla w, \nabla u\rangle\, dx,$$

to get

$$\mathrm{MS}(u, \Omega) \le \mathrm{MS}(w, \Omega) \iff \int_\Omega \langle \nabla u, \nabla(u - w)\rangle\, dx \le \mathcal{H}^{n-1}(S_w)$$

$$\iff \int_\Omega \nabla u \cdot dD(u-w) - \int_{S_w} \langle \nabla u, \nu_w\rangle (w^+ - w^-) d\mathcal{H}^{n-1} \le \mathcal{H}^{n-1}(S_w).$$

[1] The finite energy condition is actually not needed due to the local character of the notion introduced, it is assumed only for the sake of simplicity.

An integration by parts, the harmonicity of u and the equality $w = u$ on $\Omega \setminus A$ give

$$\int_\Omega \nabla u \cdot dD(u-w) = -\int_\Omega (u-w)\Delta u \, dx = 0,$$

and therefore

$$\mathrm{MS}(u,\Omega) \leq \mathrm{MS}(w,\Omega) \Longleftrightarrow -\int_{S_w} \langle \nabla u, \nu_w \rangle (w^+ - w^-) d\mathcal{H}^{n-1} \leq \mathcal{H}^{n-1}(S_w).$$

The conclusion follows from condition (2.9) as $\inf_\Omega u \leq w \leq \sup_\Omega u$ $\mathcal{L}^n$ a.e. on Ω. □

By means of the slicing theory in SBV, *i.e.* the characterization of SBV via restrictions to lines, one can also prove that *pure jumps*, *i.e.* functions as

$$a\chi_{\{\langle x-x_o,\nu\rangle>0\}} + b\chi_{\{\langle x-x_o,\nu\rangle<0\}} \tag{2.10}$$

for a and $b \in \mathbb{R}$ and $\nu \in \mathbb{S}^{n-1}$, are local minimizers as well (*cf.* [7, Proposition 6.8]]). Further examples shall be discussed in what follows (*cf.* Subsection 2.5).

As established in [33] in all dimensions (and proved alternatively in [23] and [25] in dimension two), if $u \in \mathcal{M}(\Omega)$ then the pair $(u, \Omega\cap\overline{S_u})$ is a minimizer of $\mathscr{F}$ for $\gamma = 0$. The main point is the identity $\mathcal{H}^{n-1}\big(\Omega\cap(\overline{S_u}\setminus S_u)\big) = 0$, which holds for every $u \in \mathcal{M}(\Omega)$. The groundbreaking paper [33] proves this identity via the following *density lower bound* estimate (see [7, Theorem 7.21]).

Theorem 2.6 (De Giorgi, Carriero and Leaci [33]). *There exist dimensional constants θ, $\varrho > 0$ such that for every $u \in \mathcal{M}(\Omega)$*

$$\mathrm{MS}(u, B_r(z)) \geq \theta\, r^{n-1} \tag{2.11}$$

for all $z \in \Omega \cap \overline{S_u}$, and all $r \in (0, \varrho \wedge \mathrm{dist}(z, \partial\Omega))$.

Building upon the same ideas, in [17] it is proved a slightly more precise result (see again [7, Theorem 7.21]).

Theorem 2.7 (Carriero and Leaci [17]). *There exist dimensional constants θ_0, $\varrho_0 > 0$ such that for every $u \in \mathcal{M}(\Omega)$*

$$\mathcal{H}^{n-1}(S_u \cap B_r(z)) \geq \theta_0\, r^{n-1} \tag{2.12}$$

for all $z \in \Omega \cap \overline{S_u}$, and all $r \in (0, \varrho_0 \wedge \mathrm{dist}(z, \partial\Omega))$.

In particular, from the latter we infer the so called *elimination property* for $\Omega \cap \overline{S_u}$, *i.e.* if $\mathcal{H}^{n-1}(S_u \cap B_r(z)) < \frac{\theta_0}{2^{n-1}} r^{n-1}$ then actually $\overline{S_u} \cap B_{r/2}(z) = \emptyset$.

Given Theorem 2.6 or 2.7 for granted we can easily prove the equivalence of the strong and weak formulation of the problem by means of the ensuing *density estimates*.

Lemma 2.8. *Let μ be a Radon measure on $\mathbb{R}^n$, B be a Borel set and $s \in [0, n]$ be such that*

$$\limsup_{r \downarrow 0} \frac{\mu(B_r(x))}{\omega_s r^s} \geq t \qquad \text{for all } x \in B.$$

Then, $\mu(B) \geq t\,\mathcal{H}^s(B)$.

Proposition 2.9. *Let $u \in \mathcal{M}(\Omega)$, then $\mathcal{H}^{n-1}\big(\Omega \cap (\overline{S_u} \setminus S_u)\big) = 0$. In particular $\big(u, \Omega \cap \overline{S_u}\big)$ is a local minimizer for $\mathscr{F}$ (with $\gamma = 0$).*

Proof of Proposition 2.9. In view of Theorem 2.7 we may apply the density estimates of Lemma 2.8 to $\mu = \mathcal{H}^{n-1} \llcorner S_u$ and to the Borel set $\Omega \cap (\overline{S_u} \setminus S_u)$ with $t = \theta_0$. Therefore, we deduce that

$$\theta_0 \mathcal{H}^{n-1}\big(\Omega \cap (\overline{S_u} \setminus S_u)\big) \leq \mu\big(\Omega \cap (\overline{S_u} \setminus S_u)\big) = 0.$$

Clearly, $\mathrm{MS}(u) = \mathscr{F}(u, \Omega \cap \overline{S_u})$, and the conclusion follows at once. □

The argument for (2.11) used by De Giorgi, Carriero and Leaci in [33], and similarly in [17] for (2.12), is indirect: it relies on Ambrosio's SBV compactness theorem and Poincaré-Wirtinger type inequality in SBV established in [33] (see also [7, Theorem 4.14] and [8, Proposition 2] for a version in which boundary values are preserved) to analyze blow up limits of minimizers (see Subsection 2.6 for the definition of blow ups) with vanishing jump energy and prove that they are harmonic functions (*cf.* [7, Theorem 7.21]). A contradiction argument shows that on small balls the energy of local minimizers inherits the decay properties as that of harmonic functions. Actually, the proof holds true for much more general energies (see [43], [7, Chapter 7]).

In the paper [34] an elementary proof valid only in 2-dimensions and tailored on the MS energy is given. No Poincaré-Wirtinger inequality, nor any compactness argument are required. Moreover, it has the merit to exhibit an explicit constant. Indeed, the proof in [34] is based on an observation of geometric nature and on a direct variational comparison argument. It also differs from those exploited in [23] and [25] to derive (2.12) in the 2-dimensional case.

Theorem 2.10 (De Lellis and Focardi [34]). *Let* $u \in \mathcal{M}(\Omega)$. *Then*

$$\mathrm{MS}(u, B_r(z)) \geq r \tag{2.13}$$

for all $z \in \Omega \cap \overline{S_u}$ *and all* $r \in (0, \mathrm{dist}(z, \partial\Omega))$. *More precisely, the set* $\Omega_u := \{z \in \Omega : (2.13)\ \textit{fails}\}$ *is open and* $\Omega_u = \Omega \setminus \overline{S_u}$[2].

To the aim of establishing Theorem 2.10 we prove a consequence of (2.6), a monotonicity formula discovered independently by David and Léger in [27, Proposition 3.5] and by Maddalena and Solimini in [57]. The proof we present here is that given in [34, Lemma 2.1] (an analogous result holds true in any dimension with essentially the same proof).

Lemma 2.11. *Let* $u \in \mathcal{M}(\Omega)$, $\Omega \subset \mathbb{R}^2$, *then for every* $z \in \Omega$ *and for* $\mathcal{L}^1$ *a.e.* $r \in (0, \mathrm{dist}(z, \partial\Omega))$

$$\begin{aligned} r \int_{\partial B_r(z)} \left(\left(\frac{\partial u}{\partial \nu}\right)^2 - \left(\frac{\partial u}{\partial \tau}\right)^2 \right) d\mathcal{H}^1 + \mathcal{H}^1(S_u \cap B_z(r)) \\ = \int_{S_u \cap \partial B_r(z)} |\langle \nu_u^{\perp}(x), x\rangle| d\mathcal{H}^0(x), \end{aligned} \tag{2.14}$$

$\frac{\partial u}{\partial \nu}$ *and* $\frac{\partial u}{\partial \tau}$ *being the projections of* ∇u *in the normal and tangential directions to* $\partial B_r(z)$, *respectively.*[3]

Proof of Lemma 2.11. With fixed a point $z \in \Omega, r > 0$ with $B_r(z) \subseteq \Omega$, we consider special radial vector fields $\eta_{r,s} \in \mathrm{Lip} \cap C_c(B_r(z), \mathbb{R}^2)$, $s \in (0, r)$, in the first variation formula (2.6) (with $\gamma = 0$). Moreover, for the sake of simplicity we assume $z = 0$, and drop the subscript z in what follows. Let

$$\eta_{r,s}(x) := x\, \chi_{[0,s]}(|x|) + \frac{|x| - r}{s - r}\, x\, \chi_{(s,r]}(|x|),$$

then a routine calculation leads to

$$\nabla \eta_{r,s}(x) := \mathrm{Id}\, \chi_{[0,s]}(|x|) + \left(\frac{|x| - r}{s - r} \mathrm{Id} + \frac{1}{s - r} \frac{x}{|x|} \otimes x \right) \chi_{(s,r]}(|x|)$$

$\mathcal{L}^2$ a.e. in Ω. In turn, from the latter formula we infer for $\mathcal{L}^2$ a.e. in Ω

$$\mathrm{div}\eta_{r,s}(x) = 2\, \chi_{[0,s]}(|x|) + \left(2\frac{|x| - r}{s - r} + \frac{|x|}{s - r} \right) \chi_{(s,r]}(|x|),$$

---

[2] Actually, the very same proof shows also that $\Omega_u = \Omega \setminus \overline{J_u}$, where J_u is the subset of points of S_u for which one sided traces exist. Recall that $\mathcal{H}^{n-1}(S_v \setminus J_v) = 0$ for all $v \in BV(\Omega)$.

[3] For $\xi \in \mathbb{R}^2$, $\xi^{\perp}$ is the vector obtained by an anticlockwise rotation.

and, if $\nu_u(x)$ is a unit vector normal field in $x \in S_u$, for $\mathcal{H}^1$ a.e. $x \in S_u$

$$\operatorname{div}^{S_u}\eta_{r,s}(x)=\chi_{[0,s]}(|x|)+\left(\frac{|x|-r}{s-r}+\frac{1}{|x|(s-r)}|\langle x,\nu_u^{\perp}\rangle|^2\right)\chi_{(s,r]}(|x|).$$

Consider the set $I := \{\rho \in (0, \operatorname{dist}(0, \partial\Omega)) : \mathcal{H}^1(S_u \cap \partial B_\rho) = 0\}$, then $(0, \operatorname{dist}(0, \partial\Omega)) \setminus I$ is at most countable being $\mathcal{H}^1(S_u) < +\infty$. If ρ and $s \in I$, by inserting η_s in (2.6) we find

$$\begin{aligned}
&\frac{1}{s-r}\int_{B_r\setminus B_s}|x||\nabla u|^2dx-\frac{2}{s-r}\int_{B_r\setminus B_s}|x|\left\langle\nabla u,\left(\mathrm{Id}-\frac{x}{|x|}\otimes\frac{x}{|x|}\right)\nabla u\right\rangle dx\\
&=\mathcal{H}^1(S_u\cap B_s)+\int_{S_u\cap(B_r\setminus B_s)}\frac{|x|-r}{s-r}d\mathcal{H}^1\\
&+\frac{1}{s-r}\int_{S_u\cap(B_r\setminus B_s)}|x|\left|\left\langle\frac{x}{|x|},\nu_u^{\perp}\right\rangle\right|^2d\mathcal{H}^1.
\end{aligned}$$

Next we employ Co-Area formula and rewrite equality above as

$$\begin{aligned}
&\frac{1}{s-r}\int_s^r\rho\,d\rho\int_{\partial B_\rho}|\nabla u|^2d\mathcal{H}^1-\frac{2}{s-r}\int_s^r\rho\,d\rho\int_{\partial B_\rho}\left|\frac{\partial u}{\partial\tau}\right|^2d\mathcal{H}^1\\
&=\mathcal{H}^1(S_u\cap B_s)+\int_{S_u\cap(B_r\setminus B_s)}\frac{|x|-r}{s-r}d\mathcal{H}^1\\
&+\frac{1}{s-r}\int_s^r d\rho\int_{S_u\cap\partial B_\rho}|\langle x,\nu_u^{\perp}\rangle|d\mathcal{H}^0
\end{aligned}$$

where $\nu := x/|x|$ denotes the radial unit vector and $\tau := \nu^{\perp}$ the tangential one. Lebesgue differentiation theorem then provides a subset I' of full measure in I such that if $r \in I'$ and we let $s \uparrow t^-$ it follows

$$\begin{aligned}
&-r\int_{\partial B_r}|\nabla u|^2d\mathcal{H}^1+2r\int_{\partial B_r}\left|\frac{\partial u}{\partial\tau}\right|^2d\mathcal{H}^1\\
&=\mathcal{H}^1(S_u\cap B_r)-\int_{S_u\cap\partial B_r}|\langle x,\nu_u^{\perp}\rangle|d\mathcal{H}^0.
\end{aligned}$$

Formula (2.14) then follows straightforwardly. □

We are now ready to prove Theorem 2.10.

Proof of Theorem 2.10. Given $u \in \mathcal{M}(\Omega)$, $z \in \Omega$ and $r \in (0, \operatorname{dist}(z, \partial\Omega))$ let

$$e_z(r) := \int_{B_r(z)}|\nabla u|^2dx,\quad \ell_z(r) := \mathcal{H}^1(S_u \cap B_r(z)),$$

and

$$m_z(r) := \mathrm{MS}(u, B_r(z)), \quad h_z(r) := e_z(r) + \frac{1}{2}\ell_z(r).$$

Clearly, $m_z(r) = e_z(r) + \ell_z(r) \le 2h_z(r)$, with equality if and only if $e_z(r) = 0$.

Introduce the set $S_u^\star$ of points $x \in S_u$ for which

$$\lim_{r\downarrow 0} \frac{\mathcal{H}^1(S_u \cap B_r(x))}{2r} = 1\,. \tag{2.15}$$

Since S_u is rectifiable, $\mathcal{H}^1(S_u \setminus S_u^\star) = 0$. Next let $z \in \Omega$ be such that

$$m_z(R) < R \qquad \text{for some } R \in (0, \mathrm{dist}(z, \partial\Omega)). \tag{2.16}$$

We claim that $z \notin S_u^\star$.

W.l.o.g. we take $z = 0$ and drop the subscript z in e, ℓ, m and h.

In addition we can assume $e(R) > 0$. Otherwise, by the Co-Area formula and the trace theory of BV functions, we would find a radius $r < R$ such that $u|_{\partial B_r}$ is a constant (*cf.* the argument below). In turn, u would necessarily be constant in B_r because the energy decreases under truncations, thus implying $z \notin S_u^\star$. We can also assume $\ell(R) > 0$, since otherwise u would be harmonic in B_R and thus we would conclude $z \notin S_u^\star$.

We start next to compare the energy of u with that of an harmonic competitor on a suitable disk. The inequality $\ell(R) \le m(R) < R$ is crucial to select good radii.

Step 1*:* For any fixed $r \in (0, R - \ell(R))$, there exists a set I_r of positive length in (r, R) such that

$$\frac{h(\rho)}{\rho} \le \frac{1}{2} \cdot \frac{e(R) - e(r)}{R - r - (\ell(R) - \ell(r))} \qquad \text{for all } \rho \in I_r\,. \tag{2.17}$$

Define $J_r := \{t \in (r, R) : \mathcal{H}^0(S_u \cap \partial B_t) = 0\}$. We claim the existence of $J_r' \subseteq J_r$ with $\mathcal{L}^1(J_r') > 0$ and such that

$$\int_{\partial B_\rho} |\nabla u|^2 d\mathcal{H}^1 \le \frac{e(R) - e(r)}{R - r - (\ell(R) - \ell(r))} \qquad \text{for all } \rho \in J_r'. \tag{2.18}$$

Indeed, we use the Co-Area formula for rectifiable sets (see [7, Theorem 2.93]) to find

$$\begin{aligned}\mathcal{L}^1((r, R) \setminus J_r) &\le \int_{(r,R)\setminus J_r} \mathcal{H}^0(S_u \cap \partial B_t)dt \\ &= \int_{S_u\cap(B_R\setminus\overline{B_r})} \left|\left\langle \nu_u^\perp(x), \frac{x}{|x|}\right\rangle\right| d\mathcal{H}^1(x) \le \ell(R) - \ell(r).\end{aligned}$$

In turn, this inequality implies $\mathcal{L}^1(J_r) \geq R - r - (\ell(R) - \ell(r)) > 0$, thanks to the choice of r. Then, define J_r' to be the subset of radii $\rho \in J_r$ for which

$$\int_{\partial B_\rho} |\nabla u|^2 d\mathcal{H}^1 \leq \fint_{J_r} \left(\int_{\partial B_t} |\nabla u|^2 d\mathcal{H}^1 \right) dt.$$

Formula (2.18) follows by the Co-Area formula and the estimate $\mathcal{L}^1(J_r) \geq R - r - (\ell(R) - \ell(r))$.

We define I_r as the subset of radii $\rho \in J_r'$ satisfying both (2.14) and (2.18). Therefore,

$$\int_{\partial B_\rho} \left(\frac{\partial u}{\partial \tau}\right)^2 d\mathcal{H}^1 = \frac{1}{2} \int_{\partial B_\rho} |\nabla u|^2 d\mathcal{H}^1 + \frac{\ell(\rho)}{2\rho} \qquad \forall \rho \in I_r. \quad (2.19)$$

Clearly, I_r has full measure in J_r', so that $\mathcal{L}^1(I_r) > 0$.

For any $\rho \in I_r$, we let w be the harmonic function in B_ρ with trace u on ∂B_ρ. Then, as $\frac{\partial w}{\partial \tau} = \frac{\partial u}{\partial \tau}$ $\mathcal{H}^1$ a.e. on ∂B_ρ, the local minimality of u entails

$$\begin{aligned} m(\rho) &\leq \int_{B_\rho} |\nabla w|^2 dx \\ &\leq \rho \int_{\partial B_\rho} \left(\frac{\partial u}{\partial \tau}\right)^2 d\mathcal{H}^1 \overset{(2.19)}{=} \frac{\rho}{2} \int_{\partial B_\rho} |\nabla u|^2 d\mathcal{H}^1 + \frac{\ell(\rho)}{2}. \end{aligned}$$

The inequality (2.17) follows from the latter inequality and from (2.18):

$$h(\rho) = e(\rho) + \frac{\ell(\rho)}{2} \leq \frac{\rho}{2} \int_{\partial B_\rho} |\nabla u|^2 d\mathcal{H}^1 \leq \frac{\rho}{2} \cdot \frac{e(R) - e(r)}{R - r - (\ell(R) - \ell(r))}.$$

Step 2: We now show that $0 \notin S_u^\star$.

Let $\varepsilon \in (0,1)$ be fixed such that $m(R) \leq (1-\varepsilon)R$, and fix any radius $r \in (0, R - \ell(R) - \frac{1}{1-\varepsilon} e(R))$. Step 1 and the choice of r then imply

$$\frac{h(\rho)}{\rho} \leq \frac{1}{2} \frac{e(R) - e(r)}{R - r - (\ell(R) - \ell(r))} \leq \frac{e(R)}{2(R - \ell(R) - r)} < \frac{1-\varepsilon}{2},$$

in turn giving $m(\rho) \leq 2h(\rho) < (1-\varepsilon)\rho$. Let $\rho_\infty := \inf\{t > 0 : m(t) \leq (1-\varepsilon)t\}$, then $\rho_\infty \in [0, \rho]$. Note that if ρ_∞ were strictly positive then actually ρ_∞ would be a minimum. In such a case, we could apply the argument above and find $\widetilde{\rho} \in (r_\infty, \rho_\infty)$, with $r_\infty \in (0, \rho_\infty - \ell(\rho_\infty) - \frac{1}{1-\varepsilon} e(\rho_\infty))$, such that $m(\widetilde{\rho}) < (1-\varepsilon)\widetilde{\rho}$ contradicting the minimality of

ρ_∞. Hence, there is a sequence $\rho_k \downarrow 0^+$ with $m(\rho_k) \le (1-\varepsilon)\rho_k$. Then, clearly condition (2.15) is violated, so that $0 \notin S_u^\star$.

Conclusion: We first prove that Ω_u is open. Let $z \in \Omega_u$ and let $R > 0$ and $\varepsilon > 0$ be such that $m_z(R) \le (1-\varepsilon)R$ and $B_{\varepsilon R}(z) \subset \Omega$. Let now $x \in B_{\varepsilon R}(z)$, then

$$m_x(R - |x-z|) \le m_z(R) \le (1-\varepsilon)R < R - |x-z|,$$

therefore $x \in \Omega_u$.

As Ω_u is open and $S_u^\star \cap \Omega_u = \emptyset$ by Step 2, we infer $\Omega \cap \overline{S_u^\star} \subseteq \Omega \setminus \Omega_u$. Moreover, let $z \notin \overline{S_u^\star}$ and $r > 0$ be such that $B_r(z) \subseteq \Omega \setminus \overline{S_u^\star}$. Since $\mathcal{H}^1(S_u \setminus \overline{S_u^\star}) = 0$, $u \in W^{1,2}(B_r(z))$ and thus u is an harmonic function in $B_r(z)$ by minimality. Therefore $z \in \Omega_u$, and in conclusion $\Omega \setminus \Omega_u = \Omega \cap \overline{S_u^\star} = \Omega \cap \overline{S_u}$. □

A simple iteration of Theorem 2.10 gives a density lower bound as in (2.12) with an explicit constant θ_0 (see [34, Corollary 1.2]).

Corollary 2.12 (De Lellis and Focardi [34]). *If* $u \in \mathcal{M}(\Omega)$, *then* $\mathcal{H}^1\big(\Omega \cap (\overline{S_u} \setminus J_u)\big) = 0$ *and*

$$\mathcal{H}^1(S_u \cap B_r(z)) \ge \frac{\pi}{2^{23}} r \tag{2.20}$$

for all $z \in \Omega \cap \overline{S_u}$ *and all* $r \in (0, \operatorname{dist}(z, \partial\Omega))$.

A similar result can be established for quasi-minimizers of the Mumford and Shah energy, the most prominent examples being minimizers of the functional $\mathscr{F}$ in equation (1.1). More precisely, a *quasi-minimizer* is any function u in $SBV(\Omega)$ with $\mathrm{MS}(u) < \infty$ satisfying for some $\alpha > 0$ and $c_\alpha \ge 0$ for all balls $B_\rho(z) \subset \Omega$

$$\mathrm{MS}(u, B_\rho(z)) \le \mathrm{MS}(w, B_\rho(z)) + c_\alpha\, \rho^{1+\alpha} \quad \text{whenever } \{w \ne u\} \Subset B_\rho(z).$$

One can then prove the ensuing infinitesimal version of (2.13) (*cf.* with [34, Corollary 1.3]).

Corollary 2.13 (De Lellis and Focardi [34]). *Let* u *be a quasi-minimizers of the Mumford and Shah energy, then*

$$\Omega \cap \overline{J_u} = \Omega \cap \overline{S_u} = \left\{ z \in \Omega : \liminf_{r \downarrow 0^+} \frac{m_z(r)}{r} \ge \frac{2}{3} \right\}. \tag{2.21}$$

The proof of this corollary, though, needs a blow up analysis and a new SBV Poincaré-Wirtinger type inequality of independent interest, obtained by improving upon some ideas contained in [41] (*cf.* with [34, Theorem B.6]); it is, therefore, much more technical.

Let us remark that it is possible to improve slightly Theorem 2.10 by combining the ideas of its proof hinted to above with the SBV Poincaré-Wirtinger type inequality in [34, Theorem B.6], and show that actually

$$\Omega \setminus \overline{S_u} = \{x \in \Omega : m_x(r) \leq r \quad \text{for some } r \in \operatorname{dist}(x, \partial\Omega)\}$$

(see [34, Remark 2.3]).

A natural question is the sharpness of the estimates (2.13) and (2.20). The analysis performed by Bonnet [10] suggests that $\pi/2^{23}$ in (2.20) should be replaced by 1, and 1 in (2.13) by 2. Note that the square root function $u(\rho, \theta) = \sqrt{\frac{2}{\pi}\rho} \cdot \sin(\theta/2)$ satisfies $\mathrm{MS}(u, B_\rho(0)) = \mathcal{H}^1(S_u \cap B_\rho(0)) = \rho$ for all $\rho > 0$ (its minimality will be discussed in Subsection 2.5). Thus both the constants conjectured above would be sharp by [26, Section 62]. Unfortunately, none of them have been proven so far.

We point out that Bucur and Luckhaus [16], independently from [35], have been able to improve the ideas in Theorem 2.10 carrying on the proof without the 2-dimensional limitation via a delicate induction argument. Their approach leads to a remarkable monotonicity formula for (a truncated version of) the energy valid for a broad class of approximate minimizers that shall be the topic of the next Subsection 2.4.

To conclude this paragraph we notice that the derivation of an energy upper bound for minimizers on balls is much easier as a result of a simple comparison argument.

Proposition 2.14. *For every* $u \in \mathcal{M}(\Omega)$ *and* $B_r(z) \subseteq \Omega$

$$\mathrm{MS}(u, B_r(z)) \leq n\,\omega_n\, r^{n-1}. \tag{2.22}$$

Proof. Let $\varepsilon > 0$, consider $u_\varepsilon := u\chi_{\Omega \setminus B_{r-\varepsilon}(z)}$ and test the minimality of u with u_ε. Conclude by letting $\varepsilon \downarrow 0$. □

2.4 Bucur and Luckhaus' almost monotonicity formula

Let us start off by introducing the notion of *almost-quasi minimizers* of the MS energy.

Definition 2.15. Let $\Lambda \geq 1, \alpha > 0$ and $c_\alpha \geq 0$, a function $u \in SBV(\Omega)$ with $\mathrm{MS}(u) < \infty$ is a $(\Lambda, \alpha, c_\alpha)$-almost-quasi minimizer in Ω, we write

$u \in \mathcal{M}_{\{\Lambda,\alpha,c_\alpha\}}(\Omega)$, if for all balls $B_\rho(z) \subset \Omega$

$$\mathrm{MS}(u, B_\rho(z)) \leq \int_{B_\rho(z)} |\nabla v|^2 \, dx + \Lambda \, \mathcal{H}^{n-1}\big(S_v \cap B_\rho(z)\big) + c_\alpha \, \rho^{n-1+\alpha},$$

whenever $\{u \neq v\} \Subset \Omega$.

Clearly, minimizers are $(1, \alpha, c_\alpha)$-almost minimizers for all α and c_α as in the definition above. Almost-quasi minimizers has turned out to be useful in studying the regularity properties of solutions to free boundary - free discontinuity problems with Robin conditions (*cf.* [16, Section 4]). We are now ready to state the Bucur and Luckhaus' almost monotonicity formula.

Theorem 2.16 (Bucur and Luckhaus [16]). *There is a (small) dimensional constant $C(n) > 0$ such that for all $u \in \mathcal{M}_{\{\Lambda,\alpha,c_\alpha\}}(\Omega)$ and $z \in \Omega$ the function*

$$\begin{aligned}\big(0, \operatorname{dist}(z, \partial\Omega)\big) \ni r &\longmapsto E_z(r) \\ &:= \left(\frac{\mathrm{MS}(u, B_r(z))}{r^{n-1}} \wedge \frac{C(n)}{n-1}\Lambda^{2-n}\right) + (n-1)\,\frac{c_\alpha}{\alpha}\, r^\alpha \end{aligned} \tag{2.23}$$

is non decreasing.

To explain the strategy of proof of Theorem 2.16 we need to state two additional results on SBV functions defined on boundaries of balls. The first is related to the approximation with H^1 functions, the second to the problem of lifting. The tangential gradient on boundaries of balls shall be denoted by ∇_τ in what follows.

Lemma 2.17. *Let $n \geq 2$, then there is a (small) dimensional constant $C(n) > 0$ such that for all $v \in SBV(\partial B_r)$ with*

$$\begin{aligned}\mathcal{E}(v, \partial B_r) &:= \int_{\partial B_r} |\nabla_\tau v|^2 d\mathcal{H}^{n-1} + \mathcal{H}^{n-2}(S_v \cap \partial B_r) \\ &\leq C(n)\, \Lambda^{2-n}\, r^{n-2},\end{aligned} \tag{2.24}$$

there exists $w \in H^1(\partial B_r)$ such that

$$\mathcal{E}(w, \partial B_r) + (n-1)\,\frac{\Lambda}{r}\,\mathcal{H}^{n-1}\big(\{x \in \partial B_r : v \neq w\}\big) \leq \mathcal{E}(v, \partial B_r). \tag{2.25}$$

Lemma 2.18. *Let $n \geq 2$, suppose $v \in SBV(\partial B_r)$ satisfies* (2.24) *in Lemma* 2.17. *Then there exists $\widehat{w} \in H^1(B_r)$ harmonic in B_r such that*

$$\mathrm{MS}(\widehat{w}, B_r) + \Lambda\, \mathcal{H}^{n-1}\big(\{x \in \partial B_r : v \neq \widehat{w}\}\big) \leq \frac{r}{n-1}\, \mathcal{E}(v, \partial B_r). \tag{2.26}$$

The proof of Theorem 2.16 is based on a cyclic induction argument that starts with the validation of Lemma 2.17 in $\mathbb{R}^2$ and then runs as follows:

$$\text{Lemma 2.17 in } \mathbb{R}^n \Rightarrow \text{Lemma 2.18 in } \mathbb{R}^n$$
$$\Rightarrow \text{Theorem 2.16 in } \mathbb{R}^n \Rightarrow \text{Lemma 2.17 in } \mathbb{R}^{n+1}.$$

The first inductive step can be established by using the geometric argument in the proof of Theorem 2.10 with the constant $C(2)$ being any number in $(0, 1)$.

Before proving all the implications above we establish the essential closure of S_u for almost-quasi minimizers as a consequence of Theorem 2.16.

Theorem 2.19. *Let $u \in \mathcal{M}_{\{\Lambda,\alpha,c_\alpha\}}(\Omega)$, for some $\Lambda \geq 1$, $\alpha > 0$ and $c_\alpha \geq 0$. Then, $\mathcal{H}^{n-1}\big(\Omega \cap (\overline{S_u} \setminus S_u)\big) = 0$.*

Proof. Consider the subset $S_u^\star$ of points in S_u with density one, *i.e.* $x \in S_u$ such that

$$\lim_{\rho \downarrow 0} \frac{\mathcal{H}^{n-1}\big(S_u \cap B_\rho(x)\big)}{\omega_{n-1}\rho^{n-1}} = 1$$

(*cf.* (2.15) in Theorem 2.10), then clearly $\overline{S_u^\star} = \overline{S_u}$. Theorem 2.16 yields for $x \in S_u^\star$

$$E_x(\rho) \geq \lim_{\rho \downarrow 0} E_x(\rho) \geq \omega_{n-1} \wedge \frac{C(n)}{n-1}\Lambda^{2-n} \qquad \forall \rho \in \big(0, \operatorname{dist}(x, \partial\Omega)\big).$$

By approximation it is then easy to deduce the same estimate for all points $x \in \Omega \cap \overline{S_u}$. In turn, this implies that $\Omega \cap (\overline{S_u} \setminus S_u) \subset A_u := \{y \in \Omega \setminus S_u : \liminf_\rho E_x(\rho) > 0\}$.

In particular, the density estimates in Lemma 2.8 applied to the measure induced on Borel sets by $\mathrm{MS}(u, \cdot)$ gives that $\mathcal{H}^{n-1}(A_u) = 0$, and the conclusion $\mathcal{H}^{n-1}\big(\Omega \cap (\overline{S_u} \setminus S_u)\big) = 0$ follows at once (*cf.* the proof of Proposition 2.9). □

Density lower bounds for $\mathrm{MS}(u, \cdot)$ as in Theorem 2.6 or for $\mathcal{H}^{n-1} \llcorner S_u$ as in Theorem 2.7 can also be recovered. Either by means of Theorem 2.16 and of a decay lemma due to De Giorgi, Carriero and Leaci or in an alternative direct way that provides explicit constants (*cf.* [7, Lemma 7.14] and [16, Section 3.3], respectively).

Let us now turn to the proofs of the implications among Theorem 2.16, Lemma 2.17 and Lemma 2.18. Recall that we have already commented on the validity of Lemma 2.17 in case $n = 2$.

Proof of Lemma 2.17 *in* $\mathbb{R}^n \Rightarrow$ *Lemma* 2.18 *in* $\mathbb{R}^n$. Denote by w the function provided by Lemma 2.17 and by $\widehat{w}$ its harmonic extension to B_ρ. Note that[4]

$$\frac{n-1}{\rho}\int_{B_\rho}|\nabla\widehat{w}|^2dx \leq \int_{\partial B_\rho}|\nabla_\tau\widehat{w}|^2d\mathcal{H}^{n-1} \leq \mathcal{E}(w,\partial B_\rho).$$

Hence, by (2.25) we conclude that

$$\begin{aligned}
&\mathrm{MS}(\widehat{w},B_\rho)+\Lambda\,\mathcal{H}^{n-1}\big(\{x\in\partial B_\rho : v\neq\widehat{w}\}\big)\\
&\leq \frac{\rho}{n-1}\Big(\mathcal{E}(w,\partial B_\rho)+(n-1)\frac{\Lambda}{\rho}\,\mathcal{H}^{n-1}\big(\{x\in\partial B_\rho : v\neq\widehat{w}\}\big)\Big)\\
&\leq \frac{\rho}{n-1}\,\mathcal{E}(v,\partial B_\rho).
\end{aligned}$$

□

Proof of Lemma 2.18 *in* $\mathbb{R}^n \Rightarrow$ *Theorem* 2.16 *in* $\mathbb{R}^n$.
Set $I:=\big(0,\mathrm{dist}(0,\partial\Omega)\big)$, then the energy function $m_z(\rho):=\mathrm{MS}(u,B_\rho(z)): I\to[0,+\infty)$ is non-decreasing and thus it belongs to $BV_{\mathrm{loc}}(I)$. Note that

$$E_z(\rho)=\Big(\frac{m_z(\rho)}{\rho^{n-1}}\wedge\frac{C(n)}{n-1}\,\Lambda^{2-n}\Big)+(n-1)\,\frac{c_\alpha}{\alpha}\rho^\alpha, \qquad (2.27)$$

therefore, $E_z\in BV_{\mathrm{loc}}(I)$. Denoting by m'_z and $D^s m_z$, respectively, the density of the absolutely continuous part and the singular part of Dm_z with respect to $\mathcal{L}^1\llcorner I$ in the Radon-Nikodym decomposition, the Leibnitz rule for BV functions (*cf.* [7, Example 3.97]) yields

$$D\Big(\frac{m_z(\rho)}{\rho^{n-1}}\Big)=\Big(\frac{m'_z(\rho)}{\rho^{n-1}}-(n-1)\frac{m_z(\rho)}{\rho^n}\Big)+\frac{1}{\rho^{n-1}}D^s m_z(\rho).$$

Therefore, the latter equality, (2.27) and the Chain Rule formula for Lipschitz functions [7, Theorem 3.99] imply that E_z is non-decreasing if and only if the density E'_z of the absolutely continuous part of the distributional derivative DE_z is non-negative.

By the locality of the distributional derivative (see [7, Remark 3.93]) it holds that $E'_z(\rho)=(n-1)\,c_\alpha\rho^{\alpha-1}>0$ at $\mathcal{L}^1$ a.e. $\rho\in I$ for which

$$m_z(\rho)\geq\frac{C(n)}{n-1}\,\Lambda^{2-n}\rho^{n-1}.$$

[4] The first inequality follows, for instance, from Almgren's monotonicity formula for the frequency function (*cf.* [39, Exercise 20, page 525]) and a direct comparison argument with the one-homogeneous extension of the boundary trace. Alternatively, one can expand $\widehat{w}$ in spherical harmonics.

Instead, at $\mathcal{L}^1$ a.e. $\rho \in I$ at which

$$m_z(\rho) < \frac{C(n)}{n-1} \Lambda^{2-n} \rho^{n-1} \tag{2.28}$$

we have

$$E_z'(\rho) = \frac{m_z'(\rho)}{\rho^{n-1}} - (n-1)\frac{m_z(\rho)}{\rho^n} + (n-1)\, c_\alpha \rho^{\alpha-1}.$$

Suppose by contradiction that $E_z' < 0$ on some subset J of I of positive measure, *i.e.*

$$m_z'(\rho) < (n-1)\frac{m_z(\rho)}{\rho} - (n-1)\, c_\alpha \rho^{n-2+\alpha}. \tag{2.29}$$

Since for $\mathcal{L}^1$ a.e. in I by the slicing theory $u|_{\partial B_\rho} \in SBV(\partial B_\rho)$ (*cf.* [7, Section 3.11]) and by the Co-Area formula $\mathcal{E}(u, \partial B_\rho) \leq m_z'(\rho)$ (*cf.* [7, Theorem 2.93]), we conclude that for $\mathcal{L}^1$ a.e. $\rho \in J$

$$\mathcal{E}(u, \partial B_\rho) \leq m_z'(\rho) \overset{(2.28),(2.29)}{<} C(n)\, \Lambda^{2-n}\, \rho^{n-2} - (n-1)\, c_\alpha\, \rho^{n-2+\alpha}.$$

Hence, with fixed $\rho \in J$ as above, Lemma 2.18 provides an harmonic function $\widehat{w} \in H^1(B_\rho)$ satisfying (2.26). In turn, the latter inequality and the assumption $E_z'(\rho) < 0$ give

$$\begin{aligned}&\mathrm{MS}(\widehat{w}, B_\rho) + \Lambda\, \mathcal{H}^{n-1}(\{x \in \partial B_\rho : u \neq \widehat{w}\})\\ &\quad\overset{(2.26)}{\leq} \frac{\rho}{n-1}\mathcal{E}(u, \partial B_\rho) \leq \frac{\rho}{n-1} m_z'(\rho) \overset{(2.29)}{<} m_z(\rho) - c_\alpha \rho^{n-1+\alpha},\end{aligned}$$

leading to a contradiction to the almost-quasi minimality of u in Ω by taking the trial function $\widehat{w}\, \chi_{B_\rho(z)} + u\, \chi_{\Omega \setminus B_\rho(z)}$. □

To conclude the cyclic induction argument we set some notation: in the following proof we denote by B_ρ the $(n+1)$-dimensional ball of radius ρ and by B_ρ^n its intersection with the hyperplane $\mathbb{R}^n \times \{0\}$. Moreover, we still denote by $\mathcal{E}$ the n dimensional version of the boundary energy in (2.24).

Proof of Theorem 2.16 *in* $\mathbb{R}^n \Rightarrow$ *Lemma* 2.17 *in* $\mathbb{R}^{n+1}$. Up to a scaling argument we may assume the radius r in the statement of Lemma 2.17 to be 1.

Then, consider $v \in SBV(\partial B_1)$ such that

$$\mathcal{E}(v, \partial B_1) \leq C\, \Lambda^{1-n} \tag{2.30}$$

for some constant $C > 0$.

Claim: There exists $C(n+1) > 0$ such that if v satisfies (2.30) with $C \in (0, C(n+1)]$, every minimizer $w \in SBV(\partial B_1)$ of the problem

$$\inf_{\zeta \in SBV(\partial B_1)} F(\zeta, \partial B_1)$$

actually belongs to $H^1(\partial B_1)$, where for all open sets $A \subseteq \partial B_1$ and $\zeta \in SBV(\partial B_1)$ if $\mathcal{E}(\zeta, A)$ is defined as in (2.24) by integrating ζ on A, then

$$F(\zeta, A) := \mathcal{E}(\zeta, A) + n\,\Lambda\,\mathcal{H}^n\big(\{x \in A : \zeta \neq v\}\big). \tag{2.31}$$

Given the claim above for granted we conclude the thesis of Lemma 2.17 straightforwardly by comparing the values of the energy $F(\cdot, \partial B_1)$ in (2.31) on w and v respectively, namely

$$\int_{\partial B_1} |\nabla_\tau w|^2 d\mathcal{H}^n + n\Lambda\mathcal{H}^n\big(\{x \in \partial B_1 : w \neq v\}\big) \leq \int_{\partial B_1} |\nabla_\tau v|^2 d\mathcal{H}^n + \mathcal{H}^{n-1}(S_v).$$

We are then left with proving the claim above. Suppose by contradiction that for some constant $C > 0$ some minimizer w of (2.31) satisfies $\mathcal{H}^{n-1}(S_w) > 0$. Note that C can be taken arbitrarily small, it shall be chosen suitably in what follows. Even more, assume that the north pole $\mathcal{N} := (0, \ldots, 0, 1) \in \partial B_1$ is a point of density one for S_w. Set $\lambda := 2 - \sqrt{3}$ and denote by $\pi : \partial B_1 \cap B_\lambda(\mathcal{N}) \to B^n_{1/2}$ the orthogonal projection, then $\pi \in \mathrm{Lip}_1(\partial B_1 \cap B_\lambda(\mathcal{N}), B^n_{1/2})$ and $\pi^{-1} \in \mathrm{Lip}_\ell(B^n_{1/2}, \partial B_1 \cap B_\lambda(\mathcal{N}))$, for some $\ell > 0$. Actually, as $\pi \in C^1(\partial B_1 \cap B_\lambda(\mathcal{N}), B^n_{1/2})$ and $\pi^{-1} \in C^1(B^n_{1/2}, \partial B_1 \cap B_\lambda(\mathcal{N}))$

$$\mathrm{Lip}\big(\pi, \pi^{-1}(B^n_\rho)\big) \to 1, \quad \mathrm{Lip}\big(\pi^{-1}, B^n_\rho\big) \to 1 \qquad \text{as } \rho \downarrow 0. \tag{2.32}$$

For $\zeta \in SBV(\partial B_1 \cap B_\lambda(\mathcal{N}))$ let $\overline{\zeta} := \zeta \circ \pi^{-1}$, then $\overline{\zeta} \in SBV(B^n_{1/2})$ and $S_{\overline{\zeta}} = \pi(S_\zeta)$. Moreover, for all $\rho \in (0, 1/2)$, we have

$$\mathcal{H}^{n-1}\big(S_{\overline{\zeta}} \cap B^n_\rho\big) \leq \mathcal{H}^{n-1}\big(S_\zeta \cap \pi^{-1}(B^n_\rho)\big) \leq \ell^{n-1}\,\mathcal{H}^{n-1}(S_{\overline{\zeta}} \cap B^n_\rho), \tag{2.33}$$

and by the generalized Area Formula (see [7, Theorem 2.91])

$$\int_{B^n_\rho} |\nabla \overline{\zeta}|^2 dy \leq k_1(\rho) \int_{\pi^{-1}(B^n_\rho)} |\nabla_\tau \zeta|^2 d\mathcal{H}^n, \tag{2.34}$$

and

$$\int_{\pi^{-1}(B^n_\rho)} |\nabla_\tau \zeta|^2 d\mathcal{H}^n \leq k_2(\rho) \int_{B^n_\rho} |\nabla \overline{\zeta}|^2 dy, \tag{2.35}$$

for some $k_1(\rho)$ and $k_2(\rho) > 0$, with $k_1(\rho), k_2(\rho) \downarrow 1$ as $\rho \downarrow 0$ by (2.32). In particular, for some $\tau_n > 0, 0 \leq k_1(\rho) \vee k_2(\rho) \leq 1 + \tau_n \rho$ as $\rho \downarrow 0$.

We next prove that $\overline{w}$ is a $(\Lambda_1, 1, c_1)$-almost-quasi minimizer on $B^n_{1/2}$ of the n-dimensional Mumford and Shah energy for $\Lambda_1 := \ell^{n-1}$ and a suitable $c_1 = c_1(n, \ell, \Lambda) > 0$. Indeed, recalling that $\Lambda \geq 1$, if $\overline{\zeta}$ is a test function for $\overline{w}$, *i.e.* $\{y \in B^n_{1/2} : \overline{\zeta} \neq \overline{w}\} \Subset B^n_\rho$, $\rho \in (0, 1/2)$, we deduce from (2.33), (2.34), (2.35) and the minimality of w for the energy in (2.31) that

$$
\begin{aligned}
\mathrm{MS}(\overline{w}, B^n_\rho) &\leq \int_{B^n_\rho} |\nabla \overline{w}|^2 dx + \mathcal{H}^{n-1}(S_{\overline{w}} \cap B^n_\rho) + \mathcal{H}^n(\{y \in B^n_\rho : \overline{w} \neq \overline{v}\}) \\
&\leq k_1(\rho) \int_{\pi^{-1}(B^n_\rho)} |\nabla_\tau w|^2 d\mathcal{H}^n + \mathcal{H}^{n-1}\big(S_w \cap \pi^{-1}(B^n_\rho)\big) \\
&\quad + \mathcal{H}^n\big(\{x \in \pi^{-1}(B^n_\rho) : w \neq v\}\big) \\
&\leq k_1(\rho) F(w, \pi^{-1}(B^n_\rho)) \leq k_1(\rho) F(\zeta, \pi^{-1}(B^n_\rho)) \qquad (2.36)\\
&\leq k_1(\rho)\Bigg(k_2(\rho) \int_{B^n_\rho} |\nabla \overline{\zeta}|^2 dx + \Lambda_1 \mathcal{H}^{n-1}(S_{\overline{\zeta}} \cap B^n_\rho) \\
&\qquad + n\,\Lambda\,\Lambda_1^{\frac{n}{n-1}} \mathcal{H}^n(\{y \in B^n_\rho : \overline{\zeta} \neq \overline{v}\})\Bigg) \\
&\leq (1 + \tau_n\,\rho)^2 \left(\int_{B^n_\rho} |\nabla \overline{\zeta}|^2 dx + \Lambda_1 \mathcal{H}^{n-1}(S_{\overline{\zeta}} \cap B^n_\rho)\right) \\
&\quad + (1 + \tau_n) n \omega_n\,\Lambda\,\Lambda_1^{\frac{n}{n-1}} \rho^n, \qquad (2.37)
\end{aligned}
$$

being $\mathcal{H}^n(\{y \in B^n_\rho : \overline{\zeta} \neq \overline{v}\}) \leq \omega_n \rho^n$.

Next we note that w satisfies the energy upper bound $F(w, \pi^{-1}(B^n_\rho)) \leq k_n \rho^{n-1}$, for all $\rho \in (0, 1/2)$ and for some $k_n > 0$, by a direct comparison argument similar to that in Proposition 2.14. Therefore, (2.36) yields

$$
\mathrm{MS}(\overline{w}, B^n_\rho) \leq (1 + \tau_n) k_n \rho^{n-1}.
$$

Hence, we may assume $F(\zeta, \pi^{-1}(B^n_\rho)) \leq 2(1 + \tau_n) k_n \rho^{n-1}$ being otherwise the conclusion obvious. The latter condition and (2.37) then imply $\overline{w} \in \mathcal{M}_{\{\Lambda_1, 1, c_1\}}\big(B^n_{1/2}\big)$ with $c_1 := (1 + \tau_n)\big(n\,\omega_n\,\Lambda\,\Lambda_1^{\frac{n}{n-1}} + 2(1 + \tau_n) k_n\big)$.

By inductive assumption, Theorem 2.16 implies in particular that

$$
\begin{aligned}
&\left(\frac{\mathrm{MS}(\overline{w}, B^n_\rho)}{\rho^{n-1}} \wedge \frac{C(n)}{n-1} \Lambda_1^{2-n}\right) + (n-1) c_1 \rho \\
&\geq \lim_{\rho \downarrow 0} \left(\left(\frac{\mathrm{MS}(\overline{w}, B^n_\rho)}{\rho^{n-1}} \wedge \frac{C(n)}{n-1} \Lambda_1^{2-n}\right) + (n-1) c_1 \rho\right) \qquad (2.38)\\
&\geq \omega_{n-1} \wedge \frac{C(n)}{n-1} \Lambda_1^{2-n} =: \beta_n.
\end{aligned}
$$

In the last inequality we have used that $\mathcal{N}$ is a point of density one for S_w and (2.32).

Given any $\rho \in (0, \frac{\beta_n}{2(n-1)c_1} \wedge 1/2)$, inequality (2.38) yields

$$\frac{\mathrm{MS}(\overline{w}, B^n_\rho)}{\rho^{n-1}} \geq \frac{\beta_n}{2}.$$

On the other hand, repeating the argument leading to the first and the second inequality in (2.36) imply for any $\rho \in (0, \frac{\beta_n}{2(n-1)c_1} \wedge 1/2)$

$$\frac{\mathrm{MS}(\overline{w}, B^n_\rho)}{\rho^{n-1}} \leq (1+\tau_n\rho)\frac{\mathcal{E}(v, \partial B_1)}{\rho^{n-1}} \overset{(2.30)}{\leq} C\,(1+\tau_n\rho)\,(\rho\,\Lambda)^{1-n}.$$

Thus, with fixed $\overline{\rho} \in (0, \frac{\beta_n}{2(n-1)c_1} \wedge 1/2)$, we infer a contradiction from the last two estimates by choosing the constant $C = C(\overline{\rho}) > 0$ in (2.30) so that $C\,(1+\tau_n)\,(\overline{\rho}\,\Lambda)^{1-n} < \beta_n/2$. □

2.5 The Mumford-Shah Conjecture

Having established the existence of strong (local) minimizers, it is elementary to infer that u is harmonic on $\Omega \setminus \overline{S_u}$. Hence, we will focus in the rest of the note on the regularity properties of the set $\Omega \cap \overline{S_u}$ that will be instrumental also to gain further regularity on u itself (*cf.* Theorem 3.3).

The interest of the researchers in this problem is motivated by the Mumford and Shah conjecture (in 2-dimensions) that we recall below for the readers' convenience.

Conjecture 2.20 (Mumford and Shah [69]). If $u \in \mathcal{M}(\Omega)$, $\Omega \subseteq \mathbb{R}^2$, then $\Omega \cap \overline{S_u}$ is the union of (at most) countably many injective C^1 arcs $\gamma_i : [a_i, b_i] \to \Omega$ with the following properties:

(c1) Any compact $K \subset \Omega$ intersects at most finitely many arcs;
(c2) Two arcs can have at most an endpoint p in common, and if this is the case, then p is in fact the endpoint of three arcs, forming equal angles of $2\pi/3$.

So according to this conjecture only two possible singular configurations occur: either three arcs meet in an end forming angles equal to $2\pi/3$, or an arc has a free end in Ω. In what follows, we shall call *triple junction* the first configuration and *crack-tip* the second.

A suitable theory of calibrations for free discontinuity problems established by Alberti, Bouchitté and Dal Maso [1] shows that the model case of triple junction functions,

$$a\,\chi_{\{\vartheta\in(-\pi/6,\pi/2]\}} + b\,\chi_{\{\vartheta\in(\pi/2,7\pi/6]\}} + c\,\chi_{\{\vartheta\in(7\pi/6,11\pi/6]\}} \qquad (2.39)$$

with $|a-b|\cdot|a-c|\cdot|b-c|>0$, is indeed a local minimizer (for more results on calibrations in the setting of free discontinuity problems see [22,65–68]).

Instead, Bonnet and David [11] have shown that the model crack-tip functions, *i.e.* functions that up to rigid motions can be written as

$$C \pm \sqrt{\frac{2}{\pi}\rho} \cdot \sin(\theta/2) \tag{2.40}$$

for $\theta \in (-\pi, \pi)$, $\rho > 0$ and some constant $C \in \mathbb{R}$, are *global* minimizers of the Mumford and Shah energy[5].

More recently, second order sufficient conditions for minimality have been investigated. More precisely, Bonacini and Morini in [9] for suitable critical points, strictly stable and regular in their terminology, have proved that strict local minimality is implied by strict positivity of the second variation. This approach in the case of triple junctions is currently under investigation [19].

Conjecture 2.20 has been first proven in some particular cases in the ensuing weaker form.

Conjecture 2.21. If $u \in \mathcal{M}(\Omega)$, $\Omega \subseteq \mathbb{R}^2$, then $\Omega \cap \overline{S_u}$ is the union of (at most) countably many injective C^0 arcs $\gamma_i : [a_i, b_i] \to \Omega$ which are C^1 on $]a_i, b_i[$ and satisfy the two conditions of Conjecture 2.20.

The subtle difference between Conjecture 2.20 and Conjecture 2.21 above is in the following point: assuming Conjecture 2.21 holds, if $y_0 = \gamma_i(a_i)$ is a "loose end" of the arc γ_i, *i.e.* it does not belong to any other arc, then the techniques in [10] show that any *blow up limit*, *i.e.* any limit of subsequences of the family $(u_{y_0,\rho})_\rho$ as in (2.8), is a crack-tip but do not ensure the uniqueness of the limit itself (for more details on the notion of blow up see the proof of Proposition 4.11).

2.6 Blow up analysis and the Mumford and Shah conjecture

The Mumford and Shah conjecture, in the form stated in Conjecture 2.21, has been attacked first in the contribution by Bonnet [10]. Bonnet's approach is based on a weaker notion of minimality for the strong formulation of the problem, that includes a topological condition to be satisfied by the competitors, though still sufficient to develop a regularity theory.

---

[5] The prefactor $\sqrt{\frac{2}{\pi}}$ results from a simple calculation to ensure stationarity for a crack-tip function by plugging it in the Euler-Lagrange equation (2.6).

Definition 2.22. Let $\Omega \subseteq \mathbb{R}^2$ be an open set, a pair (u, K), $K \subset \Omega$ closed and $u \in W^{1,2}_{\mathrm{loc}}(\Omega \setminus K)$, is a Bonnet minimizer of $\mathscr{F}$ in Ω if $\mathscr{F}(u, K, \Omega) \leq \mathscr{F}(v, L, \Omega)$ among all couples (v, L), $L \subset \Omega$ closed and $v \in W^{1,2}_{\mathrm{loc}}(\Omega \setminus L)$, such that there is a ball $B_\rho(x) \subset \Omega$ for which

(i) $u = v$ and $K = L$ on $\Omega \setminus \overline{B_\rho(x)}$,
(ii) any pair of points in $\Omega \setminus \left(K \cup \overline{B_\rho(x)}\right)$ that lie in different connected components of $\Omega \setminus K$ are also in different connected components of $\Omega \setminus L$.

Moreover, in case $\Omega = \mathbb{R}^2$, we say that (u, K) is a global minimizer of $\mathscr{F}$.

In particular, under the assumption that $\Omega \cap \overline{S_u}$ has a finite number of connected components Bonnet has classified all the blow up limits of minimizers as in Definition 2.22, and then also of local minimizers, as

(i) constant functions,
(ii) pure jumps (*cf.* (2.10)),
(iii) triple junction functions (*cf.* (2.39)),
(iv) crack-tip functions (*cf.* (2.40)),

establishing Conjecture 2.21 in such a restricted framework. In particular, as already mentioned, Bonnet's result does not deal with the stronger Conjecture 2.20 as it cannot exclude the possibility that γ_i "spirals" around y_0 infinitely many times (compare with the discussion at the end of [10, Section 1]). The method of Bonnet relies on a key monotonicity formula for the rescaled Dirichlet energy that so far has no counterpart in general. Recently, Lemenant [52] has exhibited another monotone quantity that plays the role of the rescaled Dirichlet energy in higher dimensions in case $\Omega \cap \overline{S_u}$ is contained in a sufficiently smooth cone. In view of this, a rigidity result for $\Omega \cap \overline{S_u}$ in the 3-dimensional case can be deduced. Let us also point out that the almost monotonicity formula established by Bucur and Luckhaus in Theorem 2.16 is of little use in the blow up analysis due to the truncation with the constant $C(n)$.

The contributions of Léger [50] and of David and Léger [27] improve upon Bonnet's results: the former addressing the case of $\Omega \cap \overline{S_u}$ satisfying a suitable flatness assumption, the latter identifying pure jumps and triple junction functions as the only minimizers in the sense of Bonnet for which $\mathbb{R}^2 \setminus \overline{S_u}$ is not connected. All these efforts are directed to push forward Bonnet's ideas. Indeed, [26, Proposition 71.2] shows in general, *i.e.* with no extra connectedness assumptions on $\Omega \cap \overline{S_u}$, that the complete classification of the blow up limits of local minimizers as in the list above turns out to be a viable strategy to establish Conjecture 2.20 (in this respect see Proposition 4.11). More precisely, coupling the latter piece of information with a detailed local description of the geometry of $\Omega \cap \overline{S_u}$, that is the topic of the ensuing Subsection 3.1, would yield the conclusion. A further interesting consequence of the analysis in the paper [27] is that the Mumford and Shah conjecture turns out to be equivalent to the uniqueness (up to rotations and translations) of crack-tips as global minimizers of the MS energy as conjectured by De Giorgi in [30].

However, we shall not enter here into this streamline of results but rather refer for more details on them to the monograph [26] and to the recent review paper [53].

Instead, a different (but related) perspective to the goal of understanding Conjecture 2.20 is taken in what follows. We shall link the latter to a sharp higher integrability property of the approximate gradient following Ambrosio, Fusco and Hutchinson [4]. In order to do this, in the next section we review the state of the art about the regularity properties of $\Omega \cap \overline{S_u}$.

3 Regularity of the jump set

The aim of this section is to survey on the regularity of $\Omega \cap \overline{S_u}$. In Subsection 3.1 we shall first recall classical and more recent ε-regularity results, and then state an estimate on the size of the subset of singular points in $\Omega \cap \overline{S_u}$ that will be dealt with in details in Section 4. In particular, in Subsection 3.2 the links of the higher integrability of the gradient with the Mumford and Shah Conjecture 2.20 will be highlighted following the approach of Ambrosio, Fusco and Hutchinson [4]. A slight improvement of the latter ideas leads to an energetic characterization of the Conjecture 2.21 as exposed in Subsection 3.3.

3.1 ε-regularity theorems

The starting point to address the regularity of $\Omega \cap \overline{S_u}$ for local minimizers is the ensuing ε-regularity result.

Theorem 3.1 (Ambrosio, Fusco and Pallara [5]). *Let $u \in \mathcal{M}(\Omega)$, then there exists $\Sigma_u \subset \Omega \cap \overline{S_u}$ relatively closed in Ω with $\mathcal{H}^{n-1}(\Sigma_u) = 0$, and such that $(\Omega \cap \overline{S_u}) \setminus \Sigma_u$ is locally a $C^{1,\gamma}$ hypersurface for all $\gamma \in (0, 1)$ and $C^{1,1}$ if $n = 2$.*

More precisely, there exist $\varepsilon_0 = \varepsilon_0(n)$, $\rho_0 = \rho_0(n) > 0$ such that

$$\Sigma_u = \{z \in \Omega \cap \overline{S_u} : \mathscr{D}(z,\rho) + \mathscr{A}(z,\rho) \geq \varepsilon_0 \ \forall \rho \in (0, \rho_0 \wedge \mathrm{dist}(z, \partial\Omega))\} \quad (3.1)$$

where

$$\mathscr{D}_u(z,\rho) := \rho^{1-n} \int_{B_\rho(z)} |\nabla u|^2 dy, \quad \textit{(scaled Dirichlet energy)}$$

$$\mathscr{A}_u(z,\rho) := \rho^{-n-1} \min_{T\in\Pi} \int_{S_u \cap B_\rho(z)} \mathrm{dist}^2(y,T) d\mathcal{H}^{n-1}(y), \quad \textit{(scaled mean flatness)},$$

with Π *the class of* $(n-1)$*-affine planes.*

For more details on Theorem 3.1 we refer to [7, Chapter 8], that is entirely devoted to the proof of it, and to [44] for a hint of the strategy of proof. Here we shall only comment on the quantities involved in (3.1).

First note that the affine change of variables mapping $B_\rho(x)$ into B_1 shows that $\mathscr{D}_u(x,\cdot)$ and $\mathscr{A}_u(x,\cdot)$ are equal to the Dirichlet energy and the mean flatness on B_1 of the rescaled maps $u_{x,\rho}$ in (2.8), respectively.

Further, the scaled mean flatness measures in an average sense the deviation of $\Omega\cap\overline{S_u}$ from being flat in z. For instance, if $\Omega\cap\overline{S_u}$ is a C^1 hypersurface in a neighborhood of z it is easy to check that $\mathscr{A}_u(z,\rho) = o(1)$ as $\rho \downarrow 0$. Actually, the density lower bound in Theorem 2.7 and the density upper bound in Proposition 2.14 allow us to show that the rescaled mean flatness $\mathscr{A}_u$ is equivalent to its L^∞ version, namely

$$\mathscr{A}_{u,\infty}(z,\rho) := \rho^{-1} \min_{T\in\Pi} \sup_{y\in S_u\cap B_\rho(z)} \mathrm{dist}(y,T).$$

Proposition 3.2. *Let* $u \in \mathcal{M}(\Omega)$*, then for all* $z \in \overline{S_u}$ *and* $B_\rho(z) \subset \Omega$ *we have*

$$\frac{\theta_0}{2^{n+1}} \mathscr{A}_{u,\infty}^{n+1}(z,\rho/2) \le \mathscr{A}_u(z,\rho) \le n\,\omega_n\, \mathscr{A}_{u,\infty}^2(z,\rho).$$

where θ_0 *is the constant in Theorem* 2.7*.*

Proof. The estimate from above easily follows by the very definitions of $\mathscr{A}_u$ and $\mathscr{A}_{u,\infty}$ by taking into account (2.22).

Let then $\bar T$ be the affine hyperplane through z giving the minimum in the definition of $\mathscr{A}_u(z,\rho)$. If $\bar y \in S_u \cap B_{\rho/2}(z)$ is a point of almost maximum distance from $\bar T$, *i.e.* for a fixed $\delta \in (0,1)$

$$d := \mathrm{dist}(\bar y, \bar T) \ge (1-\delta) \sup_{y\in S_u\cap B_{\rho/2}(z)} \mathrm{dist}(y,\bar T),$$

then we can estimate as follows thanks to (2.12)

$$\begin{aligned}\mathscr{A}_u(z,\rho) &\geq \rho^{-n-1}\int_{S_u\cap B_{d/2}(\bar{y})} \mathrm{dist}^2(y,\overline{T})d\mathcal{H}^{n-1}(y)\\ &\geq \rho^{-n-1}\frac{d^2}{4}\mathcal{H}^{n-1}\big(S_u\cap B_{d/2}(\bar{y})\big)\\ &\geq \rho^{-n-1}\theta_0\left(\frac{d}{2}\right)^{n+1}\\ &\geq (1-\delta)^{n+1}\frac{\theta_0}{2^{n+1}}\mathscr{A}^{n+1}_{u,\infty}(z,\rho/2).\end{aligned}$$

The conclusion follows at once as $\delta\downarrow 0$. □

The local graph property of $\Omega\cap\overline{S_u}$ established in Theorem 3.1, the Euler-Lagrange condition and the regularity theory for elliptic PDEs with Neumann boundary conditions determine the regularity of u close to $\Omega\cap\overline{S_u}$ (see [7, Theorem 7.49] or [26, Proposition 17.15] if $n=2$).

Theorem 3.3 (Ambrosio, Fusco and Pallara [6]). *Let $u\in\mathcal{M}(\Omega)$ and $A\cap\overline{S_u}$, $A\subset\Omega$ open, be the graph of a $C^{1,\gamma}$ function ϕ, $\gamma\in(0,1)$. Then, $\phi\in C^\infty$ and u has C^∞ extension on each side of $A\cap\overline{S_u}$.*

Actually, Koch, Leoni and Morini [46] proved that if $\overline{S_u}\cap A$ is $C^{1,\gamma}$ then it is actually analytic, as conjectured by De Giorgi (*cf.* [29,31]).

Going back to the regularity issue for $\overline{S_u}$ we resume below the outcomes of a different approach developed by David in the 2-dimensional case. In this setting it is also possible to address the situation in which $\overline{S_u}$ is close in the Hausdorff distance to a triple-junction (*cf.* Section 6 for more comments in this respect).

Theorem 3.4 (David [25], Corollary 51.17 and Theorem 53.4 [26]). *There exists $\varepsilon>0$ and an absolute constant $c\in(0,1)$ with the following properties. If $u\in\mathcal{M}(\Omega)$, $z\in\overline{S_u}$, $B_r(z)\subset\Omega$ and $\mathscr{C}$ is either a line or a triple junction such that*

$$\int_{B_r(z)}|\nabla u|^2\,dx+\mathrm{dist}_{\mathcal{H}}(\overline{S_u}\cap B_r(z),\mathscr{C}\cap B_r(z))\leq\varepsilon r, \qquad (3.2)$$

then there exists a C^1-diffeomorphism ϕ of $B_r(z)$ onto its image with

$$\overline{S_u}\cap B_{cr}(z)=\phi(\mathscr{C})\cap B_{cr}(z).$$

In addition, for any given $\delta\in(0,1/2)$, there is $\varepsilon>0$ such that, if (3.2) holds, then $\overline{S_u}\cap(B_{(1-\delta)r}(z)\setminus B_{\delta r}(z))$ is δ-close, in the C^1 norm, to $\mathscr{C}\cap(B_{(1-\delta)r}(z)\setminus B_{\delta r}(z))$.

Remark 3.5. The last sentence of Theorem 3.4 is not contained in [26, Corollary 51.17, Theorem 53.4]. However it is a simple consequence of the theory developed in there. By scaling, we can assume $r = 1$ and $x = 0$. Fix a cone $\mathscr{C}$, a $\delta > 0$ and a sequence $\{u_k\} \subset \mathcal{M}(B_1)$ for which the left hand side of (3.2) goes to 0. If $\mathscr{C}$ is a segment, then it follows from [26] (or [7]) that there are uniform $C^{1,\alpha}$ bounds on $\overline{S_{u_k}} \cap B_{1-\delta}$. We can then use the Ascoli-Arzelà Theorem to conclude that $\overline{S_{u_k}}$ is converging in C^1 to $\mathscr{C}$.

In case the minimal cone $\mathscr{C}$ is a triple junction, then observe that $\mathscr{C} \cap (B_1 \setminus B_{\delta/2})$ consists of a three distinct segments at distance $\delta/2$ from each other. Covering each of these segments with balls of radius comparable to δ and centered in a point belonging to the segment itself, we can argue as above and conclude that, for k large enough, $\overline{S_{u_k}} \cap (B_{1-\delta} \setminus B_\delta)$ consist of three arcs, with uniform $C^{1,\alpha}$ estimates. Once again the Ascoli-Arzelà Theorem shows that $\overline{S_{u_k}} \cap (B_{1-\delta} \setminus B_\delta)$ is converging in C^1 to $\mathscr{C} \cap (B_{1-\delta} \setminus B_\delta)$.

Remark 3.6. Actually assumption (3.2) can be relaxed to

$$\int_{B_r(x)} |\nabla u|^2 \, dx \leq \varepsilon r$$

(see [26, Proposition 60.1] or Theorem 4.3 below).

Remarkably, Lemenant [51] extended such a result to the 3-dimensional case with suitable changes in the statement (see also [53] for a sketch of the proof).

The techniques developed by David are also capable to describe in details the structure of $\Omega \cap \overline{S_u}$ around points that corresponds to the model case of crack-tips despite uniqueness of blow ups is not ensured.

Theorem 3.7 (David, Theorem 69.29 [26]). *For all $\varepsilon_0 > 0$ there exists $\varepsilon > 0$ such that if $u \in \mathcal{M}(\Omega)$ and*

$$\operatorname{dist}_{\mathcal{H}}(\overline{S_u} \cap B_r(z), \sigma) < \varepsilon r$$

for some radius σ of $B_r(z) \subset \Omega$, then $\overline{S_u} \cap B_{r/2}(z)$ consists of a single connected arc which joins some point $y_0 \in B_{r/4}(z)$ with $\partial B_{r/2}(z)$ and which is smooth in $B_{r/2}(z) \setminus \{y_0\}$.

More precisely, there is a point $y_0 \in B_{r/4}(z)$ such that

$$\overline{S_u} \cap B_{r/2}(z) = \big\{ y_0 + \rho(\cos\alpha(\rho), \sin\alpha(\rho)) \big\}$$

for some smooth function $\alpha : (0, r/2) \to \mathbb{R}$ which satisfies

$$\rho|\alpha'(\rho)| \leq \varepsilon_0 \quad \text{for all } \rho \in (0, r/2), \qquad \lim_{\rho \downarrow 0} \rho|\alpha'(\rho)| = 0. \tag{3.3}$$

In addition, there is a constant C such that, up to a change of sign,

$$u\big(y_0+\rho(\cos\theta,\sin\theta)\big)=\sqrt{\frac{2}{\pi}\rho}\,\sin\left(\frac{\theta-\alpha(\rho)}{2}\right)+C+\rho^{1/2}\omega(\rho,\theta)$$

for all $\rho\in(0,r/2)$ *and* $\theta\in(\alpha(\rho)-\pi,\alpha(\rho)+\pi)$, *with* $\lim_{\rho\downarrow 0}\sup_\theta|\omega(\rho,\theta)|=0$. *Finally,*

$$\lim_{\rho\downarrow 0}\frac{1}{\rho}\int_{B_\rho(y_0)}|\nabla u|^2dx=\lim_{\rho\downarrow 0}\frac{1}{\rho}\mathcal{H}^1(\overline{S_u}\cap B_\rho(y_0))=1.$$

As remarked above, the latter theorem does not guarantee that such arc is C^1 *up to the loose end* y_0: in particular it leaves the possibility that the arc spirals infinitely many times around it. Hence, it does not establish the uniqueness of the blowup in the point y_0.

3.2 Higher integrability of the gradient and the Mumford and Shah conjecture

Theorem 3.1, or better the characterization of the *singular set* Σ_u in (3.1), can be employed to subdivide Σ_u as follows: $\Sigma_u=\Sigma_u^{(1)}\cup\Sigma_u^{(2)}\cup\Sigma_u^{(3)}$, where

$$\Sigma_u^{(1)}:=\{x\in\Sigma_u:\lim_{\rho\downarrow 0}\mathscr{D}_u(x,\rho)=0\},$$
$$\Sigma_u^{(2)}:=\{x\in\Sigma_u:\lim_{\rho\downarrow 0}\mathscr{A}_u(x,\rho)=0\},$$
$$\Sigma_u^{(3)}:=\{x\in\Sigma_u:\liminf_{\rho\downarrow 0}\mathscr{D}_u(x,\rho)>0,\ \liminf_{\rho\downarrow 0}\mathscr{A}_u(x,\rho)>0\}.$$

According to the Mumford and Shah Conjecture 2.20 we should have $\Sigma_u^{(3)}=\emptyset$ if $n=2$. Furthermore, being inspired by the 2d case, we shall refer to $\Sigma_u^{(1)}$ as the set of triple junctions, and to $\Sigma_u^{(2)}$ as the set of crack-tips. Actually, in 2-dimensions we will fully justify the latter terminology in Proposition 4.11 (see also Remark 4.10).

In the general n-dimensional setting De Giorgi conjectured that $\mathcal{H}^{n-2}(\Sigma_u\cap\Omega')<\infty$ for all $\Omega'\Subset\Omega$ (*cf.* [31, conjecture 6]). In particular, the validity of the latter conjecture would imply $\dim_{\mathcal{H}}\Sigma_u\le n-2$.

A first breakthrough in this direction has been obtained by Ambrosio, Fusco and Hutchinson in [4].

Theorem 3.8 (Ambrosio, Fusco and Hutchinson, [4]). *For every* $u\in\mathcal{M}(\Omega)$

$$\dim_{\mathcal{H}}\Sigma_u^{(1)}\le n-2.$$

Actually, the set $\Sigma_u^{(1)}$ turns out to be countable in 2-dimensions (see Remark 4.10 below).

In the same paper [4], Ambrosio, Fusco and Hutchinson investigated the connection between the higher integrability of ∇u and the Mumford and Shah conjecture.

If Conjecture 2.20 does hold, then $\nabla u \in L^p_{\mathrm{loc}}$ for all $p < 4$ (*cf.* with [4, Proposition 6.3] under $C^{1,1}$ regularity assumptions on $\overline{S_u}$, see also Theorem 3.11 below). It was indeed conjectured by De Giorgi in all space dimensions that $\nabla u \in L^p_{\mathrm{loc}}$ for all $p < 4$ (*cf.* with [31, Conjecture 1]). So far only a first step into this direction has been established.

Theorem 3.9. *There is $p > 2$ such that $\nabla u \in L^p_{\mathrm{loc}}(\Omega)$ for all $u \in \mathcal{M}(\Omega)$ and for all open sets $\Omega \subseteq \mathbb{R}^n$.*

A proof of Theorem 3.9 in 2-dimension has been given by De Lellis and Focardi in [35], shortly after De Philippis and Figalli established the result without any dimensional limitation in [37]. The exponent p in both papers is not explicitly computed despite some suggestions to do that are also proposed.

The higher integrability can be translated into an estimate for the size of the singular set Σ_u of $\overline{S_u}$ (see [4, Corollary 5.7]) that improves upon the conclusion of Theorem 3.1 (*cf.* [47,61,62] and the survey [63] for related issues for minima of variational integrals and for solutions to nonlinear elliptic systems in divergence form).

Theorem 3.10 (Ambrosio, Fusco and Hutchinson [4]). *If $u \in \mathcal{M}(\Omega)$ and $|\nabla u| \in L^p_{\mathrm{loc}}(\Omega)$ for some $p > 2$, then*

$$\dim_{\mathcal{H}} \Sigma_u \leq \max\{n-2, n-p/2\} \in (0, n-1). \tag{3.4}$$

The estimate $\dim_{\mathcal{H}} \Sigma_u < n-1$ has also been established by David [25] for $n = 2$, and lately by Rigot [71] and by Maddalena and Solimini [55] in general by establishing the porosity of $\Omega \cap \overline{S_u}$. Despite this, it was not related to the higher integrability property of the gradient. In Section 6 below we shall comment more in details on how porosity implies such an estimate and moreover on how porosity can be employed to prove the higher integrability of the gradients of minimizers following De Philippis and Figalli [37] (see Section 7).

For the time being few remarks are in order:

(i) the upper bound $p < 4$ is motivated not only because we need the rhs in the estimate (3.4) to be positive if $n = 2$, but also because explicit examples show that it is the best exponent one can hope for: consider in 2 dimensions a crack-tip minimizer (Bonnet and

David [11]), *i.e.* a function that up to a rigid motion can be written as

$$u(\rho, \theta) = C \pm \sqrt{\frac{2}{\pi}\rho} \cdot \sin(\theta/2)$$

for $\theta \in (-\pi, \pi)$ and $\rho > 0$, and some constant $C \in \mathbb{R}$. Simple calculations imply that crack-tip minimizers satisfy

$$|\nabla u| \in L^p_{\mathrm{loc}} \setminus L^4_{\mathrm{loc}}(\mathbb{R}^2) \quad \text{for all } p < 4;$$

(ii) If we were able to prove the higher integrability property for every $p < 4$ then we would infer that $\dim_{\mathcal{H}} \Sigma_u \leq n-2$, and actually in 2-dimensions $\dim_{\mathcal{H}} \Sigma_u = 0$. Clearly, this would be a big step towards the solution in positive of the Mumford and Shah conjecture. For further progress in this direction see Theorem 3.11 below.

Given Theorems 3.8 and 3.9 for granted, Theorem 3.10 is a simple consequence of soft measure theoretic arguments. We shall establish Theorem 3.8 in Section 4. Instead, here we prove Theorem 3.10 to underline the role of the higher integrability that, in turn, shall be established in Sections 5 and 7 following two different paths.

Proof of Theorem 3.10. Suppose that $|\nabla u| \in L^p_{\mathrm{loc}}(\Omega)$ for some $p > 2$, then for all $s \in (n - p/2, n-1)$ the set

$$\Lambda_s := \left\{ x \in \Omega : \limsup_{\rho} \rho^{-s} \int_{B_\rho(x)} |\nabla u|^p dy = \infty \right\}$$

satisfies $\mathcal{H}^s(\Lambda_s) = 0$ by elementary density estimates for the Radon measure

$$\mu(A) := \int_A |\nabla u|^p dy \quad A \subseteq \Omega \text{ open subset.}$$

Indeed, for all $\delta > 0$, Proposition 2.9 gives that

$$\delta\, \mathcal{H}^s(\Lambda_s) \leq \mu(\Lambda_s) \leq \mu(\Omega) < \infty.$$

Therefore, $\mathcal{H}^s(\Lambda_s) = 0$.

Hence, if we rewrite Σ_u as the disjoint union of $\Sigma_u \cap \Lambda_s$ and of $\Sigma_u \setminus \Lambda_s$, we deduce $\mathcal{H}^s(\Sigma_u \cap \Lambda_s) = 0$ and thus the estimate $\dim_{\mathcal{H}}(\Sigma_u \cap \Lambda_s) \leq s$.

Furthermore, it is easy to prove that $\Sigma_u \setminus \Lambda_s \subseteq \Sigma_u^{(1)}$. If $x \in \Sigma_u \setminus \Lambda_s$ by the higher integrability and Hölder inequality it follows that

$$\begin{aligned} \mathscr{D}_u(x, \rho) &= \rho^{1-n} \int_{B_\rho(x)} |\nabla u|^2 dy \\ &\leq \omega_n^{1-\frac{2}{p}} \rho^{\frac{2}{p}\left(s-n+\frac{p}{2}\right)} \left(\rho^{-s} \int_{B_\rho(x)} |\nabla u|^p dy \right)^{\frac{2}{p}} \overset{\rho \downarrow 0^+}{\longrightarrow} 0, \end{aligned}$$

since $s > n - p/2$. By taking into account Theorem 3.8 we have that $\dim_{\mathcal{H}}(\Sigma_u \setminus \Lambda_s) \leq n - 2$.

In conclusion, we infer that for all $s \in (n - p/2, n - 1)$

$$\dim_{\mathcal{H}} \Sigma_u = \max\{\dim_{\mathcal{H}}(\Sigma_u \cap \Lambda_s), \dim_{\mathcal{H}}(\Sigma_u \setminus \Lambda_s)\} \leq \max\{n - 2, s\},$$

by letting $s \downarrow (n - p/2)$ we are done. □

3.3 An energetic characterization of the Mumford and Shah Conjecture 2.21

Let us go back to the upper bound $p < 4$ in the higher integrability result. We consider again the crack-tip function in item (i) after Theorem 3.10. A simple calculation shows that its gradient belongs to $L^p_{\text{loc}} \setminus L^4_{\text{loc}}(\mathbb{R}^2)$ for all $p < 4$. Beyond the scale of L^p spaces something better holds true: $|\nabla u| \in L^{4,\infty}_{\text{loc}}(\mathbb{R}^2)$. The latter is a weak-Lebesgue space, *i.e.* if $U \subseteq \mathbb{R}^2$ is open then $f \in L^{4,\infty}_{\text{loc}}(U)$ if and only if for all $U' \Subset U$ there exists $K = K(U') > 0$ such that

$$\mathcal{L}^2\big(\{x \in U' : |f(x)| > \lambda\}\big) \leq K\lambda^{-4} \quad \text{for all } \lambda > 0.$$

As a side effect of the considerations in [35] one deduces an energetic characterization of the modified Mumford and Shah Conjecture 2.21 (see [35, Proposition 5]). Indeed, the validity of the latter is equivalent to a sharp integrability property of the gradient of the minimizers.

Theorem 3.11 (De Lellis and Focardi [35]). *If $\Omega \subseteq \mathbb{R}^2$, Conjecture 2.21 is true for $u \in \mathcal{M}(\Omega)$ if and only if $\nabla u \in L^{4,\infty}_{\text{loc}}(\Omega)$.*

The characterization above would hold for the original Mumford and Shah Conjecture 2.20 if C^1 regularity of $\overline{S_u}$ up to crack-tip points had been established.

To prove Theorem 3.11 we need the following preliminary observation.

Lemma 3.12. *Let $f \in L^{4,\infty}_{\text{loc}}(\Omega)$, $\Omega \subseteq \mathbb{R}^2$, then for all $\varepsilon > 0$ the set*

$$D_\varepsilon := \left\{x \in \Omega : \liminf_r \frac{1}{r} \int_{B_r(x)} f^2(y)\, dy \geq \varepsilon\right\} \tag{3.5}$$

is locally finite.

Proof. We shall show in what follows that if $f \in L^{4,\infty}(\Omega)$ then D_ε is finite, an obvious localization argument then proves the general case.

Let $\varepsilon > 0$ and consider the set D_ε in (3.5) above. First note that, for any $B_r(x) \subset \Omega$ and any $\lambda > 0$ we have the estimate

$$\begin{aligned}\int_{\{y\in B_r(x):|f(y)|\geq\lambda\}} f^2(y)\,dy &\leq \int_{\{y\in\Omega:|f(y)|\geq\lambda\}} f^2(y)\,dy \\ &= 2\int_\lambda^{+\infty} t\,\mathcal{L}^2\big(\{y\in\Omega : |f(y)|\geq t\}\big)dt \quad (3.6)\\ &\leq \int_\lambda^{+\infty}\frac{2K}{t^3}dt = \frac{K}{\lambda^2}.\end{aligned}$$

If $x \in D_\varepsilon$ and $r > 0$ satisfy

$$\int_{B_r(x)} f^2(y)\,dy \geq \frac{\varepsilon}{2}r, \tag{3.7}$$

choosing $\lambda = 2(K/r\varepsilon)^{1/2}$ in (3.6) we conclude

$$\int_{\{y\in B_r(x):|f(y)|<2(\frac{K}{r\varepsilon})^{1/2}\}} f^2(y)\,dy \geq \frac{\varepsilon}{4}r. \tag{3.8}$$

Furthermore, the trivial estimate

$$\int_{\{y\in B_r(x):|f(y)|<\lambda\}} f^2(y)\,dy < \pi\lambda^2 r^2,$$

implies for $\lambda = (\varepsilon/8\pi r)^{1/2}$

$$\int_{\{y\in B_r(x):|f(y)|<(\frac{\varepsilon}{8\pi r})^{1/2}\}} f^2(y)\,dy < \frac{\varepsilon}{8}r. \tag{3.9}$$

By collecting (3.8) and (3.9) we infer

$$\int_{\{y\in B_r(x):(\frac{\varepsilon}{8\pi r})^{1/2}\leq|f(y)|<2(\frac{K}{r\varepsilon})^{1/2}\}} f^2(y)\,dy \geq \frac{\varepsilon}{8}r,$$

that in turn implies

$$\mathcal{L}^2\left(\left\{y\in B_r(x) : |f(y)| \geq \left(\frac{\varepsilon}{8\pi r}\right)^{1/2}\right\}\right) \geq \frac{\varepsilon^2 r^2}{32K}. \tag{3.10}$$

Let $\{x_1,\ldots,x_N\} \subseteq D_\varepsilon$ and $r > 0$ be a radius such that the balls $B_r(x_i) \subseteq \Omega$ are disjoint and (3.7) holds for each x_i. Then, from (3.10) and the fact that $f \in L^{4,\infty}(\Omega)$, we infer

$$\begin{aligned}N\frac{\varepsilon^2 r^2}{32K} &\leq \mathcal{L}^2\left(\left\{y\in\Omega : |f(y)| \geq \left(\frac{\varepsilon}{8\pi r}\right)^{1/2}\right\}\right)\\ &\leq \frac{K(8\pi r)^2}{\varepsilon^2} \Longrightarrow N \leq \frac{2^{11}K^2\pi^2}{\varepsilon^4},\end{aligned}$$

and the conclusion follows at once. □

We are now ready to give the proof of Theorem 3.11. The "if" direction is achieved by first proving that $\overline{S_u}$ has locally finitely many connected components and then invoking the regularity theory developed by Bonnet [10]. In turn, the proof that the connected components are locally finite is a fairly simple application of David's ε-regularity theory (see Theorem 3.4). Vice versa, the "only if" direction is proved by means of Bonnet blow up analysis and standard elliptic regularity theory.

Proof of Theorem 3.11. To prove the direct implication we assume without loss of generality that $\Omega = B_R$ for some $R > 1$, being the result local. In addition, we may also suppose that $\overline{S_u} \cap \partial B_1 = \{y_1, \ldots, y_M\}$. Theorem 3.4 and Theorem 4.3 in Section 4 below yield that there exists some $\varepsilon_0 > 0$ such that for all points $x \in B_R \setminus D_{\varepsilon_0}$ the set $\overline{S_u} \cap B_r(x)$ is either empty or diffeomorphic to a minimal cone, for some $r > 0$. In particular, in the latter event $B_r(x) \setminus \overline{S_u}$ is not connected.

Supposing that $D_{\varepsilon_0} \cap B_1 = \{x_1, \ldots, x_N\}$, and setting

$$\Omega_k := B_{1-1/k} \setminus \bigcup_{i=1}^{N} B_{1/k}(x_i),$$

a covering argument and the last remark give that for every $x \in \Omega_k \cap \overline{S_u}$ there is a continuous arc $\gamma_k : [0,1] \to \overline{S_u}$ with $\gamma_k(0) = x$ and $\gamma_k(1) = y \in \partial\Omega_k$. Then, the sequence $(\widetilde{\gamma}_k)_{k\in\mathbb{N}}$ of reparametrizations of the γ_k's by arc length converges to some arc $\gamma : [0,1] \to \overline{S_u}$ with $\gamma(0) = x$ and $\gamma(1) \in \{x_1, \ldots, x_N, y_1, \ldots, y_M\}$.

From this, we deduce that $\overline{B_1} \cap \overline{S_u}$ has a finite number of connected components. Bonnet's regularity results [10, Theorems 1.1 and 1.3] then provide the thesis.

To conclude we prove the opposite implication. To this aim we consider $\Omega' \Subset \Omega'' \Subset \Omega$ and suppose that $\overline{S_u} \cap \Omega''$ is a finite union of C^1 arcs of finite length. Denote by $\{x_1, \ldots, x_N\}$ the end points of the arcs in Ω' and let $r > 0$ be such that $B_{4r}(x_i) \subseteq \Omega'$ for all i, and $B_{4r}(x_i) \cap B_{4r}(x_j) = \emptyset$ if $i \neq j$. Theorem 3.3 implies that ∇u has a $C^{0,\alpha}$ extension on both sides of $(\Omega'' \cap \overline{S_u}) \setminus \cup_i \overline{B_r(x_i)}$ for all $\alpha < 1$. In particular, ∇u is bounded on $\overline{\Omega'} \setminus \cup_i B_{2r}(x_i)$.

Next consider the sequence $r_k = r/2^{k-1}$, $k \geq 0$, and fix $i \in \{1, \ldots, N\}$. Then, by [26, Proposition 37.8] (or [10, Theorem 2.2]) we can extract a subsequence $k_j \uparrow \infty$ along which the blow up functions $u_j(x) := r_{k_j}^{-1/2}\big(u(x_i + r_{k_j}x) - c_j(x)\big)$ converge to some w in $W^{1,2}_{\mathrm{loc}}(B_4 \setminus K)$, for some piecewise constant function $c_j : \Omega \setminus \overline{S_{u_j}} \to \mathbb{R}$, and $(\overline{S_{u_j}})_{j\in\mathbb{N}}$ converges to some set K in the Hausdorff metric.

By Bonnet's blow up theorem [10, Theorem 4.1] only two possibilities occur: either x_i is a triple junction point, *i.e.*, K is a triple junction and w is locally constant on $B_4 \setminus K$, or x_i is a crack-tip, *i.e.*, up to a rotation $K = \{(x,0) : x \leq 0\}$ and $w(\rho,\theta) = C \pm \sqrt{\frac{2}{\pi}\rho} \cdot \sin(\theta/2)$ for $\theta \in (-\pi,\pi)$, $\rho > 0$ and some constant $C \in \mathbb{R}$ (note that in this argument we do not need to know that the blow up limit is unique).

In both cases, we claim that ∇u_j has a $C^{0,\alpha}$ extension on the closure of each connected component of $U_j := (B_3 \setminus \overline{B_1}) \setminus \overline{S_{u_j}}$ with $\sup_j \|\nabla u_j\|_{L^\infty(U_j)} \leq C$. This follows as in [7, Theorem 7.49] (or [26, Proposition 17.15], see also Remark 3.5) locally straightening $\overline{S_{u_j}} \cap (B_4 \setminus \overline{B_{1/2}})$ onto $K \cap (B_4 \setminus \overline{B_{1/2}})$ via a $C^{1,\alpha}$ conformal map, a reflection argument and standard Schauder estimates for the laplacian. Scaling back the previous estimate gives

$$|\nabla u(x)| \leq C\,|x - x_i|^{-1/2} \qquad \text{for } x \in \cup_{j \in \mathbb{N}} (\overline{B_{3r_{k_j}}(x_i)} \setminus B_{r_{k_j}}(x_i)),$$

in turn from this, the maximum principle and Hopf's lemma we infer

$$|\nabla u(x)| \leq C\, r_k^{-1/2} \qquad \text{for } x \in B_{2r}(x_i) \setminus B_{r_k}(x_i).$$

The latter inequality finally implies $\nabla u \in L^{4,\infty}(B_{2r}(x_i))$.

Eventually, we are able to conclude $\nabla u \in L^{4,\infty}(\Omega')$, being on one hand ∇u bounded on $\overline{\Omega'} \setminus \cup_i B_{2r}(x_i)$, and on the other hand belonging to $L^{4,\infty}(\cup_i B_{2r}(x_i))$. □

4 Hausdorff dimension of the set of triple-junctions

In order to prove Theorem 3.8 we need to analyze the asymptotic behavior of MS minimizer in points of vanishing Dirichlet energy. This issue has been first investigated in [4, Proposition 5.3, Theorem 5.4]. Those results hinge upon the notion of *Almgren's area minimizing sets*, *i.e.* a $\mathcal{H}^{n-1}$ rectifiable set $S \subset B_1$ such that

$$\mathcal{H}^{n-1}(S) \leq \mathcal{H}^{n-1}(\varphi(S)), \quad \forall \varphi \in \mathrm{Lip}(\mathbb{R}^n, \mathbb{R}^n), \ \{\varphi \neq \mathrm{Id}\} \Subset B_1.$$

Following this approach to infer Theorem 3.8 requires a delicate study of the behavior of the composition of SBV functions with Lipschitz deformations that are not necessarily one-to-one, and some specifications on the regularity theory for Almgren's area minimizing sets are needed (*cf.* [4]). Therefore, following Ambrosio, Fusco and Hutchinson, Theorem 3.10 is a straightforward corollary of a much deeper and technically demanding result (given the higher integrability for granted).

Instead, in Theorem 4.3 below (*cf.* [35, Proposition 5.1]) we set the analysis into the more natural framework of Caccioppoli partitions.

Definition 4.1. A Caccioppoli partition of Ω is a countable partition $\mathscr{E} = \{E_i\}_{i=1}^{\infty}$ of Ω in sets of (positive Lebesgue measure and) finite perimeter with $\sum_{i=1}^{\infty} \mathrm{Per}(E_i, \Omega) < \infty$.

For each Caccioppoli partition $\mathscr{E}$ the *set of interfaces* is given by $J_{\mathscr{E}} := \bigcup_i \partial^* E_i$.

The partition $\mathscr{E}$ is said to be *minimal* if

$$\mathcal{H}^{n-1}(J_{\mathscr{E}}) \leq \mathcal{H}^{n-1}(J_{\mathscr{F}})$$

for all Caccioppoli partitions $\mathscr{F}$ for which $\sum_{i=1}^{\infty} \mathcal{L}^n\big((F_i \triangle E_i) \cap (\Omega \setminus \Omega')\big) = 0$, for some open subset $\Omega' \Subset \Omega$.

There is an important correspondence between Caccioppoli partitions and the subspace SBV_0 of "piecewise constant" SBV functions recalled as prototype example in Subsection 2.1, in such a way that minimizing the Mumford and Shah energy over SBV_0 corresponds exactly to the minimal area problem for Caccioppoli partitions (see [7, Theorems 4.23, 4.25 and 4.39]).

Existence of minimal Caccioppoli partitions is guaranteed by Ambrosio's SBV closure and compactness Theorem 2.3 without imposing any L^∞ bound simply by composition with $\arctan(t)$, provided the partitions are either equi-finite or ordered, *i.e.* if $\mathscr{E} = \{E_i\}_{i=1}^{\infty}$ then $\mathcal{L}^n(E_i) \geq \mathcal{L}^n(E_j)$ for $j \geq i$.

A regularity theory for minimal Caccioppoli partitions has been established by Massari and Tamanini [58]. We limit ourselves here to the ensuing statement.

Theorem 4.2 (Massari and Tamanini [58]). *Let $\mathscr{E}$ be a minimal Caccioppoli partition in Ω,*

$$\omega_{n-1} \leq \liminf_{r \downarrow 0} \frac{\mathcal{H}^{n-1}(J_{\mathscr{E}} \cap B_r(x))}{r^{n-1}} \leq \limsup_{r \downarrow 0} \frac{\mathcal{H}^{n-1}(J_{\mathscr{E}} \cap B_r(x))}{r^{n-1}} \leq n\omega_n.$$

In particular, $J_{\mathscr{E}}$ is essentially closed, i.e. $\mathcal{H}^{n-1}\big((\Omega \cap \overline{J_{\mathscr{E}}}) \setminus J_{\mathscr{E}}\big) = 0$.

Moreover, there exists a relatively closed subset $\Sigma_{\mathscr{E}}$ of $J_{\mathscr{E}}$ such that $J_{\mathscr{E}} \setminus \Sigma_{\mathscr{E}}$ is a $C^{1,1/2}$ hypersurface and $\dim_{\mathcal{H}} \Sigma_{\mathscr{E}} \leq n-2$. If, in addition, $n = 2$, then $\Sigma_{\mathscr{E}}$ is locally finite.

We are now ready to prove a compactness result for sequences of MS-minimizers with vanishing L^1-gradient energy.

Theorem 4.3 (De Lellis and Focardi [35]). *Let $(u_k)_{k \in \mathbb{N}} \subset \mathcal{M}(B_1)$ be such that*

$$\lim_k \|\nabla u_k\|_{L^1(B_1)} = 0. \tag{4.1}$$

Then, (up to the extraction of a subsequence not relabeled for convenience) there exists a minimal Caccioppoli partition $\mathscr{E} = \{E_i\}_{i\in\mathbb{N}}$ such that $(\overline{S_{u_k}})_{k\in\mathbb{N}}$ converges locally in the Hausdorff distance on $\overline{B}_1$ to $\overline{J_{\mathscr{E}}}$ and for all open sets $A \subseteq B_1$

$$\lim_k \mathrm{MS}(u_k, A) = \lim_k \mathcal{H}^{n-1}(S_{u_k} \cap A) = \mathcal{H}^{n-1}(J_{\mathscr{E}} \cap A). \tag{4.2}$$

Though this last statement is, intuitively, quite clear, it is technically demanding, because we do not have any a priori control of the norms $\|u_k\|_{L^1}$, thus preventing the use of Ambrosio's $(G)SBV$ compactness theorem. We can not even expect to gain pre-compactness via De Giorgi's SBV Poincaré-Wirtinger type inequality, since the latter holds true in a regime of small jumps rather than of small gradients as the current one.

Proof of Theorem 4.3. The sequence $(u_k)_{k\in\mathbb{N}}$ does not satisfy, a priori, any L^p bound, thus in order to gain some insight on the asymptotic behavior of the corresponding jump sets we first construct a new sequence $(w_k)_{k\in\mathbb{N}}$ with null gradients introducing an infinitesimal error on the length of the jump set of w_k with respect to that of u_k. Then, we investigate the limit behavior of the corresponding Caccioppoli partitions.
Step 1. *There exists a sequence $(w_k)_{k\in\mathbb{N}} \subseteq SBV(B_1)$ satisfying*

(i) $\nabla w_k = 0$ $\mathcal{L}^n$ *a.e. on* B_1,
(ii) $\|u_k - w_k\|_{L^\infty(B_1)} \leq 2\|\nabla u_k\|^{1/2}_{L^1(B_1)}$,
(iii) $\mathcal{H}^{n-1}\left(S_{w_k} \setminus (S_{u_k} \cup H_k)\right) = 0$ *for some Borel measurable set* H_k, *with* $\mathcal{H}^{n-1}(H_k) = o(1)$ *as* $k \uparrow \infty$.

Note that in turn item (iii) implies that

$$\mathrm{MS}(w_k) = \mathcal{H}^{n-1}(S_{w_k}) \leq \mathcal{H}^{n-1}(S_{u_k}) + o(1) \leq \mathrm{MS}(u_k) + o(1). \tag{4.3}$$

In Step 2 below we shall eventually show that $|\mathrm{MS}(w_k) - \mathrm{MS}(u_k)| \leq o(1)$.

Recall that the BV Co-Area formula (see [7, Theorem 3.40]) establishes

$$\int_{B_1} |\nabla u_k| dx = |Du_k|(B_1 \setminus S_{u_k}) = \int_{\mathbb{R}} \mathrm{Per}\left(\{u_k \geq t\} \setminus S_{u_k}\right) dt. \tag{4.4}$$

Denote by I_i^k a partition of $\mathbb{R}$ of intervals of equal length $\|\nabla u_k\|^{1/2}_{L^1(B_1)}$. Equation (4.4) and the Mean value Theorem provide the existence of levels $t_i^k \in I_i^k$ satisfying

$$\sum_{i=1}^{\infty} \mathrm{Per}\left(\{u_k \geq t_i^k\} \setminus S_{u_k}\right) \leq \|\nabla u_k\|^{1/2}_{L^1(B_1)}. \tag{4.5}$$

Then define the functions w_k to be equal to t_i^k on $\{u_k \geq t_i^k\} \setminus \{u_k \geq t_{i+1}^k\}$. The choice of the I_i^k's, (4.5) and the very definition yield that w_k belongs to $SBV(B_1)$ and that it satisfies properties (i) and (ii). To conclude, note that $\mathcal{H}^{n-1}\left(S_{w_k} \setminus (\cup_i \partial^*\{u_k \geq t_i^k\} \cup S_{u_k})\right) = 0$ by construction, thus item (iii) follows at once from (4.5).

Step 2. *Compactness for the jump sets.*
Each function w_k determines a Caccioppoli partition $\mathscr{E}_k = \{E_i^k\}_{i\in\mathbb{N}}$ of B_1 (see [18, Lemma 1.11]). In addition, upon reordering the sets E_i^k's, we may assume that $\mathcal{L}^n(E_i^k) \geq \mathcal{L}^n(E_j^k)$ if $i < j$. Then, the compactness theorem for Caccioppoli partitions (see [54, Theorem 4.1, Proposition 3.7] and [7, Theorem 4.19]) provides us with a subsequence (not relabeled) and a Caccioppoli partition $\mathscr{E} := \{E_i\}_{i\in\mathbb{N}}$ such that

$$\begin{aligned} &\lim_j \sum_{i=1}^{\infty} \mathcal{L}^n(E_i^k \triangle E_i) = 0, \\ &\text{and } \sum_{i=1}^{\infty} \operatorname{Per}(E_i, A) \leq \liminf_k \sum_{i=1}^{\infty} \operatorname{Per}(E_i^k, A) \end{aligned} \tag{4.6}$$

for all open subsets A in B_1. We claim that $\mathscr{E}$ determines a minimal Caccioppoli partition and in proving this we will also establish (4.2).

We start off observing that the first identity (4.6) and the Co-Area formula yield the existence of a set $I \subset (0,1)$ of full measure such that

$$\liminf_k \sum_{i=1}^{\infty} \mathcal{H}^{n-1}\left((E_i^k \triangle E_i) \cap \partial B_\rho\right) = 0 \qquad \forall \rho \in I. \tag{4.7}$$

Define the measures μ_k as $\mu_k(A) := \operatorname{MS}(u_k, A) + \operatorname{MS}(w_k, A)$ (A being an arbitrary Borel subset of B_1). Proposition 2.14 and item (iii) in Step 1 ensure that, upon the extraction of a further subsequence, μ_k converges weakly* to a finite measure μ on B_1. W.l.o.g. we may assume that for all $\rho \in I$ we have, in addition, $\mu(\partial B_\rho) = 0$.

Let us now fix a Caccioppoli partition $\mathscr{F} := \{F_i\}_{i\in\mathbb{N}}$ suitable to test the minimality of $\mathscr{E}$, *i.e.* $\sum_{i=1}^{\infty} \mathcal{L}^n\left((F_i \triangle E_i) \cap (B_1 \setminus \overline{B_t})\right) = 0$ for some $t \in (0,1)$. Moreover, we may also suppose that $\sum_{i=1}^{\infty} \mathcal{H}^{n-1}\left((F_i \triangle E_i) \cap \partial B_\rho\right) = 0$ for all $\rho \in I \cap (t,1)$. Let then ρ and r be radii in $I \cap (t,1)$ with $\rho < r$ and assume, after passing to a subsequence (not relabeled) that the inferior limit in (4.7) is actually a limit for these two radii. We define

$$\omega_k := \begin{cases} w_k & \text{on } B_1 \setminus \overline{B_\rho} \\ t_i^k & \text{on } F_i \cap B_\rho. \end{cases}$$

Note that $\omega_k \in SBV(B_1)$ with $\nabla\omega_k = 0$ $\mathcal{L}^n$ a.e. on B_1, and since $t < \rho \in I$ it follows

$$\mathcal{H}^{n-1}\Big(S_{\omega_k}\setminus\big((J_{\mathscr{F}}\cap B_\rho)\cup(\cup_{i\in\mathbb{N}}(E_i^k\Delta E_i)\cap\partial B_\rho)\cup(S_{w_k}\cap(B_1\setminus\overline{B_\rho}))\big)\Big)=0.$$

Consider $\varphi \in \mathrm{Lip}\cap C_c(B_1,[0,1])$ with $\varphi|_{B_r}\equiv 1$, and $|\nabla\varphi|\le(1-r)^{-1}$ on B_1, and set $v_k := \varphi\,\omega_k + (1-\varphi)\,u_k$. Clearly, v_k is admissible to test the minimality of u_k. Then, simple calculations lead to

$$\begin{aligned}
&\mathrm{MS}(u_k)\le \mathrm{MS}(v_k)\\
&\le \mathrm{MS}(\omega_k)+2\mathrm{MS}(u_k,B_1\setminus\overline{B_r})+\frac{2}{(1-r)^2}\|u_k-\omega_k\|^2_{L^2(B_1\setminus\overline{B_r})}\\
&\le \mathcal{H}^{n-1}\left(J_{\mathscr{F}}\right)+\sum_{i\in\mathbb{N}}\mathcal{H}^{n-1}\left((E_i^k\Delta E_i)\cap\partial B_\rho\right)+\mathcal{H}^{n-1}\left(S_{w_k}\setminus\overline{B_\rho}\right)\\
&\quad+2\mathrm{MS}(u_k,B_1\setminus\overline{B_r})+\frac{2}{(1-r)^2}\|u_k-w_k\|^2_{L^2(B_1\setminus\overline{B_r})}\\
&\le \mathcal{H}^{n-1}\left(J_{\mathscr{F}}\right)+\sum_{i\in\mathbb{N}}\mathcal{H}^{n-1}\left((E_i^k\Delta E_i)\cap\partial B_\rho\right)+3\mu_k(B_1\setminus\overline{B_\rho})\\
&\quad+\frac{2}{(1-r)^2}\|u_k-w_k\|^2_{L^\infty(B_1)}.
\end{aligned}\tag{4.8}$$

Note that in the third inequality we have used that ω_k and w_k coincide on $B_1\setminus\overline{B_\rho}$, and that $\rho<r$. By letting $k\uparrow\infty$ in (4.8), we infer

$$\begin{aligned}
\mathcal{H}^{n-1}(J_{\mathscr{E}})&\le\liminf_j\mathcal{H}^{n-1}(S_{u_k})\le\liminf_j\mathrm{MS}(u_k)\le\limsup_j\mathrm{MS}(u_k)\\
&\le\limsup_k\mathrm{MS}(v_k)\le\mathcal{H}^{n-1}\left(J_{\mathscr{F}}\right)+3\mu(B_1\setminus\overline{B_\rho}),
\end{aligned}$$

where we have used that r and ρ belong to I, inequality (4.3), the convergence $\mu_k \rightharpoonup^* \mu$, the estimate $\sup_k \mathrm{MS}(\mu_k B_1\setminus B_t)\le n\omega_n(1-t^{n-1})$ $\forall t\in(0,1)$, that is derived as inequality (2.22) in Proposition 2.14, and the limit (4.7). Finally, by letting $\rho\in I$ tend to 1^- we conclude

$$\begin{aligned}
\mathcal{H}^{n-1}(J_{\mathscr{E}})&\le\liminf_k\mathcal{H}^{n-1}(S_{u_k})\le\liminf_k\mathrm{MS}(u_k)\\
&\le\limsup_k\mathrm{MS}(u_k)\le\mathcal{H}^{n-1}\left(J_{\mathscr{F}}\right),
\end{aligned}\tag{4.9}$$

which proves the minimality of $\mathscr{E}$. In addition, choosing $\mathscr{E}=\mathscr{F}$, we infer (4.2) for $A=B_1$. Actually, for $\mathscr{E}=\mathscr{F}$ the same same argument employed above gives (4.2) (it suffices to take $v_k=\varphi w_k+(1-\varphi)u_k$).

In particular, $J_{\mathscr{E}}$ is essentially closed (by Theorem 4.2) and it satisfies a density lower bound estimate. Using this and the De Giorgi, Carriero, Leaci density lower bound in formula (2.20) we conclude that $(\overline{S_{u_k}})_{k\in\mathbb{N}}$ converges to $\overline{J_{\mathscr{E}}}$ in the local Hausdorff topology on $\overline{B_1}$. □

Interesting (immediate) consequences of Theorem 4.3 are contained in the ensuing two statements.

Corollary 4.4. *Let $(u_k)_{k\in\mathbb{N}} \subset \mathcal{M}(B_1)$ be as in the statement of Theorem* 4.3*, then*

$$\lim_k \|\nabla u_k\|_{L^2(B_1)} = 0, \quad \textit{and} \quad \mathcal{H}^{n-1} \llcorner S_{u_k} \stackrel{*}{\to} \mathcal{H}^{n-1} \llcorner J_{\mathscr{E}}.$$

Actually, $\mathcal{H}^{n-1} \llcorner S_{u_k} \to \mathcal{H}^{n-1} \llcorner J_{\mathscr{E}}$ in the narrow convergence of measures, i.e. *in the duality with $C_b(\Omega)$.*

Proof. It is an easy consequence of the equalities in (4.2). □

Corollary 4.5. *Let $x \in \Sigma_u^{(1)}$ and $\rho_k \downarrow 0$, then (up to subsequences not relabeled) there exists a minimal Caccioppoli partition $\mathscr{E}$ such that $(S_{u_{x,\rho_k}})_{k\in\mathbb{N}}$, u_{x,ρ_k} defined in* (2.8)*, converges locally in the Hausdorff distance to $\overline{J_{\mathscr{E}}}$ and*

$$\mathcal{H}^{n-1} \llcorner S_{u_{x,\rho_k}} \stackrel{*}{\to} \mathcal{H}^{n-1} \llcorner J_{\mathscr{E}}.$$

Remark 4.6. Under the assumptions of Corollary 4.5, it is natural to expect the limit partition $\mathscr{E}$ to be *conical, i.e.* $\mathscr{E} = \{E_i\}_{i=1}^{\infty}$ with the E_i's cones with vertices in the origin, as a result of the blow up procedure. In general this latter property can be proven only for suitable sequences $\rho_k \downarrow 0$ by combining a blow up argument and Theorem 4.3 (*cf.* [4, Proposition 5.8]).

Actually, in 2-dimensions the result is true for every sequence $\rho_k \downarrow 0$ since a structure theorem for minimal Caccioppoli partitions assures that (locally) they are minimal connections (*cf.* Proposition 5.3). The lack of monotonicity formulas for the Mumford and Shah problem prevents the derivation of such a statement in the general case.

A more precise result in the 2d case will be established in Proposition 4.11. Note that no uniqueness is ensured for the limits except for 2d in view of David's ε-regularity result (see Theorem 3.4), and in 3d in view of the analogous result established by Lemenant in [51].

By means of Theorems 4.2, 4.3 and standard blow up arguments we are able to establish Theorem 3.8. Let us first recall a technical lemma.

Lemma 4.7 (Section 3.6 [74]). *Let $s \geq 0$, then*

(i) $\mathcal{H}^s(\Sigma) = 0 \iff \mathcal{H}^{s,\infty}(\Sigma) = 0$, *for all sets $\Sigma \subseteq \mathbb{R}^n$;*
(ii) *if Σ_j and Σ are compact sets such that* $\sup_{\Sigma_j} \mathrm{dist}(\cdot, \Sigma) \downarrow 0$ *as $j \uparrow \infty$, then*

$$\mathcal{H}^{s,\infty}(\Sigma) \geq \limsup_j \mathcal{H}^{s,\infty}(\Sigma_j);$$

(iii) *if* $\mathcal{H}^s(\Sigma) > 0$, *then for* $\mathcal{H}^s$*-a.e.* $x \in \Sigma$

$$\limsup_{\rho\downarrow 0+} \frac{\mathcal{H}^{s,\infty}(\Sigma \cap B_\rho(x))}{\omega_s \rho^s} \geq 2^{-s}.$$

The strategy of proof below is essentially that by Ambrosio, Fusco and Hutchinson as reworked by De Lellis, Focardi and Ruffini in light of [36, Theorem 4.2].

Proof of Theorem 3.8. We argue by contradiction: suppose that there exists $s > n-2$ such that $\mathcal{H}^s\left(\Sigma_u^{(1)}\right) > 0$. From this we infer that $\mathcal{H}^{s,\infty}\left(\Sigma_u^{(1)}\right) > 0$, and moreover that for $\mathcal{H}^s$-a.e. $x \in \Sigma_u^{(1)}$ it holds

$$\limsup_{\rho\downarrow 0+} \frac{\mathcal{H}^{s,\infty}(\Sigma_u^{(1)} \cap B_\rho(x))}{\rho^s} \geq \frac{\omega_s}{2^s} \tag{4.10}$$

(see for instance [7, Theorem 2.56 and formula (2.43)]). Without loss of generality, suppose that (4.10) holds at $x = 0$, and consider a sequence $\rho_k \downarrow 0$ for which

$$\mathcal{H}^{s,\infty}(\Sigma_u^{(1)} \cap B_{\rho_k}) \geq \frac{\omega_s}{2^{s+1}} \rho_k^s \qquad \text{for all } k \in \mathbb{N}. \tag{4.11}$$

Theorem 4.3 provides a subsequence, not relabeled for convenience, and a minimal Caccioppoli partition $\mathscr{E}$ such that

$$\lim_k \mathcal{H}^{n-1}(S_{u_k} \cap A) = \mathcal{H}^{n-1}(J_{\mathscr{E}} \cap A) \quad \text{for all open sets } A \subseteq B_1. \tag{4.12}$$

and that

$$\rho_k^{-1}\overline{S_u} \to \overline{J_{\mathscr{E}}} \text{ locally Hausdorff.} \tag{4.13}$$

In turn, from the latter we claim that if $\mathcal{F}$ is any open cover of $\Sigma_{\mathscr{E}} \cap \overline{B}_1$, then for some $h_0 \in \mathbb{N}$

$$\rho_k^{-1}\Sigma_u^{(1)} \cap \overline{B}_1 \subseteq \cup_{F\in\mathcal{F}} F \qquad \text{for all } k \geq k_0. \tag{4.14}$$

Indeed, if this is not the case we can find a sequence $x_{k_j} \in \rho_{k_j}^{-1}\Sigma_u^{(1)} \cap \overline{B}_1$ converging to some point $x_0 \notin \Sigma_{\mathscr{E}}$. If $T_{x_0}^{\mathscr{E}}$ is the tangent plane to $J_{\mathscr{E}}$ at x_0 (which exists by the property of $\Sigma_{\mathscr{E}}$ in Theorem 4.2), then for some ρ_0 we have

$$\rho^{-1-n} \int_{B_\rho(x_0)\cap J_{\mathscr{E}}} \operatorname{dist}^2(y, T_{x_0}^{\mathscr{E}}) d\mathcal{H}^{n-1} < \varepsilon_0, \qquad \text{for all } \rho \in (0, \rho_0).$$

In turn, from the latter inequality and the convergence in (4.12), it follows that, for $\rho \in (0, \rho_0 \wedge 1)$,

$$\limsup_{j\uparrow\infty} \rho^{-1-n} \int_{B_\rho(x_{k_j})\cap\rho_{k_j}^{-1}S_u} \mathrm{dist}^2(y, T_{x_0}^{\mathscr{E}})\, d\mathcal{H}^{n-1} < \varepsilon_0.$$

Therefore, as $x_{k_j} \in \rho_{k_j}^{-1}\Sigma_u^{(1)}$, we get for j large enough

$$\limsup_{\rho\downarrow 0} \big(\mathscr{D}(x_{k_j}, \rho) + \mathscr{A}(x_{k_j}, \rho)\big) < \varepsilon_0,$$

a contradiction in view of the characterization of the singular set in (3.1).

To conclude, we note that by (4.14) we get

$$\mathcal{H}^{s,\infty}(\Sigma_{\mathscr{E}} \cap \overline{B_1}) \geq \limsup_{k\uparrow\infty} \mathcal{H}^{s,\infty}(\rho_k^{-1}\Sigma_u^{(1)} \cap \overline{B_1});$$

given this, (4.11) and (4.13) yield that

$$\mathcal{H}^s(\Sigma_{\mathscr{E}} \cap \overline{B_1}) \geq \mathcal{H}^{s,\infty}(\Sigma_{\mathscr{E}} \cap \overline{B_1}) \geq \limsup_{k\uparrow\infty} \mathcal{H}^{s,\infty}(\rho_k^{-1}\Sigma_u^{(1)} \cap \overline{B_1}) \geq \frac{\omega_s}{2^{s+1}},$$

thus contradicting Theorem 4.2. □

Corollary 4.8. *If $\Omega \subseteq \mathbb{R}^2$ and $u \in \mathcal{M}(\Omega)$, then $\Sigma_u^{(1)}$ is at most countable.*

Proof. This claim follows straightforwardly from the compactness result Theorem 4.3, David's ε-regularity Theorem 3.4, and a direct application of Moore's triod theorem showing that in the plane every system of disjoint triods, *i.e.* unions of three Jordan arcs that have all one endpoint in common and otherwise disjoint, is at most countable (see [64, Theorem 1] and [70, Proposition 2.18]). □

Remark 4.9. Analogously, in 3-dimensions the set of points with blow-up a $\mathbb{T}$ cone, *i.e.* a cone with vertex the origin constructed upon the 1-skeleton of a regular tetrahedron, is at most countable. The latter claim follows thanks to Theorem 4.3, the 3d extension of David's ε-regularity result by Lemenant in [51, Theorem 8], and a suitable extension of Moore's theorem on triods established by Young in [76].

Let us point out that we employ topological arguments to compensate for the lack of monotonicity formulas. The latter would allow one to exploit Almgren's stratification type results and get, actually, a more precise picture of the set $\Sigma_u^{(1)}$ (*cf.* with [75, Theorem 3.2] and [42]).

Remark 4.10. In 2d Theorem 4.2 provides the local finiteness of the singular set for minimal Caccioppoli partitions, the blow up limits of $\rho^{-1}(\overline{S_u} - x)$ in points $x \in \Sigma_u^{(1)}$. This conclusion is far from being established for the set $\Sigma_u^{(1)}$ itself. With the results at hand one can prove that every convergent sequence $(x_j)_{j\in\mathbb{N}} \subset \Sigma_u^{(1)}$ has a limit $x_0 \notin \big(\Sigma_u^{(1)} \cup \Sigma_u^{(2)}\big)$. To show this, first note that $x_0 \notin \Sigma_u^{(1)}$ thanks to item (iii) in Proposition 5.3 or Theorem 3.4; moreover $x_0 \notin \Sigma_u^{(2)}$ thanks to item (ii) in Proposition 4.11 below and Theorem 3.7. Similarly, any converging sequence $(x_j)_{j\in\mathbb{N}} \subset \Sigma_u^{(2)}$ has a limit $x_0 \notin \big(\Sigma_u^{(1)} \cup \Sigma_u^{(2)}\big)$. Therefore, in both instances, it might happen that the limit point x_0 belong to

$$\Sigma_u^{(3)} = \{x \in \Sigma_u : \liminf_{\rho\downarrow 0} \mathscr{D}_u(x,\rho) > 0, \ \liminf_{\rho\downarrow 0} \mathscr{A}_u(x,\rho) > 0\}.$$

We conclude the section by justifying the denomination used for the sets $\Sigma_u^{(1)}$ and $\Sigma_u^{(2)}$ in 2-dimensions.

Proposition 4.11. *Let* $\Omega \subseteq \mathbb{R}^2$ *and* $u \in \mathcal{M}(\Omega)$, *then*

(i) $x \in \Sigma_u^{(1)}$ *if and only if every blow up of* u *in* x *is a triple junction function;*
(ii) $x \in \Sigma_u^{(2)}$ *if and only if every blow up of* u *in* x *is a crack-tip function.*

Proof. We start off recalling that $(u, \Omega \cap \overline{S_u})$ is an *essential pair*, *i.e.* $\mathcal{H}^1(\overline{S_u} \cap B_r(x)) > 0$ for all $x \in \overline{S_u}$ and $B_r(x) \subseteq \Omega$ (see Theorem 2.6). Then the existence and several properties of blow up limits are guaranteed by [26, Propositions 37.8, 40.9, Corollary 38.48]. More precisely, with fixed a point $x \in \Omega$ and a sequence $\rho_k \downarrow 0$, up to subsequences not relabeled, we may assume that the sets $K_k := \rho_k^{-1}(\overline{S_u} - x)$ locally Hausdorff converges in $\mathbb{R}^2$ to some closed set K as $k \uparrow \infty$. Then there is a subsequence (not relabeled for convenience) and continuous piecewise constant functions c_k on $\mathbb{R}^2 \setminus K_k$, such that the pairs (u_{x,ρ_k}, K_k) with $u_{x,\rho_k}(y) := \rho_k^{-1/2}\big(u(x+\rho_k y) - c_k(y)\big)$ satisfy:

(a) $(u_{x,\rho_k})_k$ converges to some w in $W^{1,2}_{\text{loc}}(\mathbb{R}^2 \setminus K)^6$,
(b) (w, K) is a global Bonnet minimizer according to Definition 2.22, and an essential pair (see [26, Remark 54.8]),
(c) for $\mathcal{L}^1$ a.e. $r > 0$

$$\lim_k \int_{B_r\setminus K_k} |\nabla u_{x,\rho_k}|^2 dy = \int_{B_r\setminus K} |\nabla w|^2 dy,$$
$$\lim_k \mathcal{H}^1\big(B_r \setminus K_k\big) = \mathcal{H}^1\big(B_r \setminus K\big).$$

[6] As $(K_k)_k$ locally Hausdorff converges to K, every $O \Subset \mathbb{R}^2 \setminus K$ is contained for k sufficiently big in $\rho_k^{-1}(\Omega - x) \setminus K$, so that the convergence of (u_{x,ρ_k}) in $W^{1,2}(O)$ is well defined.

To prove the direct implication in case (i) note that by Corollary 4.5, the jump set of any blow up limit in a point $x \in \Sigma_u^{(1)}$ is a minimal Caccioppoli partition $\mathscr{E}$, and that 0 is a singular point for it (*cf.* the argument leading to (4.14) in the proof of Theorem 3.8). Finally, Proposition 5.3 below ensures then that $J_{\mathscr{E}}$ is (locally) a triple junction around 0.

For the direct implication in item (ii), as by the very definition of $\Sigma_u^{(2)}$ for $\mathcal{L}^1$ a.e. $r > 0$

$$\lim_k \mathscr{A}_u(x, r\,\rho_k) = \lim_k \mathscr{A}_{u_{x,\rho_k}}(0, r) = \mathscr{A}_w(0, r) = 0,$$

$\overline{S_w}$ is actually contained in a 1-dimensional vector space. In this case a result by Léger [50] ensures that $\overline{S_w}$ is either empty or a line or a half line (*cf.* [26, Theorem 64.1]). Therefore, the energy upper bound in (2.22) and item (iii) above yield for $\mathcal{L}^1$ a.e. $r > 0$

$$\lim_k \mathscr{D}_u(x, r\,\rho_k) = \lim_k \mathscr{D}_{u_{x,\rho_k}}(0, r) = \mathscr{D}_w(0, r) \in [\varepsilon_0, 2\pi\, r]. \qquad (4.15)$$

The possibility that $\overline{S_w} = \emptyset$ is ruled out as follows: in such a case $|\nabla w|^2$ would be subharmonic on $\mathbb{R}^2$, being w harmonic there, and thus we would deduce that

$$\sup_{B_{r/2}} |\nabla w|^2 \le \frac{4}{\pi r^2} \int_{B_r} |\nabla w|^2 dx \overset{(2.22)}{\le} \frac{8}{r}.$$

By letting $r \uparrow \infty$ we would conclude w to be constant, in contrast to (4.15). Analogously, if $\overline{S_w}$ would be a line, w would be harmonic in $\mathbb{R}^2 \setminus \overline{S_w}$. Considering the restriction of w to one of the two half-spaces forming $\mathbb{R}^2 \setminus \overline{S_w}$ and performing an even reflection, since $\frac{\partial w}{\partial \nu} = 0$ on $\overline{S_w}$ we would get an harmonic function $\widetilde{w}$ on $\mathbb{R}^2$ satisfying for all $r > 0$

$$\int_{B_r} |\nabla \widetilde{w}|^2 dx \le 4\pi r.$$

Arguing as before $\widetilde{w}$ would be constant. Hence, w would be locally constant on $\mathbb{R}^2 \setminus \overline{S_w}$, leading again to a contradiction to (4.15). Therefore, $\overline{S_w}$ is a half-line, that up to a rotation can be written as $\overline{S_w} = \{(x, 0) : x \le 0\}$. Then the map $\widetilde{w} : \{z \in \mathbb{C} : \mathrm{Re} z \ge 0\} \to \mathbb{R}$ defined by $\widetilde{w}(z) := w(z^2)$ is harmonic, $\frac{\partial \widetilde{w}}{\partial \nu} = 0$ on $\{\mathrm{Re} z = 0\}$ and it satisfies

$$\int_0^r \int_{-\pi/2}^{\pi/2} \left(\rho \left|\frac{\partial \widetilde{w}}{\partial r}\right|^2 + \frac{1}{\rho}\left|\frac{\partial \widetilde{w}}{\partial \theta}\right|^2\right) d\rho\, d\theta = \int_{B_{r^2}} |\nabla w|^2 \le 2\pi r^2. \qquad (4.16)$$

In view of (4.16), the even extension of $\widetilde{w}$ on B_1, that we still denote by $\widetilde{w}$, has Fourier decomposition

$$\widetilde{w}(r,\theta) = \alpha_0 + \beta_1 r \sin\theta.$$

Changing back coordinates and taking into account the minimality of w we get

$$w(r\cos\theta, r\sin\theta) = \alpha_0 \pm \sqrt{\frac{2}{\pi}} r \sin\frac{\theta}{2},$$

the conclusion follows at once.

The reverse implications in both cases are easily concluded. Indeed, in case (i) if $(\rho_k)_k$ satisfies

$$\lim_k \mathscr{D}_u(x,\rho_k) = \limsup_{\rho\downarrow 0} \mathscr{D}_u(x,\rho),$$

then the blow up limit w of $(u_{x,\rho_k})_k$ is a triple junction function by assumption. By taking into account item (c) above we infer for $\mathcal{L}^1$ a.e. $r > 0$

$$\lim_k \mathscr{D}_u(x, r\rho_k) = \lim_k \mathscr{D}_{u_{x,\rho_k}}(0,r) = \mathscr{D}_w(0,r) = 0,$$

thus implying that $x \in \Sigma_u^{(1)}$.

Similarly, one can prove the reverse implication in case (ii). □

Actually, in the first instance of Proposition 4.11 uniqueness of the blow up limit is ensured by Theorem 3.4. As already remarked, case (ii) is still open.

5 Higher integrability of the gradient in dimension 2

The higher integrability of the gradient has been first established by De Lellis and Focardi [35] in dimension 2. Following a classical path, the key ingredient to establish Theorem 3.9 is a reverse Hölder inequality for the gradient, which we state independently (see [35, Theorem 1.3]).

Theorem 5.1 (De Lellis and Focardi [35]). *For all $q \in (1,2)$ there exist $\rho \in (0,1)$ and $C > 0$ such that*

$$\|\nabla u\|_{L^2(B_\rho)} \le C \|\nabla u\|_{L^q(B_1)} \qquad \text{for any } u \in \mathcal{M}(B_1). \tag{5.1}$$

Using the obvious scaling invariance of (2.3), Theorem 5.1 yields a corresponding reverse Hölder inequality for balls of arbitrary radius. Theorem 3.9 is then a consequence of a by now classical result.

Theorem 5.2 (Giaquinta and Modica [49]). *Let $v \in L^q_{\rm loc}(\Omega)$, $q > 1$, be nonnegative such that for some constants $\beta > 0$, $\lambda \geq 1$ and $R_0 > 0$*

$$\left(\fint_{B_r(z)} v^q\,dy\right)^{1/q} \leq \beta \fint_{B_{\lambda r}(z)} v\,dy$$

for all $z \in \Omega$, $r \in \big(0, R_0 \wedge \mathrm{dist}(z, \partial\Omega)\big)$.

Then $v \in L^p_{\rm loc}(\Omega)$ for some $p > q$ and there is $C = C(\beta, n, q, p, \lambda) > 0$ such that

$$\left(\fint_{B_r(z)} v^p\,dy\right)^{1/p} \leq C\left(\fint_{B_{2r}(z)} v^q\,dy\right)^{1/q}.$$

The exponent p could be explicitly estimated in terms of q, C and ρ. However, since our argument for Theorem 5.1 is indirect, we do not have any explicit estimate for C (ρ can instead be computed). Hence, combining Theorem 3.9 with [4] we can only conclude that the dimension of the singular set of $\overline{S_u}$ is strictly smaller than 1. Guy David pointed out that the corresponding dimension estimate could be made explicit. In fact, he suggested to the Authors of [35] that also the constant C in Theorem 5.1 might be estimated: a viable strategy would combine the core argument of this paper with some ideas from [26] (the proof of Theorem 5.1 given here makes already a fundamental use of the paper [26], but depends only on the ε-regularity theorems for "triple junctions" and "segments" stated in Section 3). However, the resulting estimate would give an extremely small number, whereas the proof would very likely become much more complicated.

In spite of the dimensional restriction, the indirect proof has as interesting side results Theorem 4.3 and its related consequences highlighted in Section 4. No dimensional limitation is present in Theorem 4.3, instead dimension 2 enters dramatically in the proof of Theorem 5.1 as the structure of minimal Caccioppoli partitions in $\mathbb{R}^2$ can be described precisely via minimal connections. Recall that a *minimal connection* of $\{q_1, \ldots, q_N\} \subset \mathbb{R}^2$ is any minimizer of the Steiner problem

$$\min\left\{\mathcal{H}^1(\Gamma) : \Gamma \text{ closed and connected, and } q_1, \ldots, q_N \in \Gamma\right\}.$$

Proposition 5.3 (Proposition 11, Lemma 12 [35]). *Let $\mathscr{E}$ be a minimal Caccioppoli partition in $\Omega \subset \mathbb{R}^2$, then*

(i) $\mathcal{H}^0(\overline{J_{\mathscr{E}}} \cap \partial B_\rho(x)) < +\infty$ *if* $B_\rho(x) \Subset \Omega$;
(ii) $\mathcal{H}^0(K \cap \partial B_\rho(x)) \geq 2$ *for each connected component* K *of* $\overline{J_{\mathscr{E}}} \cap \overline{B}_\rho(x)$, *and it is a minimal connection of* $K \cap \partial B_\rho(x)$*;*

(iii) *if* $\Omega = B_1$, *then there exists* $\rho_0 \in (0, 1)$ *such that for all* $t \in (0, \rho_0)$

$$\mathcal{H}^0(\overline{J_{\mathcal{E}}} \cap \partial B_t) \leq 3, \quad \textit{and} \quad \mathcal{H}^1(\overline{J_{\mathcal{E}}} \cap B_t) \leq 3t.$$

We are now ready to sketch the proof of Theorem 5.1 in 2-dimensions following De Lellis and Focardi [35].

Proof of Theorem 5.1. We fix an exponent $q \in (1, 2)$ and a suitable radius ρ (whose choice will be specified later) for which (5.1) is false, that is

$$\|\nabla u_k\|_{L^2(B_\rho)} \geq k\|\nabla u_k\|_{L^q(B_1)} \text{ for a sequence } (u_k)_{k\in\mathbb{N}} \in \mathcal{M}(B_1). \quad (5.2)$$

Since the Mumford and Shah energy of any $u \in \mathcal{M}(B_1)$ can be easily bounded a priori by 2π (*cf.* Proposition 2.14), we have $\|\nabla u_k\|_{L^q(B_1)} \to 0$. Theorem 4.3 and Proposition 5.3 then show that:

(i) The L^2 energy of the gradients of u_k converge to 0;
(ii) $\overline{S_{u_k}}$ converge in the local Hausdorff metric to the (closure of) set of interfaces of a minimal Caccioppoli partition $\overline{J_{\mathcal{E}}}$;
(iii) $\mathcal{H}^0(\overline{J_{\mathcal{E}}} \cap \partial B_t) \leq 3$ for $t \in (0, \rho_0)$.

An elementary argument shows the existence of $t \geq \rho_0/4$ such that

(a) either $\overline{J_{\mathcal{E}}} \cap B_t = \emptyset$;
(b) or $\overline{J_{\mathcal{E}}} \cap B_t$ is a segment and $\partial B_t \setminus \overline{J_{\mathcal{E}}}$ is the union of two arcs each with length $< \frac{4\pi}{3}t$;
(c) or $\overline{J_{\mathcal{E}}} \cap B_t$ is a triple junction and $\partial B_t \setminus \overline{J_{\mathcal{E}}}$ the union of three arcs each with length $< (2\pi - \frac{1}{8})t$.

In any case we set $\rho := \rho_0/9$ (*cf.* with (5.2)). By Theorem 3.4 (we keep the notation introduced there), we may find a constant $\beta \in (0, 1/3)$ such that for all k sufficiently big one of the following alternatives happens

(a_1) $\overline{S_{u_k}} \cap B_t = \emptyset$;
(b_1) For each $s \in ((1-\beta)t, t)$, $\partial B_s \setminus \overline{S_{u_k}}$ is the union of two arcs γ_1^k and γ_2^k each with length $< (2\pi - \frac{1}{9})s$, whereas $\overline{S_{u_k}} \cap B_s$ is connected and divides B_s in two components B_1^k, B_2^k with $\partial B_i^k = \gamma_i^k \cup (\overline{S_{u_k}} \cap \overline{B_s})$;
(c_1) For each $s \in ((1-\beta)t, t)$, $\partial B_s \setminus \overline{S_{u_k}}$ is the union of three arcs γ_1^k, γ_2^k and γ_3^k each with length $< (2\pi - \frac{1}{9})s$, whereas $\overline{S_{u_k}} \cap B_s$ is connected and divides B_s in three connected components B_1^k, B_2^k and B_3^k with $\partial B_i^k \subset \gamma_i^k \cup (\overline{S_{u_k}} \cap \overline{B_s})$.

Choose then $r \in (2/3t, t)$ and a subsequence (not relabeled) such that

$$g_k := u_k|_{\partial B_r} \in W^{1,q}(\gamma, \mathcal{H}^1) \text{ for any connected component of } \partial B_r \setminus S_{u_k}$$

and

$$\int_{\partial B_r \setminus S_{u_k}} |g_k'|^q d\mathcal{H}^1 \leq \frac{3}{t} \int_{B_t} |\nabla u_k|^q \, dx \leq \frac{12}{\rho_0} \int_{B_1} |\nabla u_k|^q \, dx.$$

Let us first deal with the (easier) case (a). By compactness, as $\overline{J_{\mathscr{E}}} \cap B_t = \emptyset$ then

$$\overline{S_{u_k}} \cap B_t = \emptyset \quad \text{for } k \gg 1.$$

Hence, being $u_k \in \mathcal{M}(B_1)$ we get that u_k is the harmonic extension of its trace in B_t. In conclusion, as $\rho = \rho_0/9 < 2/3\,t < r$ we have

$$\int_{B_{\rho_0/9}} |\nabla u_k|^2 dx \leq \int_{B_r} |\nabla u_k|^2 dx \leq C \min_{\lambda} \|g_k - \lambda\|^2_{H^{1/2}(\partial B_r)}$$
$$\overset{W^{1,q} \hookrightarrow H^{1/2}}{\leq} C \left(\int_{\partial B_r} |g_k'|^q \, d\mathcal{H}^1 \right)^{2/q} \leq C \left(\frac{12}{\rho_0} \int_{B_1} |\nabla u_k|^q \, dx \right)^{2/q},$$

for some $C > 0$ (independent of k), contradicting (5.2).

In case (b) or (c) hold the construction is similar. Denote by K_k the minimal connection relative to $\overline{S_{u_k}} \cap \partial B_r$. Then K_k splits $\overline{B_r}$ into two (case (b$_1$)) or three (case (c$_1$)) regions denoted by B_r^i. Let γ^i be the arc of ∂B_r contained in the boundary of B_r^i. Having all the arcs length uniformly bounded from below, it is easy to check that for all i we can find a function $w_k^i \in W^{1,2}(B_r)$ with boundary trace g_k and satisfying for some absolute constant $C > 0$

$$\int_{B_r} |\nabla w_k^i|^2 \, dx \leq C \left(\int_{\gamma^i} |g_k'|^q \, d\mathcal{H}^1 \right)^{2/q} \tag{5.3}$$

(*cf.* [35, Lemma 7]). Denote by w_k the function equal to w_k^i on B_k^i, then $w_k \in SBV(B_r)$ and $S_{w_k} \subseteq K_k$. The minimality of u_k implies then that

$$\int_{B_{\rho_0/9}} |\nabla u_k|^2 \leq \int_{B_r} |\nabla u_k|^2 \leq \int_{B_r} |\nabla w_k|^2 + \mathcal{H}^1(K_k) - \mathcal{H}^1(S_{u_k} \cap B_r)$$
$$\leq \int_{B_r} |\nabla w_k|^2 \overset{(5.3)}{\leq} C \left(\int_{\partial B_r \setminus S_{u_k}} |g_k'|^q \, d\mathcal{H}^1 \right)^{2/q} \leq C \left(\frac{12}{\rho_0} \int_{B_1} |\nabla u_k|^q \, dx \right)^{2/q},$$

contradicting (5.2). □

6 Higher integrability of the gradient in any dimension: Porosity of the Jump set

A central role in establishing the higher integrability of the gradient in any dimension shall be played by the following improvement of Theorem 3.1.

Theorem 6.1 (Rigot [71], Maddalena and Solimini [55]). *There are dimensional constants $\varepsilon(n)$, $C_0(n) > 0$ such that for every $\varepsilon \in (0, \varepsilon(n))$ there exists $\alpha_\varepsilon \in (0, 1/2)$ such that if $u \in \mathcal{M}(B_2)$ and $B_\rho(x) \subset \Omega$, with $x \in \overline{S_u}$ and $\rho \in (0, 1)$, then there exists a ball $B_r(y) \subset B_\rho(x)$ with radius $r \in (\alpha_\varepsilon \rho, \rho)$ such that*

(i) $\mathscr{D}_u(y,r) + \mathscr{A}_u(y,r) < \varepsilon_0$, *$\varepsilon_0 > 0$ the constant in Theorem* 3.1*;*
(ii) $\overline{S_u} \cap B_r(y)$ *is a $C^{1,\gamma}$ graph, for all $\gamma \in (0, 1)$, containing y;*
(iii)
$$r\,\|\nabla u\|^2_{L^\infty(B_r(y))} \leq C_0\,\varepsilon. \tag{6.1}$$

We can restate the result above by saying that Σ_u is $(\alpha_\varepsilon, 1)$-*porous* in $\overline{S_u}$ according to the following definition.

Definition 6.2. Given a metric space (X, d_X), a subset K is (α, δ)-*porous* in X, with $\alpha \in (0, 1/2)$ and $\delta > 0$, if for every $x \in X$ and $\rho \in (0, \delta)$ we can find $y \in B_\rho(x)$ and $r \in (\alpha\rho, \rho)$ such that

$$B_r(y) \subset B_\rho(x) \setminus K.$$

Clearly, in our case $X = \overline{S_u}$, $K = \Sigma_u$ and d_X is the metric induced by the Euclidean one. The Hausdorff dimension estimate in the papers by David [25], Rigot [71] and Maddalena and Solimini [55] follows from the porosity property in Theorem 6.1 and Theorem 6.4 below.

To this aim we recall that, given an Alfhors regular metric space (X,d_X) of dimension ℓ, *i.e.* (X, d_X) is complete and there is $\Lambda > 0$ such that

$$\Lambda^{-1} r^\ell \leq \mathcal{H}^\ell(B_r(z)) \leq \Lambda r^\ell \quad \text{for all } z \in X \text{ and } r > 0,$$

the *lower/upper Minkowski dimension* of K is defined as

$$\underline{\dim}_{\mathcal{M}} K := \inf\{s \in (0, \ell] : \mathcal{M}^s_*(K) = 0\},$$
$$\overline{\dim}_{\mathcal{M}} K := \inf\{s \in (0, \ell] : \mathcal{M}^{*s}(K) = 0\},$$

where the *lower/upper Minkowski content* is given by

$$\mathcal{M}^s_*(K) := \liminf_{r\downarrow 0} \frac{\mathcal{H}^\ell\big((K)_r\big)}{r^{\ell-s}}, \qquad \mathcal{M}_{*s}(K) := \limsup_{r\downarrow 0} \frac{\mathcal{H}^\ell\big((K)_r\big)}{r^{\ell-s}}.$$

and

$$(K)_r := \{x \in X : d_X(x, K) < r\}. \tag{6.2}$$

Let us first relate the 2-dimensions introduced above.

Lemma 6.3. *For all sets $K \subset X$*

$$\dim_{\mathcal{H}} K \leq \dim_{\mathcal{M}} K.$$

Proof. We may assume $\dim_{\mathcal{M}} K < \ell$ since otherwise the inequality is trivial.

Let $N(K, r)$ be the minimal number of balls of radius r covering K, and $P(K, r)$ be the maximal number of disjoint balls with centers belonging to K, then

$$N(K, 2r) \leq P(K, r)^7.$$

Indeed, set $N := N(K, 2r)$, $P := P(K, r)$ and let $\mathscr{B} = \{B_r(x_i)\}_{i=1}^P$ be the corresponding maximal family of disjoint balls with $x_i \in K$. If there exists $x \in K \setminus \cup_{i=1}^P B_{2r}(x_i)$, then $\{B_r(x)\} \cup \mathscr{B}$ is a disjoint family of balls with centers on K, a contradiction.

Therefore, for all $s \in (\dim_{\mathcal{M}} K, \ell)$

$$N\omega_\ell r^\ell \leq P\omega_\ell r^\ell \leq \mathcal{H}^\ell\big((K)_r\big) \Longrightarrow N\omega_\ell r^s \leq \frac{\mathcal{H}^\ell\big((K)_r\big)}{r^{\ell-s}},$$

from which one easily conclude that $2^{-s}\frac{\omega_\ell}{\omega_s}\mathcal{H}^s(K) \leq \mathcal{M}_*^s(K) = 0$. □

We are now ready to state an estimate on the Hausdorff dimension of porous sets.

Theorem 6.4 (David and Semmes [28]). *If (X, d_X) is Alfhors regular of dimension ℓ, then every (α, δ)-porous subset K of X of diameter d satisfies*

$$\mathcal{H}^\ell\big((K)_r\big) \leq C\, r^{\ell-\eta} \quad \forall r \in (0, d),$$

for some constant $C = C(\ell, \delta, d) > 0$ and $\eta = \eta(\alpha, d, \Lambda) \in [0, \ell)$. Hence,

$$\dim_{\mathcal{H}} K \leq \dim_{\mathcal{M}} K \leq \eta.$$

The higher integrability property of the gradient for MS-minimizers will be (essentially) a consequence of the result above. Actually, we cannot take advantage directly of Theorem 6.4 since we are not able to relate $(\Sigma_u)_r$ and $\{|\nabla u|^2 \geq r^{-1}\}$. Therefore, we shall prove a suitable version of Theorem 6.4 and establish its links with the higher integrability property in the Section 7 following the approach by De Philippis and Figalli [37].

[7] One can also prove that $P(K, r) \leq N(K, r/2)$

In passing, we mention that recently porosity has been employed in several instances to estimate the Hausdorff dimension of singular sets of solutions to variational problems (*cf.* [38,45,48]).

In the rest of the present section we shall comment on porosity in the more standard Euclidean setting, *i.e.* $X = \mathbb{R}^n$, and prove analogous results to those of interest for us. This is done to get more acquainted with porosity and it is intended as a warm up to the proof of Theorem 3.9 by De Philippis and Figalli.

Let then $K \subseteq \mathbb{R}^n$ be a (α, δ)-porous set. Few remarks are in order:

(i) By Lebesgue's differentiation theorem clearly $\mathcal{L}^n(K) = 0$;
(ii) K is nowhere dense, i.e. $\mathrm{int}\overline{K} = \emptyset$, since the latter is equivalent to: for every $x \in X$ and $\rho > 0$ there exists $y \in X$ and $r > 0$ such that $B_r(y) \subset B_\rho(x) \setminus K$.
(iii) Zajíček [77] actually proved that there are non-porous sets which are nowhere dense and with zero Lebesgue measure.

An elementary covering argument actually provides an estimate on the Hausdorff dimension of K and therefore improves item (i) above.

Proposition 6.5 (Salli [72]). *Suppose that K is a bounded (α, δ)-porous set in $\mathbb{R}^n$ with* $\mathrm{diam}K \leq d$*, then*

$$\mathcal{L}^n\big((K)_r\big) \leq C\, r^{n-\gamma} \qquad \forall r \in (0, d). \tag{6.3}$$

for some constants $C = C(n, \delta, d) > 0$ and $\gamma = \gamma(\alpha, n) < n$. In particular,

$$\dim_{\mathcal{H}} K \leq \dim_{\mathcal{M}} K \leq \gamma. \tag{6.4}$$

Proof. The building step of the argument goes as follow: consider a cube Q of $\mathrm{diam}Q < \delta$ and center x_Q, then by taking into account the (α, δ)-porosity of K, we find a point y_Q such that $B_{\alpha\,\mathrm{diam}Q/2\sqrt{n}}(y_Q) \subset B_{\mathrm{diam}Q/2\sqrt{n}}(x_Q) \setminus K$. If $\{Q_i\}_i$ is a covering of Q of k^n sub cubes with $\mathrm{diam}Q_i = \mathrm{diam}Q/k$, $k \in \mathbb{N}$, then at least one of those cubes does not intersect K if k is sufficiently big. Indeed, it suffices to impose

$$\alpha\,\mathrm{diam}Q > 2\frac{\mathrm{diam}Q}{k} \Longleftrightarrow \alpha\, k > 2.$$

Hence, we may choose $k = k(\alpha)$ for which the previous condition is satisfied.

Therefore, given a covering of K of $m = m(\delta, d)$ cubes with diameter $\delta/2$, we can construct another covering made of $m(k^n - 1) = mk^\gamma$ cubes of diameter $\delta/2k$, where $\gamma = \gamma(\alpha, n) \in (0, n)$ is such that $k^\gamma = k^n - 1$.

Clearly, we can iterate this procedure in each of the new cubes, so that for all $N \in \mathbb{N}$ we may find a covering of K made of $mk^{N\gamma}$ cubes $\{Q_i^k(x_i^k)\}_{i=1}^{mk^{N\gamma}}$ of diameter $\delta/2k^N$. In particular, each ball $B_{\delta/4k^N}(x_i^k)$ contains Q_i^k, and their union covers K. Thus,

$$(K)_{\delta/4k^N} \subset \cup_{i=1}^{mk^{N\gamma}} B_{\delta/2k^N}(x_i^k).$$

Hence

$$\mathcal{L}^n\left((K)_{\delta/4k^N}\right) \leq \omega_n mk^{N\gamma}\left(\frac{\delta}{2k^N}\right)^n = o(1) \quad N \uparrow \infty.$$

Estimate (6.3) follows at once by a simple dyadic argument on the radii, *i.e.* given $r > 0$ choosing k such that $r \in [\frac{\delta}{2k^{N+1}}, \frac{\delta}{2k^N})$.

Instead, estimate (6.4) is an easy consequence of Lemma 6.3. □

Remark 6.6. Theorem 6.4 can be proved exactly as Proposition 6.5 once the existence of a family of dyadic cubes in (X, d_X) has been established (*cf.* [28, Lemma 5.8]). By this, we mean a collection $\{\Delta_j\}_{\mathbb{Z}\ni j<j_0}$ of families of measurable subsets of X, $j_0 = \infty$ if $\mathrm{diam}X = \infty$ and otherwise $j_0 \in \mathbb{Z}$ such that $2^{j_0} \leq \mathrm{diam}X < 2^{j_0+1}$, having the following properties:

(i) each Δ_j is a partition of X, *i.e.* $X = \cup_{Q\in\Delta_j} Q$ for any j as above;
(ii) $Q \cap Q' = \emptyset$ whenever $Q, Q' \in \Delta_j$ and $Q \neq Q'$;
(iii) if $Q \in \Delta_j$ and $Q' \in \Delta_k$ for $k \geq j$, then either $Q \subseteq Q'$ or $Q \cap Q' = \emptyset$;
(iv) $\lambda^{-1}2^j \leq \mathrm{diam}Q \leq \lambda 2^j$ and $\lambda^{-1}2^{j\ell} \leq \mathcal{H}^\ell(Q) \leq \lambda 2^{j\ell}$ for all j and all $Q \in \Delta_j$;
(v) for all j and all $Q \in \Delta_j$, and $\tau > 0$

$$\mathcal{H}^\ell\left(\{x \in Q : \mathrm{dist}(x, X \setminus Q) \leq \tau 2^j\}\right) + \mathcal{H}^\ell\left(\{x \in X \setminus Q : \mathrm{dist}(x, Q) \leq \tau 2^j\}\right) \leq \lambda\tau^{1/\lambda}\mathcal{H}^\ell(Q).$$

with $\lambda = \lambda(\ell, \Lambda)$ (for the existence of such families see [24] and [73]).

With fixed a given porosity $\alpha \in (0, 1/2)$, we are then interested in analyzing the worst case, *i.e.*

$$D(\alpha, n) := \sup\left\{\dim_{\mathcal{H}} K : K \subset \mathbb{R}^n \text{ is } (\alpha, \delta)\text{-porous for some } \delta > 0\right\}.$$

We can easily deduce the estimate

$$n - 1 \leq D(\alpha, n) \leq \gamma(\alpha, n) < n,$$

as $(n-1)$-dimensional vector spaces are (α,δ)-porous for all $\delta > 0$ and $\alpha \in (0,1/2)$. Furthermore, Mattila (see, for instance, [59, Theorem 11.14]) has shown that

$$\lim_{\alpha\uparrow 1/2} D(\alpha,n) = n-1. \tag{6.5}$$

Remark 6.7. Salli [72] has actually improved upon the previous result by showing that

$$n-1+\frac{B(n)}{|\ln(1-2\alpha)|} \le D(\alpha,n) \le n-1+\frac{A(n)}{|\ln(1-2\alpha)|} \quad \text{for all } \alpha\in(0,1/2),$$

for some strictly positive dimensional constants A and B.

Finally, we note that the analogous property in (6.5) in the case of interest for us, if true, would then let us conclude another characterization of the conjectured estimate on the Hausdorff dimension of Σ_u: *If Σ_u is (α,δ)-porous in $\overline{S_u}$ for all $\alpha \in (0,1/2)$ and some $\delta = \delta(\alpha) > 0$, then $\dim_{\mathcal{H}}\Sigma_u \le n-2$.*

7 Higher integrability of the gradient in any dimension: the proof

In this section we shall prove the higher integrability property of the gradient following De Philippis and Figalli [37]. We shall first establish in Proposition 7.3 below a particular case of Theorem 6.4 that is sufficient for our purposes.

To this aim we recall that the conclusions of Theorem 2.7 and Proposition 2.14 show the Alfhors regularity of $\Omega \cap \overline{S_u}$: for some constants $C_0 = C_0(n) > 0$, $\rho_0 = \rho_0(n) > 0$

$$C_0^{-1} r^{n-1} \le \mathcal{H}^{n-1}(\overline{S_u} \cap B_r(z)) \le C_0 r^{n-1} \tag{7.1}$$

for all $z \in \Omega \cap \overline{S_u}$, and all $r \in (0, \rho_0 \wedge \operatorname{dist}(z,\partial\Omega))$, $u \in \mathcal{M}(B_2)$.

To prove Proposition 7.3 we need two technical lemmas. The first one is obtained via De Giorgi's slicing/averaging principle.

Lemma 7.1. *There are dimensional constants M_1, C_1 such that if $M \ge M_1$ for every $u \in \mathcal{M}(B_2)$ we can find three decreasing sequences of radii such that*

(i) $1 \ge R_h \ge S_h \ge T_h \ge R_{h+1}$;
(ii) $8M^{-(h+1)} \le R_h - R_{h+1} \le M^{-(h+1)/2}$, *and* $S_h - T_h = T_h - R_{h+1} = 4M^{-(h+1)}$;
(iii) $\mathcal{H}^{n-1}\big(\overline{S_u} \cap (\overline{B}_{S_h} \setminus \overline{B}_{R_{h+1}})\big) \le C_1 M^{-(h+1)/2}$;
(iv) $R_\infty = S_\infty = T_\infty \ge 1/2$.

Proof. Let $R_1 = 1$, given R_h we construct S_h, T_h and R_{h+1} as follows.

Set $N_h := \lfloor M^{(h+1)/2}/8 \rfloor \in \mathbb{N}$ and fix $M_1 \in \mathbb{N}$ such that $N_h \geq \lfloor M^{(h+1)/2}/16 \rfloor$ for $M \geq M_1$. Here, $\lfloor \alpha \rfloor$ denotes the integer part of $\alpha \in \mathbb{R}$.

The annulus $B_{R_h} \setminus \overline{B}_{R_h - 8M^{-(h+1)/2}}$ contains the N_h disjoint sub annuli $\overline{B}_{R_h - 8(i-1)M^{-(h+1)}} \setminus \overline{B}_{R_h - 8i\,M^{-(h+1)}}$, $i \in \{1, \ldots, N_h\}$, of equal width $8\, M^{-(h+1)}$. By averaging we can find an index $i_h \in \{1, \ldots, N_h\}$ such that

$$\begin{aligned}
&\mathcal{H}^{n-1}\big(K \cap (\overline{B}_{R_h - 8(i_h-1)M^{-(h+1)/2}} \setminus \overline{B}_{R_h - 8i_h\, M^{-(h+1)/2}})\big) \\
&\leq \frac{1}{N_h}\, \mathcal{H}^{n-1}\big(K \cap (\overline{B}_{R_h} \setminus \overline{B}_{R_h - 8M^{-(h+1)/2}})\big) \\
&\overset{\text{d.u.b. in (7.1)}}{\leq} C_0\, \frac{R_h^{n-1}}{N_h} \leq C_1\, M^{-(h+1)/2},
\end{aligned}$$

so that (iii) is established. Finally, set

$$\begin{aligned}
S_h &:= R_h - 8(i_h - 1)M^{-(h+1)}, \\
R_{h+1} &:= R_h - 8i_h\, M^{-(h+1)}, \\
T_h &:= \frac{1}{2}(S_h + R_{h+1}),
\end{aligned}$$

then items (i) and (ii) follow by the very definition, and item (iv) from (ii) if M_1 is sufficiently big. □

The second lemma has a geometric flavor.

Lemma 7.2. *Let* $f : \mathbb{R}^{n-1} \to \mathbb{R}$ *be Lipschitz with*

$$f(0) = 0, \quad \|\nabla f\|_{L^\infty} \leq \eta. \tag{7.2}$$

If $G :=$ graph$(f) \cap B_2$ *and* $\eta \in (0, 1/15]$, *then for all* $\delta \in (0, 1/2)$ *and* $x \in (\overline{B}_{1+\delta} \setminus B_1) \cap G$

$$\operatorname{dist}(x, (\overline{B}_{1+2\delta} \setminus B_{1+\delta}) \cap G) \leq \frac{3}{2}\delta.$$

Proof. Clearly by (7.2) we get

$$\|f\|_{W^{1,\infty}(B_2)} \leq 3\eta.$$

Let $x = (y, f(y)) \in (\overline{B}_{1+\delta} \setminus B_1) \cap G$ and $\hat{x} := (\lambda\, y, f(\lambda\, y))$, with λ to be chosen suitably. Note that as $|x| \geq 1$ we have

$$\begin{aligned}
|f(\lambda\, y) - \lambda\, f(y)| &\leq |f(\lambda\, y) - f(y)| + |\lambda - 1|\, |f(y)| \\
&\leq |\lambda - 1| \big(\|\nabla f\|_{L^\infty}\, |y| + \|f\|_{L^\infty}\big) \leq 3\eta|\lambda - 1|\, |x|.
\end{aligned}$$

Hence,

$$|\hat{x} - x| \leq |\hat{x} - \lambda x| + |\lambda - 1||x| \leq \big(3\eta + 1\big)|\lambda - 1|\,|x|.$$

It is easy to check that the choice $\lambda = 1 + \frac{5}{4}\delta|x|^{-1}$ gives the conclusion. □

We are now ready to prove the version of Theorem 6.4 of interest for our purposes.

Proposition 7.3 (De Philippis and Figalli [37]). *Let C_0, C_1, M_1 be the constants in* (7.1) *and Lemma* 7.1, *respectively.*

There exist dimensional constants C_2, $M_2 > 0$ and $\alpha \in (0, 1/4)$, $\beta \in (0, 1/4)$, with $M_2 \geq M_1$, such that for every $M \geq M_2$, $u \in \mathcal{M}(B_2)$, we can find families $\mathcal{F}_j$ of disjoint balls

$$\mathcal{F}_j = \big\{B_{\alpha M^{-j}}(y_i) : \; y_i \in \overline{S_u}, \; 1 \leq j \leq N_j\big\}$$

such that for all $h \in \mathbb{N}$

(i) *$B, B' \in \cup_{j=1}^{h} \mathcal{F}_j$ are distinct balls, then $(B)_{4M^{-(h+1)}} \cap (B')_{4M^{-(h+1)}} = \emptyset$;*

(ii) *if $B_{\alpha M^{-j}}(y_i) \subset \mathcal{F}_j$, then $\overline{S_u} \cap B_{2\alpha M^{-j}}(y_i)$ is a $C^{1,\gamma}$ graph, $\gamma \in (0, 1)$ any, containing y_i,*

$$\mathscr{D}_u\big(y_i, 2\alpha M^{-j}\big) + \mathscr{A}_u\big(y_i, 2\alpha M^{-j}\big) < \varepsilon_0;$$

$$\|\nabla u\|_{L^\infty\left(B_{2\alpha M^{-j}}(y_i)\right)} < M^{j+1}; \tag{7.3}$$

(iii) *let $\{R_h\}, \{S_h\}, \{T_h\}$ be the sequences of radii in Lemma 7.1, and let*

$$K_h := (\overline{S_u} \cap \overline{B}_{S_h}) \setminus \Big(\cup_{j=1}^{h} \cup_{\mathcal{F}_j} B\Big),$$

(note that by construction $K_{h+1} \subset K_h \setminus \cup_{\mathcal{F}_{h+1}} B$), and

$$\widetilde{K}_h := (\overline{S_u} \cap \overline{B}_{T_h}) \setminus \Big(\cup_{j=1}^{h} \cup_{\mathcal{F}_j} (B)_{2M^{-(h+1)}}\Big) \subset K_h.$$

Then, there exists a finite set of points $\mathcal{C}_h := \{x_i\}_{i \in I_h} \subset \widetilde{K}_h$ such that

$$|x_j - x_k| \geq 3M^{-(h+1)} \quad \forall j, k \in I_h, \; j \neq k; \tag{7.4}$$

$$(K_h \cap \overline{B}_{R_{h+1}})_{M^{-(h+1)}} \subset \cup_{i \in I_h} B_{8M^{-(h+1)}}(x_i); \tag{7.5}$$

$$\mathcal{H}^{n-1}(K_h) \leq C_1\, h\, M^{-2h\beta}; \tag{7.6}$$

$$\mathcal{L}^n\left((K_h \cap \overline{B}_{R_{h+1}})_{M^{-(h+1)}}\right) \leq C_2\, h\, M^{-h(1+2\beta)-1}. \tag{7.7}$$

(iv) $\Sigma_u \cap B_{1/2} \subset K_h$ *for all* $h \in \mathbb{N}$ *and*

$$\mathcal{L}^n\big((\Sigma_u \cap B_{1/2})_r\big) \leq C_2 r^{1+\beta} \qquad \forall r \in (0, 1/2]. \qquad (7.8)$$

In particular, $\dim_{\mathcal{M}}(\Sigma_u \cap B_{1/2}) \leq n - 1 - \beta$.

Proof. For notational convenience we set $K = \overline{S_u}$. In what follows we shall repeatedly use Theorem 6.1 with $\varepsilon \in (0, 1)$ fixed and sufficiently small.

We split the proof in several steps.

Step 1. *Inductive definition of the families* $\mathcal{F}_j$.
For $h = 1$ we define

$$\mathcal{F}_1 := \emptyset, \; K_1 = K \cap \overline{B}_{S_1}, \; \widetilde{K_1} = K \cap \overline{B}_{T_1},$$

and choose $\mathcal{C}_1$ to be a maximal family of points at distance $3M^{-2}$ from each other. Of course, properties (i) and (ii) and (7.4) are satisfied. To check the others, one can argue as in the verification below.

Suppose that we have built the families $\{\mathcal{F}_j\}_{j=1}^h$ as in the statement, to construct $\mathcal{F}_{h+1}$ we argue as follows. Let $\mathcal{C}_h = \{x_i\}_{i \in I_h} \subset \widetilde{K_h}$ be a family of points satisfying (7.4), *i.e.* $|x_i - x_k| \geq 3M^{-(h+1)}$ for all $j, k \in I_h$ with $j \neq k$, and consider

$$\mathcal{G}_{h+1} := \{B_{M^{-(h+1)}}(x_i)\}_{i \in I_h}.$$

By the porosity assumption on K for every ball $B_{M^{-(h+1)}}(x_i) \in \mathcal{G}_{h+1}$ we can find a sub-ball $B_{2\alpha M^{-(h+1)}}(y_i) \subset B_{M^{-(h+1)}}(x_i) \setminus K$, $\alpha \in (0, 1/4)$ for which the theses of Theorem 6.1 are satisfied. Then, define

$$\mathcal{F}_{h+1} := \{B_{\alpha M^{-(h+1)}}(y_i)\}_{i \in I_h}.$$

By condition (7.4), the balls $B_{\frac{3}{2}M^{-(h+1)}}(x_i)$ are disjoint and do not intersect

$$\cup_{j=1}^h \cup_{\mathcal{F}_j} (B)_{\frac{1}{2}M^{-(h+1)}}$$

by the very definition of $\widetilde{K_h}$. Thus, item (i) and (ii) are satisfied.

Hence, we can define K_{h+1}, $\widetilde{K}_{h+1}$ and $\mathcal{C}_{h+1}$ as in the statement.

Step 2. *Proof of* (7.5).
Let $x \in (K_h \cap \overline{B}_{R_{h+1}})_{M^{-(h+2)}}$ and let z be a point of minimal distance from $K_h \cap \overline{B}_{R_{h+1}}$. In case $z \in \widetilde{K}_{h+1}$, by maximality there is $x_i \in \mathcal{C}_{h+1}$ such that $|z - x_i| \leq 3M^{-(h+2)}$ and thus we conclude $x \in B_{5M^{-(h+2)}}(x_i)$. Instead, if $z \in (K_h \cap \overline{B}_{R_{h+1}}) \setminus \widetilde{K}_{h+1}$, the definitions of K_{h+1} and $\widetilde{K}_{h+1}$ yield the existence of a ball $\widetilde{B} \in \cup_{j=1}^{h+1} \mathcal{F}_j$ for which $z \in (K \cap (\widetilde{B})_{2M^{-(h+2)}}) \setminus \widetilde{B}$.

In view of property (ii), a rescaled version of Lemma 7.2 gives a point y satisfying

$$y \in \big(K \cap (\widetilde{B})_{4M^{-(h+2)}}\big) \setminus (\widetilde{B})_{2M^{-(h+2)}}, \quad \text{and} \quad |z-y| \le 3M^{-(h+2)}.$$

Therefore, as $z \in \overline{B}_{R_{h+2}}$ and $T_{R_{h+1}} = R_{h+2}+4M^{-(h+2)}$ we get by property (i) and the definition of $\widetilde{K}_{h+1}$

$$y \in \big(K \cap (\widetilde{B})_{4M^{-(h+2)}} \cap \widetilde{B})_{4M^{-(h+2)}}\big) \setminus (\widetilde{B})_{2M^{-(h+2)}}.$$

Finally, by maximality we may find $x_i \in \mathcal{C}_{h+1}$ such that $|y - x_i| \le 3M^{-(h+2)}$. In conclusion, we have

$$|x-x_i| \le |x-z| + |z-y| + |y-x_i| \le 7M^{-(h+2)},$$

so that (7.5) follows at once.

Step 3. *the K_h's satisfy a suitable d.l.b. as that of K in* (7.1).
We claim that for every $h \in \mathbb{N}$

$$K_h \cap B_{M^{-(h+1)}}(x_i) = K \cap B_{M^{-(h+1)}}(x_i) \qquad \text{for all } x_i \in \mathcal{C}_h. \tag{7.9}$$

In particular, from the latter we infer the conclusion of this step.

The equality above is proven by contradiction: assume we can find $x_i \in \mathcal{C}_h$ and

$$x \in (K \setminus K_h) \cap B_{M^{-(h+1)}}(x_i).$$

As $x_i \in \widetilde{K}_h$ then $x_i \in B_{T_h}$, in turn implying $x \in B_{S_h}$ since $S_h - T_h = 4M^{-(h+1)}$. Therefore $x \in \big(K \setminus K_h\big) \cap \overline{B}_{S_h}$, and by definition of K_h we can find a ball $B \in \mathcal{F}_j$, $j \le h$, such that $x \in B$. We conclude that

$$\operatorname{dist}(x_i, B) \le |x-x_i| \le M^{-(h+1)},$$

contradicting that $x_i \in \widetilde{K}_h$.

Step 4. *Proof of* (7.6).
We get first a lower bound for $\#(I_h)$: use (7.5) and the d.u.b in (7.1) to get

$$\begin{aligned}\mathcal{H}^{n-1}\big(K_h \cap \overline{B}_{R_{h+1}}\big) &= \mathcal{H}^{n-1}\big(K_h \cap \overline{B}_{R_{h+1}} \cap \cup_{i\in I_h} B_{8M^{-(h+1)}}(x_i)\big)\\ &\le C_0\,\#(I_h)\big(8M^{-(h+1)}\big)^{n-1}\end{aligned}$$

that is

$$\#(I_h) M^{-(h+1)(n-1)} \ge 8^{1-n} C_0^{-1} \mathcal{H}^{n-1}\big(K_h \cap \overline{B}_{R_{h+1}}\big). \tag{7.10}$$

Thus, we estimate as follows

$$
\begin{aligned}
\mathcal{H}^{n-1}(K_{h+1}) &\leq \mathcal{H}^{n-1}\left(K_h \setminus \cup_{\mathcal{F}_{h+1}} B\right) \\
&\overset{\text{disjoint balls}}{=} \mathcal{H}^{n-1}(K_h) - \sum_{\mathcal{F}_{h+1}} \mathcal{H}^{n-1}(K_h \cap B) \\
&\overset{\text{d.l.b. in (7.1), (7.9)}}{\leq} \mathcal{H}^{n-1}(K_h) - \frac{\alpha^{n-1}}{C_0} \#(I_h) M^{-(h+1)(n-1)} \\
&\overset{(7.10)}{\leq} \mathcal{H}^{n-1}(K_h) - \frac{8^{1-n}\alpha^{n-1}}{C_0^2} \mathcal{H}^{n-1}\big(K_h \cap \overline{B}_{R_{h+1}}\big) \qquad (7.11)\\
&= (1-\eta)\mathcal{H}^{n-1}(K_h) + \eta\left(\mathcal{H}^{n-1}(K_h) - \mathcal{H}^{n-1}\big(K_h \cap \overline{B}_{R_{h+1}}\big)\right) \\
&\overset{\text{def. of } K_h}{\leq} (1-\eta)\mathcal{H}^{n-1}(K_h) + \eta\, \mathcal{H}^{n-1}\big(K \cap (\overline{B}_{S_h} \setminus \overline{B}_{R_{h+1}})\big) \\
&\overset{\text{(iii) Lemma 7.1}}{\leq} (1-\eta)\mathcal{H}^{n-1}(K_h) + C_1 M^{-\frac{h+1}{2}},
\end{aligned}
$$

where we have set $\eta := 8^{1-n}/\alpha^{n-1}C_0^2$.

By iteration of (7.11), we find by Young inequality

$$
\mathcal{H}^{n-1}(K_h) \leq C_1 \sum_{i=0}^{h} (1-\eta)^{h-i} M^{-i/2} \leq C_1\, h\, \max\left\{(1-\eta)^h, M^{-h/2}\right\}.
$$

Choose $\beta \in (0, 1/4)$ such that $(1-\eta) \leq M^{-2\beta}$, the previous estimate then yields (7.6),

$$
\mathcal{H}^{n-1}(K_h) \leq C_1\, h\, \max\{M^{-2h\beta}, M^{-h/2}\} = C_1\, h\, M^{-2h\beta}.
$$

Step 5. *Proof of* (7.7).
Then, we exploit (7.5) to get

$$
\begin{aligned}
&\mathcal{L}^n\left((K_{h+1} \cap \overline{B}_{R_{h+2}})_{M^{-(h+2)}}\right) \\
&\leq \mathcal{L}^n\left(\cup_{i \in I_{h+1}} B_{8M^{-(h+2)}}(x_i)\right) \\
&\leq \#(I_{h+1})\big(8M^{-(h+2)}\big)^n \\
&\overset{\text{d.l.b. in (7.1), (7.9)}}{\leq} \frac{8^n}{C_0} M^{-(h+2)} \sum_{i \in I_{h+1}} \mathcal{H}^{n-1}\big(K_{h+1} \cap B_{M^{-(h+2)}}(x_i)\big) \qquad (7.12)\\
&\overset{\text{disjoint balls}}{\leq} \frac{8^n}{C_0} M^{-(h+2)} \mathcal{H}^{n-1}\big(K_{h+1}\big) \\
&\overset{(7.6)}{\leq} \frac{8^n C_1}{C_0} (h+1)\, M^{-2(h+1)\beta-(h+2)}.
\end{aligned}
$$

Step 6. *Proof of* (7.8)

By construction we have that $\Sigma_u \cap B_{1/2} \subseteq K_h$. Therefore, (7.7) gives as $R_h \geq R_\infty \geq 1/2$

$$\mathcal{L}^n\left((\Sigma_u \cap B_{1/2})_{M^{-(h+1)}}\right) \leq \mathcal{L}^n\left((K_h \cap \overline{B}_{R_{h+1}})_{M^{-(h+1)}}\right) \leq C_2\, h\, M^{-h(1+2\beta)-1}.$$

Hence, if $r \in (M^{-(h+2)}, M^{-(h+1)}]$ we get

$$\mathcal{L}^n\left((\Sigma_u \cap B_{1/2})_r\right) \leq C_2\, h\, M^{-h(1+2\beta)-1} \leq C_2\, M^{-h(1+\beta)-1} \leq C_2\, r^{1+\beta}. \quad \square$$

Remark 7.4. Apart from Step 2, all the arguments in the other steps of Proposition 7.3 employ only the Alfhors regularity of $\Omega \cap \overline{S_u}$ and its consequence Lemma 7.1.

In addition, note that one can easily infer the (more) intrinsic estimates

$$\mathcal{H}^{n-1}\left(\overline{S_u} \cap (K_h \cap \overline{B}_{R_{h+1}})_{M^{-(h+1)}}\right) \leq C_2\, h\, M^{-2h\beta},$$

and

$$\mathcal{H}^{n-1}\left(\overline{S_u} \cap (\Sigma_u \cap B_{1/2})_r\right) \leq C_2\, r^\beta \qquad \forall r \in (0, 1/2].$$

Indeed, by arguing as in (7.12) we get

$$\begin{aligned}
&\mathcal{H}^{n-1}\left(\overline{S_u} \cap (K_{h+1} \cap \overline{B}_{R_{h+2}})_{M^{-(h+2)}}\right) \\
&\leq \mathcal{H}^{n-1}\left(\overline{S_u} \cap \cup_{i \in I_{h+1}} B_{8M^{-(h+2)}}(x_i)\right) \\
&\leq \#(I_{h+1})\left(8M^{-(h+2)}\right)^{n-1} \\
&\overset{\text{d.l.b. in (7.1), (7.9)}}{\leq} \frac{8^{n-1}}{C_0} \sum_{i \in I_{h+1}} \mathcal{H}^{n-1}\left(K_{h+1} \cap B_{M^{-(h+2)}}(x_i)\right) \\
&\overset{\text{disjoint balls}}{\leq} \frac{8^{n-1}}{C_0} \mathcal{H}^{n-1}\left(K_{h+1}\right) \\
&\overset{(7.6)}{\leq} \frac{8^{n-1}\, C_1}{C_0}\, (h+1)\, M^{-2(h+1)\beta}.
\end{aligned}$$

As outlined in Section 6 the former result leads to the higher integrability of the gradient for MS-minimizers in any dimension.

Theorem 7.5 (De Philippis and Figalli [37]). *There is $p > 2$ such that $\nabla u \in L^p_{\mathrm{loc}}(\Omega)$ for all $u \in \mathcal{M}(\Omega)$ and for all open sets $\Omega \subseteq \mathbb{R}^n$.*

Proof. Clearly, it is sufficient for our purposes to prove a localized estimate. Hence, for the sake of simplicity we suppose that $\Omega = B_2$.

We keep the notation of Proposition 7.3 and furthermore denote for all $h \in \mathbb{N}$

$$A_h := \left\{x \in B_2 \setminus K : |\nabla u(x)|^2 > M^{h+1}\right\}. \tag{7.13}$$

We claim that

$$A_{h+2} \cap B_{R_{h+2}} \subset (K_h \cap B_{R_{h+1}})_{M^{-(h+1)}}. \tag{7.14}$$

Given this for granted we conclude as follows: we use (7.7) to deduce that

$$\begin{aligned} \mathcal{L}^n(A_{h+2} \cap B_{R_{h+2}}) &\leq \mathcal{L}^n\big((K_h \cap B_{R_{h+1}})_{M^{-(h+1)}}\big) \\ &\leq C_2\, h\, M^{-h(1+2\beta)-1}. \end{aligned} \tag{7.15}$$

Therefore, recalling that $1/2 \leq R_\infty \leq R_h$, in view of (7.15) and Cavalieri's formula for $q > 1$ we get that

$$\begin{aligned} \int_{B_{1/2}} |\nabla u|^{2q} dx &= q \int_0^\infty t^{q-1} \mathcal{L}^n\big(\big\{x \in B_{1/2} \setminus K : |\nabla u(x)|^2 > t\big\}\big) dt \\ &\leq q \sum_{h\geq 3} \int_{M^h}^{M^{h+1}} t^{q-1} \mathcal{L}^n\big(\big\{x \in B_{1/2} \setminus K : |\nabla u(x)|^2 > t\big\}\big) dt + M^{3q} \mathcal{L}^n(B_{1/2}) \\ &\leq \sum_{h\geq 0} M^{(h+4)q} \mathcal{L}^n\big(A_{h+2} \cap B_{1/2}\big) + M^{3q} \mathcal{L}^n(B_{1/2}) \\ &\leq C_2 \sum_{h\geq 0} h\, M^{(h+4)q-h(1+2\beta)-1} + M^{3q} \mathcal{L}^n(B_{1/2}). \end{aligned}$$

The conclusion follows at once by taking $q \in (1, 1+2\beta)$ and $p = 2q$.

Let us now prove formula (7.14) in two steps.

Step 1. *For all $M > n$ and $R \in (0, 1]$ we have that*

$$A_h \cap B_{R-2M^{-h}} \subset \big(K \cap B_R\big)_{M^{-h}} \qquad \text{for all } h \in \mathbb{N}. \tag{7.16}$$

Indeed, for $x \in A_h \cap B_{R-2M^{-h}}$ let $z \in K$ be such that $\mathrm{dist}(x, K) = |x - z|$. If $|x-z| > M^{-h}$ then $B_{M^{-h}}(x) \cap K = \emptyset$ so that u is harmonic on $B_{M^{-h}}(x)$. Therefore, by subharmonicity of $|\nabla u|^2$ on the same set and the d.u.b. in (7.1) we infer that

$$M^{h+1} \overset{x\in A_h}{\leq} |\nabla u(x)|^2 \leq \fint_{B_{M^{-h}}(x)} |\nabla u|^2 \leq n\, M^h,$$

that is clearly impossible for $M > n$.

Finally, as $x \in B_{R-2M^{-h}}$ and $|x - z| \leq M^{-h}$ we conclude that $z \in B_R$.

Step 2. *Proof of* (7.14).

Since $R_{h+1} - R_{h+2} \geq 8M^{-(h+2)}$ (*cf.* (i) Lemma 7.1), we apply Step 1 to A_{h+2} and $R = R_{h+1}$ and then (7.16) implies that

$$A_{h+2} \cap B_{R_{h+2}} \subset \big(K \cap B_{R_{h+1}}\big)_{M^{-(h+1)}}.$$

Let $x \in A_{h+2} \cap B_{R_{h+2}}$, $z \in K \cap B_{R_{h+1}}$ be a point of minimal distance, and suppose that $z \in K \setminus K_h$.

Since $R_{h+1} \leq S_h$, by the very definition of K_h we find a ball $B \in \cup_{j=1}^h \mathcal{F}_j$ such that $z \in B$. Since $B = B_t(y)$ for some y and with the radius $t \geq \alpha M^{-h}$, then $x \in B_{2t}(y)$ as $|x - z| \leq M^{-(h+1)}$ for M sufficiently large. Thus, estimate $|\nabla u(x)|^2 < M^{h+1}$ follows from (7.3) in item (ii) of Proposition 7.3. This is in contradiction with $x \in A_{h+2}$. □

References

[1] G. ALBERTI, G. BOUCHITTÉ and G. DAL MASO, *The calibration method for the Mumford-Shah functional and free discontinuity problems*, Calc. Var. Partial Differential Equations **16** (2003), 299–333.

[2] L. AMBROSIO, *A compactness theorem for a new class of functions of bounded variation*, Boll. Un. Mat. Ital. B (7) **3** (1989), 857–881.

[3] L. AMBROSIO and D. PALLARA, *Partial regularity of free discontinuity sets. I*, Ann. Scuola Norm. Sup. Pisa Cl. Sci. (4) **24** (1997), 1–38.

[4] L. AMBROSIO, N. FUSCO and J. E. HUTCHINSON, *Higher integrability of the gradient and dimension of the singular set for minimizers of the Mumford-Shah functional*, Calc. Var. Partial Differential Equations **16** (2003), 187–215.

[5] L. AMBROSIO, N. FUSCO and D. PALLARA, *Partial regularity of free discontinuity sets. II*, Ann. Scuola Norm. Sup. Pisa Cl. Sci. (4) **24** (1997), 39–62.

[6] L. AMBROSIO, N. FUSCO and D. PALLARA, *Higher regularity of solutions of free discontinuity problems*, Diff. Int. Eqs. **12** (1999), 499–520.

[7] L. AMBROSIO, N. FUSCO and D. PALLARA, "Functions of Bounded Variation and Free Discontinuity Problems", in the Oxford Mathematical Monographs, The Clarendon Press Oxford University Press, New York, 2000.

[8] M. BARCHIESI and M. FOCARDI, *Homogenization of the Neumann problem in perforated domains: an alternative approach*, Calc. Var. Partial Differential Equations **42** (2011), 257–288.

[9] M. BONACINI and M. MORINI, *Stable regular critical points of the Mumford-Shah functional are local minimizers*, Ann. Inst. H. Poincaré Anal. Non Linéaire, to appear.

[10] A. BONNET, *On the regularity of edges in image segmentation*, Ann. Inst. H. Poincaré Analyse Non Linéaire, **13** (1996), 485–528.

[11] A. BONNET and G. DAVID, "Cracktip is a Global Mumford-Shah Minimizer", Astérisque No. 274, 2001.

[12] B. BOURDIN, G. FRANCFORT and J. J. MARIGO, *The variational approach to fracture*, J. Elasticity **91** (2008), 5–148.

[13] A. BRAIDES, "Approximation of Free Discontinuity Problems", Lecture Notes in Mathematics, Vol. 1694, Springer-Verlag, Berlin, 1998.

[14] A. BRAIDES, "Γ-Convergence for Beginners", Oxford University Press, Oxford, 2002.

[15] A. BRAIDES, "A handbook of Γ-Convergence", In: Handbook of Differential Equations. Stationary Partial Differential Equations, Volume 3, M. Chipot and P. Quittner (eds.), Elsevier, 2006.

[16] D. BUCUR and S. LUCKHAUS, *Monotonicity formula and regularity for general free discontinuity problems*, Arch. Ration. Mech. Anal. **211** (2014), 489–511.

[17] M. CARRIERO and A. LEACI, *Existence theorem for a Dirichlet problem with free discontinuity set*, Nonlinear Anal. **15** (1990), 661–677.

[18] G. CONGEDO and I. TAMANINI, *On the existence of solutions to a problem in multidimensional segmentation*, Ann. Inst. Henry Poincaré **8** (1991), 175–195.

[19] R. CRISTOFERI, *A local minimality criterion for the triple point of the Mumford-Shah functional*, in preparation.

[20] G. DAL MASO, "An Introduction to Γ-Convergence", Progress in Nonlinear Differential Equations and their Applications, Vol. 8, Birkhäuser Boston, Inc., Boston, MA, 1993.

[21] G. DAL MASO, G. FRANCFORT and R. TOADER, *Quasistatic crack growth in nonlinear elasticity*, Arch. Ration. Mech. Anal. **176** (2005), 165–225.

[22] G. DAL MASO, M. G. MORA and M. MORINI, *Local calibrations for minimizers of the Mumford-Shah functional with rectilinear discontinuity sets*, J. Math. Pures Appl. (9) **79** (2000), 141–162.

[23] G. DAL MASO, J. M. MOREL and S. SOLIMINI, *A variational method in image segmentation: existence and approximation results*, Acta Math. **168** (1992), 89–151.

[24] G. DAVID, "Wavelets and Singular Integrals on Curves and Surfaces", Lecture Notes in Mathematics, Vol. 1465. Springer-Verlag, Berlin, 1991.

[25] G. DAVID, *C^1-arcs for minimizers of the Mumford-Shah functional*, SIAM J. Appl. Math. **56** (1996), 783–888.

[26] G. DAVID, "Singular Sets of Minimizers for the Mumford-Shah Functional", Progress in Mathematics, Vol. 233, Birkhäuser Verlag, Basel, 2005.

[27] G. DAVID and J. C. LÉGER, *Monotonicity and separation for the Mumford-Shah problem*, Ann. Inst. H. Poincaré Anal. Non Linéaire **19** (2002), 631–682.

[28] G. DAVID and S. SEMMES, "Fractured Fractals and Broken Dreams. Self-similar Geometry through Metric and Measure", Oxford Lecture Series in Mathematics and its Applications, Vol. 7, The Clarendon Press, Oxford University Press, New York, 1997.

[29] E. DE GIORGI, *Variational free discontinuity problems*, International Conference in Memory of Vito Volterra (Rome, 1990), Atti Convegni Lincei, Accad. Naz. Lincei, Roma **92** (1992), 133–150.

[30] E. DE GIORGI, *Problemi con discontinuità libera*, Int. Symp. "Renato Caccioppoli" (Napoli 1989), Ricerche Mat., suppl. **40** (1991), 203–214.

[31] E. DE GIORGI, *Free discontinuity problems in calculus of variations*, Frontiers in Pure and Applied Mathemathics, North Holland, Amsterdam (1991), 55–62.

[32] E. DE GIORGI and L. AMBROSIO, *Un nuovo funzionale del calcolo delle variazioni*, Atti Accad. Naz. Lincei Rend. Cl. Sci. Fis. Mat. Natur. **82** (1988), 199–210.

[33] E. DE GIORGI, M. CARRIERO and A. LEACI, *Existence theorem for a minimum problem with free discontinuity set*, Arch. Ration. Mech. Anal. **108** (1989), 195–218.

[34] C. DE LELLIS and M. FOCARDI, *Density lower bound estimates for local minimizers of the 2d Mumford-Shah energy*, Manuscripta Math. **142** (2013), 215–232.

[35] C. DE LELLIS and M. FOCARDI, *Higher integrability of the gradient for minimizers of the 2d Mumford-Shah energy*, J. Math. Pures Appl. (9) **100** (2013), 391–409.

[36] C. DE LELLIS, M. FOCARDI and B. RUFFINI, *A note on the Hausdorff dimension of the singular set for minimizers of the Mumford-Shah energy*, Adv. Calc. Var. **7** (2014), 539–545.

[37] G. DE PHILIPPIS and A. FIGALLI, *Higher Integrability for Minimizers of the Mumford-Shah Functional*, Arch. Ration. Mech. Anal. **213** (2014), 491–502.

[38] G. DE PHILIPPIS and A. FIGALLI, *A note on the dimension of the singular set in free interface problems*, Differential Integral Equations **28** (2015), 523–536.

[39] L. C. EVANS, "Partial Differential Equations", Second edition. Graduate Studies in Mathematics, Vol. 19, American Mathematical Society, Providence, RI, 2010.

[40] M. FOCARDI, *Γ-convergence: a tool to investigate physical phenomena across scales*, Math. Methods Appl. Sci. **35** (2012), 1613–1658.

[41] M. FOCARDI, M. S. GELLI and M. PONSIGLIONE, *Fracture mechanics in perforated domains: a variational model for brittle porous media*, Math. Models Methods Appl. Sci. **19** (2009), 2065–2100.

[42] M. FOCARDI, A. MARCHESE and E. SPADARO, *Improved estimate of the singular set of Dir-minimizing Q-valued functions via an abstract regularity result*, J. Funct. Anal. **268** (2015), 3290-3325.

[43] I. FONSECA and N. FUSCO, *Regularity results for anisotropic image segmentation models*, Ann. Scuola Norm. Sup. Pisa Cl. Sci. (4) **24** (1997), 463–499.

[44] N. FUSCO, *An overview of the Mumford-Shah problem*. Milan J. Math. **71** (2003), 95–119.

[45] L. KARP, T. KILPELÄINEN, A. PETROSYAN and H. SHAHGHOLIAN, *On the porosity of free boundaries in degenerate variational inequalities*, J. Differential Equations **164** (2000), 110–117.

[46] H. KOCH, G. LEONI and M. MORINI, *On optimal regularity of free boundary problems and a conjecture of De Giorgi*, Comm. Pure Appl. Math. **58** (2005), 1051–1076.

[47] J. KRISTENSEN and G. MINGIONE, *The singular set of minima of integral functionals*, Arch. Ration. Mech. Anal. **180** (2006), 331–398.

[48] J. KRISTENSEN and G. MINGIONE, *The singular set of Lipschitzian minima of multiple integrals*, Arch. Ration. Mech. Anal. **184** (2007), 341–369.

[49] M. GIAQUINTA and G. MODICA, *Regularity results for some classes of higher order nonlinear elliptic systems*, J. Reine Angew. Math., **311/312** (1979), 145–169.

[50] J. C. LÉGER, *Flatness and finiteness in the Mumford-Shah problem*, J. Math. Pures App. **78** (1999), 431–459.

[51] A. LEMENANT, *Regularity of the singular set for Mumford-Shah minimizers in $\mathbb{R}^3$ near a minimal cone*, Ann. Sc. Norm. Super. Pisa Cl. Sci. **10** (2011), 561–609.

[52] A. LEMENANT, *A rigidity result for global Mumford-Shah minimizers in dimension three*, J. Math. Pures Appl. (9) **103** (2015), 1003–1023.

[53] A. LEMENANT, *A selective review on Mumford-Shah minimizers*, Boll. Unione Mat. Ital. **9** (2016), 69–113.

[54] G. P. LEONARDI and I. TAMANINI, *Metric spaces of partitions, and Caccioppoli partitions*, Adv. Math. Sci. Appl. **12** (2002), 725–753.

[55] F. MADDALENA and S. SOLIMINI, *Regularity properties of free discontinuity sets*. Ann. Inst. H. Poincaré Anal. Non Linéaire **18** (2001), no. 6, 675–685.

[56] F. MADDALENA and S. SOLIMINI, *Lower semicontinuity properties of functionals with free discontinuities*, Arch. Rational Mech. Anal. **159** (2001), 273–294.

[57] F. MADDALENA and S. SOLIMINI, *Blow-up techniques and regularity near the boundary for free discontinuity problems*, Advanced Nonlinear Studies **1** (2001), 1–41.

[58] U. MASSARI and I. TAMANINI, *Regularity properties of optimal segmentations*, J. für Reine Angew. Math. **420** (1991), 61–84.

[59] P. MATTILA, "Geometry of Sets and Measures in Euclidean Spaces. Fractals and Rectifiability", Cambridge Studies in Advanced Mathematics, Vol. 44, Cambridge University Press, Cambridge, 1995.

[60] A. MIELKE, *Evolution in rate-independent systems* (ch. 6), In: "Handbook of Differential Equations, Evolutionary Equations", C. Dafermos and E. Feireisl, (eds.), Elsevier B.V. **2** (2005), 461–559.

[61] G. MINGIONE, *The singular set of solutions to non-differentiable elliptic systems*, Arch. Ration. Mech. Anal. **166** (2003), 287–301.

[62] G. MINGIONE, *Bounds for the singular set of solutions to non linear elliptic systems*, Calc. Var. Partial Differential Equations **18** (2003), 373–400.

[63] G. MINGIONE, *Singularities of minima: a walk on the wild side of the calculus of variations*, J. Global Optim. **40** (2008), 209–223.

[64] R. L. MOORE, *Concerning triods in the plane and the junction points of plane continua*, Proceedings of the National Academy of Sciences of the United States of America **14** (1928), 85–88.

[65] M. G. MORA, *Local calibrations for minimizers of the Mumford-Shah functional with a triple junction*, Commun. Contemp. Math. **4** (2002), 297–326.

[66] M. G. MORA, *The calibration method for free discontinuity problems on vector-valued maps*, J. Convex Anal. **9** (2002), 1–29.

[67] M. G. MORA and M. MORINI, *Local calibrations for minimizers of the Mumford-Shah functional with a regular discontinuity set*, Ann. Inst. H. Poincaré Anal. Non Linéaire **18** (2001), 403–436.

[68] M. MORINI, *Global calibrations for the non-homogeneous Mumford-Shah functional*, Ann. Sc. Norm. Super. Pisa Cl. Sci. (5) **1** (2002), 603–648.

[69] D. MUMFORD and J. SHAH, *Optimal approximations by piecewise smooth functions and associated variational problems*, Comm. Pure Appl. Math. **42** (1989), 577–685.

[70] C. POMMERENKE, "Boundary Behavior of Conformal Maps", Grundlehren der Mathematischen Wissenschaften [Fundamental Principles of Mathematical Sciences], Vol. 299, Springer-Verlag, Berlin, 1992.

[71] S. RIGOT, *Big pieces of $C^{1,\alpha}$-graphs for minimizers of the Mumford-Shah functional*, Ann. Scuola Norm. Sup. Pisa Cl. Sci. (4) **29** (2000), 329–349.

[72] A. SALLI, *On the Minkowski dimension of strongly porous fractal sets in $\mathbb{R}^n$*, Proc. London Math. Soc. **62** (1991), 353–372.

[73] S. SEMMES, *Measure-preserving quality within mappings*, Rev. Mat. Iberoamericana **16** (2000), 363–458.

[74] L. M. SIMON, "Lectures on Geometric Measure Theory", Proceedings of the Center for Mathematical Analysis, Australian National University, Vol. 3, Canberra, 1983.

[75] B. WHITE, *Stratification of minimal surfaces, mean curvature flows, and harmonic maps*, J. reine angew. Math. **488** (1997), 1–35.

[76] G. S. YOUNG, JR., *A generalization of Moore's theorem on simple triods*, Bull. Amer. Math. Soc. **50** (1944), 714.

[77] L. ZAJÍČEK, *Porosity and σ-porosity*, Real Anal. Exchange **13**, (1987/88), 314–350.

Variational models for epitaxial growth

Giovanni Leoni

1 Introduction

These lecture notes are based on a series of lectures that I gave at the ERC School on *Free Discontinuity Problems*, at the De Giorgi Center, Pisa, Italy, July 7–July 11, 2014. The school was organized by Aldo Pratelli and Nicola Fusco. Although the basic structure of the lecture notes follows the one given at the school, I am adding here more material and a lot of the details that I skipped in class.

We are interested in the deposition of a crystalline film onto a substrate. The film is called a *mechanically thin film* when its thickness is small compared to that of the substrate (see [40]).

The most common methods for transferring material, atom by atom, to the surface of a film, which is being deposited onto a substrate, are *chemical vapor deposition* (CVD) and *physical vapor deposition* (PVD). In these methods, usually carried out in a vacuum chamber, a solid is immersed in a vapor and becomes larger due to the transference of material from the vapor to the solid. If the vapor is created with a chemical reaction, the deposition is called CVD, while if it created with physical means, it is classified as PVD.

The interface between the film crystal and the substrate crystal is called *epitaxial* if the atoms of the film material occupy natural lattice positions of the substrate material. If the film and the substrate are of the same material, the epitaxial deposition is called *homoepitaxy*, while if they are of different material, it is called *eteroepitaxy*. In eteroepitaxial growth if the mismatch in lattice parameter of the film crystal and the substrate crystal is large, then the film tends to gather into islands on the surface and the substrate is exposed between islands. This is called the *Volmer-Weber (VW) growth mode*. On the other hand, if the mismatch in lattice parameter is small the atoms of the film tend to align themselves with those of the substrate, continuing its atomic structure. This is due to the fact that the energy gain associated with the chemical bonding effect is greater than the strain in the film. This layer-by-layer film growth mode is called the *Frank–van der Merwe (FM) growth mode*. However, as the film continues to grow, the stored strain energy per unit area of interface increases linearly with the film thickness. Eventually, the presence of such a strain renders a flat layer of the film morphologically unstable or metastable, after a critical value of the thickness is reached. To release some of the elastic energy due to the strain, the atoms on the free surface of the film tend to rearrange into a more favorable configuration. Typically, after entering the instability regime, the film surface becomes wavy or the material agglomerates into clusters or isolated islands on the substrate surface. Island formation in systems such as In-GaAs/GaAs or SiGe/Si turns out to be useful in the fabrication of modern semiconductor electronic and optoelectronic devices such as quantum dots laser.

This growth mode in which islands are separated by a *thin wetting layer* is known as the *Stranski-Krastanow (SK) growth mode*.

In the literature several atomistic and continuum theories for the growth of epitaxially strained solid films are available. Here we work in the context of continuum mechanics and we essentially follow the variational approach contained in the work of Spencer [75] (see also [58,76], and the references contained therein).

We now describe the model considered in these lecture notes. Both the film and the substrate are modeled as linearly elastic solids. To keep the geometry as simple as possible we restrict attention to an epitaxial layer (with variable thickness h) grown on a flat semi-infinite substrate. We further restrict attention to two-dimensional morphologies which correspond to three-dimensional configurations with planar symmetry.

We assume that the material occupies the infinite strip

$$\Omega_h := \{\boldsymbol{x} = (x, y) : 0 < x < b, \ y < h(x)\}, \tag{1.1}$$

where $h : [0, b] \to [0, \infty)$ is a Lipschitz function. Thus the graph of h represents the *free* profile of the film, the open set

$$\Omega_h^+ := \Omega_h \cap \{y > 0\} \tag{1.2}$$

is the reference configuration of the film, and the line $y = 0$ corresponds to the film/substrate interface. We work within the theory of small deformations, so that

$$\boldsymbol{E}(\boldsymbol{u}) := \frac{1}{2}\left(\nabla \boldsymbol{u} + \nabla \boldsymbol{u}^T\right)$$

represents the linearized strain, with $\boldsymbol{u} : \Omega_h \to \mathbb{R}^2$ the planar displacement. The displacement is measured from a configuration of the layer in which the lattices of the film and the layer are perfectly matched; this configuration, in which $\boldsymbol{E} \equiv \mathbf{0}$, will not correspond to a minimum energy state of the film, which we assume to occur at a strain $\boldsymbol{E}_0 = \boldsymbol{E}_0(y)$. We assume that this *mismatch strain* has the specific form

$$\boldsymbol{E}_0(y) = \begin{cases} \hat{\boldsymbol{E}}_0 & \text{if } y \geq 0, \\ 0 & \text{if } y < 0, \end{cases} \tag{1.3}$$

with $\hat{\boldsymbol{E}}_0 \neq \mathbf{0}$.

In our setting the film and the substrate have similar material properties, and so they share the same homogeneous elasticity tensor C. Hence, bearing in mind the mismatch, the elastic energy per unit area is given by $W(\boldsymbol{E} - \boldsymbol{E}_0(y))$, where

$$W(\boldsymbol{E}) := \frac{1}{2}\boldsymbol{E} \cdot C[\boldsymbol{E}], \tag{1.4}$$

with C a positive definite fourth-order tensor, that is,

$$\frac{1}{2}\boldsymbol{E} \cdot C[\boldsymbol{E}] > 0 \tag{1.5}$$

for all symmetric matrices $\boldsymbol{E} \neq \mathbf{0}$.

In the sharp interface model the interfacial energy density ψ has a step discontinuity at $y = 0$: It is $\gamma_{\text{film}} > 0$ if the film has positive thickness and $\gamma_{\text{sub}} > 0$ if the substrate is exposed, to be precise,

$$\psi(y) := \begin{cases} \gamma_{\text{film}} & \text{if } y > 0, \\ \gamma_{\text{sub}} & \text{if } y = 0. \end{cases} \tag{1.6}$$

Hence the total energy of the system is given by

$$\mathcal{F}(\boldsymbol{u}, h) := \int_{\Omega_h} W(\boldsymbol{E}(\boldsymbol{u}) - \boldsymbol{E}_0)\, d\boldsymbol{x} + \int_{\Gamma_h} \psi\, ds, \tag{1.7}$$

where Γ_h represents the free surface of the film, that is,

$$\Gamma_h := \partial\Omega_h \cap ((0, b) \times \mathbb{R}). \tag{1.8}$$

These lecture notes are organized as follows: in Section 2 we present a proof of Korn's inequality in C^1 and in Lipschitz domains. In Section 3 we discuss the (lack of) regularity of solutions of the Lamé system in domains with a corner. In Sections 4 and 5 we prove existence and regularity of equilibrium configurations for the functional (1.7).

Finally, in Section 6 we discuss recent results and open problems.

The material in Sections 4 and 5 is based on papers of Bonnetier and Chambolle [21], Chambolle and Larsen [22], Fonseca, Fusco, Morini and the author [33], and Fonseca, Fusco, Millot and the author [32].

ACKNOWLEDGEMENTS. The author warmly thanks the De Giorgi Center, Pisa, Italy, for the hospitality during the ERC School on *Free Discontinuity Problems*, July 7–July 11, 2014 as well as its organizers, Aldo Pratelli and Nicola Fusco.

The author would also like to thank all the participants of the school for many interesting comments and discussions, in particular, Massimiliano Morini and Nicola Fusco for their helpful insights. He also thanks Giovanni Gravina for carefully reading these lecture notes, and Irene Fonseca and Ian Tice for useful conversations.

The author wishes to acknowledge the Center for Nonlinear Analysis (NSF Grant No. DMS-0635983 and NSF PIRE Grant No. OISE-0967140) where part of this work was carried out. This work was partially funded by the NSF under Grant No. DMS-1412095, by the ERC Project "Geometric Measure Theory in non-Euclidean Spaces" and the ERC Project "Analysis of optimal sets and optimal constants: old questions and new results".

2 Korn's inequality

To prove existence of equilibrium configurations for the functional (1.7) we will need Korn's inequality in Lipschitz domains. In this section we present a proof for smooth and for Lipschitz domains. This section is based on the classical papers of Friedrichs [41] and Nitsche [66]. The following theorem is due to Friedrichs [41].

Theorem 2.1 (Korn's inequality in $\mathbb{R}^N$). *Assume that $\boldsymbol{u} \in H^1\left(\mathbb{R}^N;\mathbb{R}^N\right)$. Then*

$$\|\nabla \boldsymbol{u}\|_{L^2(\mathbb{R}^N)} \leq 2\,\|\boldsymbol{E}\,(\boldsymbol{u})\|_{L^2(\mathbb{R}^N)}\,.$$

Proof. **Step 1:** Assume first that $\boldsymbol{u} \in C_c^2\left(\mathbb{R}^N;\mathbb{R}^N\right)$. For every $i, j = 1, \ldots, N$ we write

$$\begin{aligned}\frac{\partial u_i}{\partial x_j} &= \frac{1}{2}\left(\frac{\partial u_i}{\partial x_j} + \frac{\partial u_j}{\partial x_i}\right) + \frac{1}{2}\left(\frac{\partial u_i}{\partial x_j} - \frac{\partial u_j}{\partial x_i}\right)\\ &=: s_{i,j} + a_{i,j}.\end{aligned}$$

Then

$$\frac{\partial u_i}{\partial x_j}\frac{\partial u_j}{\partial x_i} = \left(s_{i,j} + a_{i,j}\right)\left(s_{j,i} + a_{j,i}\right) = \left(s_{i,j} + a_{i,j}\right)\left(s_{i,j} - a_{i,j}\right)$$
$$= s_{i,j}^2 - a_{i,j}^2.$$

It follows that

$$\int_{\mathbb{R}^N}\left(\frac{\partial u_i}{\partial x_j}\frac{\partial u_j}{\partial x_i} - \frac{\partial u_i}{\partial x_i}\frac{\partial u_j}{\partial x_j}\right) d\boldsymbol{x} = \int_{\mathbb{R}^N}\left(s_{i,j}^2 - a_{i,j}^2 - \frac{\partial u_i}{\partial x_i}\frac{\partial u_j}{\partial x_j}\right) d\boldsymbol{x}.$$

Next we observe that by the Schwartz theorem,

$$\frac{\partial u_i}{\partial x_j}\frac{\partial u_j}{\partial x_i} - \frac{\partial u_i}{\partial x_i}\frac{\partial u_j}{\partial x_j} = \frac{\partial}{\partial x_j}\left(u_i\frac{\partial u_j}{\partial x_i}\right) - \frac{\partial}{\partial x_i}\left(u_i\frac{\partial u_j}{\partial x_j}\right),$$

and so by the fundamental theorem of calculus the left-hand side of the previous integral identity is zero. Hence, summing over i, j we get

$$\sum_{i,j}\int_{\mathbb{R}^N}\frac{\partial u_i}{\partial x_i}\frac{\partial u_j}{\partial x_j}\,d\boldsymbol{x} + \sum_{i,j}\int_{\mathbb{R}^N} a_{i,j}^2 d\boldsymbol{x} = \sum_{i,j}\int_{\mathbb{R}^N} s_{i,j}^2 d\boldsymbol{x},$$

which can be written as

$$\int_{\mathbb{R}^N}\left(\sum_i \frac{\partial u_i}{\partial x_i}\right)^2 d\boldsymbol{x} + \sum_{i,j}\int_{\mathbb{R}^N} a_{i,j}^2 d\boldsymbol{x} = \sum_{i,j}\int_{\mathbb{R}^N} s_{i,j}^2 d\boldsymbol{x}.$$

Thus,

$$\sum_{i,j}\int_{\mathbb{R}^N} a_{i,j}^2 d\boldsymbol{x} \leq \sum_{i,j}\int_{\mathbb{R}^N} s_{i,j}^2 d\boldsymbol{x}. \tag{2.1}$$

In turn, since $\frac{\partial u_i}{\partial x_j} = s_{i,j} + a_{i,j}$, we get $\left(\frac{\partial u_i}{\partial x_j}\right)^2 \leq 2s_{i,j}^2 + 2a_{i,j}^2$, and so

$$\int_{\mathbb{R}^N} |\nabla \boldsymbol{u}|^2\, d\boldsymbol{x} \leq 2\sum_{i,j}\int_{\mathbb{R}^N} s_{i,j}^2 d\boldsymbol{x} + 2\sum_{i,j}\int_{\mathbb{R}^N} a_{i,j}^2 d\boldsymbol{x}$$
$$\leq 4\sum_{i,j}\int_{\mathbb{R}^N} s_{i,j}^2 d\boldsymbol{x} = 4\int_{\mathbb{R}^N} |\boldsymbol{E}(\boldsymbol{u})|^2\, d\boldsymbol{x},$$

which concludes the proof in the case of $\boldsymbol{u} \in C_c^2\left(\mathbb{R}^N;\mathbb{R}^N\right)$.

Step 2: Let $\boldsymbol{u} \in C^2\left(\mathbb{R}^N;\mathbb{R}^N\right) \cap H^2\left(\mathbb{R}^N;\mathbb{R}^N\right)$. For every $j = 1, \ldots, N$ with an abuse of notation we write $\boldsymbol{x} = \left(\boldsymbol{x}_j, x_j\right) \in \mathbb{R}^{N-1} \times \mathbb{R}$, where $\boldsymbol{x}_j$ is the vector obtained from $\boldsymbol{x}$ by removing the j-th component x_j. Let

$\boldsymbol{x}_j \in \mathbb{R}^{N-1}$ be such that $u_i\left(\boldsymbol{x}_j,\cdot\right) \in H^2(\mathbb{R})$ and $\frac{\partial u_j}{\partial x_i}\left(\boldsymbol{x}_j,\cdot\right) \in H^1(\mathbb{R})$. Then

$$\liminf_{x_j\to-\infty}\left|u_i\left(\boldsymbol{x}_j,x_j\right)\frac{\partial u_j}{\partial x_i}\left(\boldsymbol{x}_j,x_j\right)\right| = 0,$$

$$\liminf_{x_j\to\infty}\left|u_i\left(\boldsymbol{x}_j,x_j\right)\frac{\partial u_j}{\partial x_i}\left(\boldsymbol{x}_j,x_j\right)\right| = 0.$$

Let $s_n \to -\infty$ and $t_n \to \infty$ be two sequences along which the above limits are zero. By the fundamental theorem of calculus

$$\left(u_i\frac{\partial u_j}{\partial x_i}\right)\left(\boldsymbol{x}_j,t_n\right) - \left(u_i\frac{\partial u_j}{\partial x_i}\right)\left(\boldsymbol{x}_j,s_n\right) = \int_{s_n}^{t_n}\frac{\partial}{\partial x_j}\left(u_i\frac{\partial u_j}{\partial x_i}\right)\left(\boldsymbol{x}_j,t\right)\,dt.$$

Since $\frac{\partial}{\partial x_j}\left(u_i\frac{\partial u_j}{\partial x_i}\right)\left(\boldsymbol{x}_j,\cdot\right)$ is integrable, by letting $n \to \infty$ it follows by the Lebesgue dominated convergence theorem that

$$0 = \int_{-\infty}^{\infty}\frac{\partial}{\partial x_j}\left(u_i\frac{\partial u_j}{\partial x_i}\right)\left(\boldsymbol{x}_j,t\right)\,dt.$$

Upon integration over $\boldsymbol{x}_j \in \mathbb{R}^{N-1}$ we obtain that

$$\int_{\mathbb{R}^N}\frac{\partial}{\partial x_j}\left(u_i\frac{\partial u_j}{\partial x_i}\right)\,d\boldsymbol{x} = 0,$$

which is what we used in Step 1.

Step 3: The general case in which $\boldsymbol{u} \in H^1\left(\mathbb{R}^N;\mathbb{R}^N\right)$ follows by mollification. We omit the details. □

Remark 2.2. The previous proof continues to work if we assume that $\boldsymbol{u} \in H^1(Q' \times \mathbb{R};\mathbb{R}^N)$, where $Q' \subset \mathbb{R}^{N-1}$ is a cube and $\boldsymbol{u}(\cdot,x_N)$ is Q'-periodic. In this case we would obtain

$$\|\nabla\boldsymbol{u}\|_{L^2(Q'\times\mathbb{R})} \le 2\,\|\boldsymbol{E}(\boldsymbol{u})\|_{L^2(Q'\times\mathbb{R})}\,.$$

The remaining of this section follows the paper of Nitsche [66]. Define the half spaces

$$\mathbb{R}^N_+ := \left\{\boldsymbol{x}\in\mathbb{R}^N :\ x_N > 0\right\},\quad \mathbb{R}^N_- := \left\{\boldsymbol{x}\in\mathbb{R}^N :\ x_N < 0\right\}.$$

Theorem 2.3 (Korn's inequality in $\mathbb{R}^N_+$). *There exists $c > 0$ such that*

$$\|\nabla\boldsymbol{u}\|_{L^2(\mathbb{R}^N_+)} \le c\,\|\boldsymbol{E}(\boldsymbol{u})\|_{L^2(\mathbb{R}^N_+)}$$

for all $\boldsymbol{u} \in H^1\left(\mathbb{R}^N_+;\mathbb{R}^N\right)$.

Proof. The idea is to extend $\boldsymbol{u}$ to a function $\boldsymbol{v} \in H^1\left(\mathbb{R}^N; \mathbb{R}^N\right)$ and such that

$$\|\boldsymbol{E}(\boldsymbol{v})\|_{L^2(\mathbb{R}^N)} \leq c\,\|\boldsymbol{E}(\boldsymbol{u})\|_{L^2(\mathbb{R}^N_+)}\,.$$

The usual reflection argument (see, *e.g.*, Theorem 12.3 in [59]) will not work, so we need to modify it appropriately. For $i = 1, \ldots, N-1$ and $\boldsymbol{x} \in \mathbb{R}^N_-$ we extend $\boldsymbol{u}$ by defining

$$\begin{aligned} v_i(\boldsymbol{x}) &:= p u_i\left(\boldsymbol{x}', -2x_N\right) + q u_i\left(\boldsymbol{x}', -x_N\right), \\ v_N(\boldsymbol{x}) &:= r u_N\left(\boldsymbol{x}', -2x_N\right) + s u_N\left(\boldsymbol{x}', -x_N\right), \end{aligned}$$

where p, q, r, s are numbers that we will choose later. Note that in order for the trace of $\boldsymbol{v}$ on $\mathbb{R}^{N-1} \times \{0\}$ to coincides with the trace of $\boldsymbol{u}$ on $\mathbb{R}^{N-1} \times \{0\}$ we need

$$p + q = 1, \quad r + s = 1.$$

For $i, j = 1, \ldots, N-1$ and $\boldsymbol{x} \in \mathbb{R}^N_-$ we have

$$\begin{aligned} \frac{\partial v_i}{\partial x_j}(\boldsymbol{x}) + \frac{\partial v_j}{\partial x_i}(\boldsymbol{x}) = p &\left(\frac{\partial u_i}{\partial x_j}\left(\boldsymbol{x}', -2x_N\right) + \frac{\partial u_j}{\partial x_i}\left(\boldsymbol{x}', -2x_N\right)\right) \\ &+ q\left(\frac{\partial u_i}{\partial x_j}\left(\boldsymbol{x}', -x_N\right) + \frac{\partial u_j}{\partial x_i}\left(\boldsymbol{x}', -x_N\right)\right), \end{aligned}$$

while

$$\frac{\partial v_N}{\partial x_N}(\boldsymbol{x}) = -2r\frac{\partial u_N}{\partial x_N}\left(\boldsymbol{x}', -2x_N\right) - s\frac{\partial u_N}{\partial x_N}\left(\boldsymbol{x}', -x_N\right).$$

Finally for $i = 1, \ldots, N-1$ and $\boldsymbol{x} \in \mathbb{R}^N_-$ we have

$$\begin{aligned} \frac{\partial v_i}{\partial x_N}(\boldsymbol{x}) + \frac{\partial v_N}{\partial x_i}(\boldsymbol{x}) &= -2p\frac{\partial u_i}{\partial x_N}\left(\boldsymbol{x}', -2x_N\right) + r\frac{\partial u_N}{\partial x_i}\left(\boldsymbol{x}', -2x_N\right) \\ &\quad - q\frac{\partial u_i}{\partial x_N}\left(\boldsymbol{x}', -x_N\right) + s\frac{\partial u_N}{\partial x_i}\left(\boldsymbol{x}', -2x_N\right) \\ &= r\left(\frac{\partial u_i}{\partial x_N}\left(\boldsymbol{x}', -2x_N\right) + \frac{\partial u_N}{\partial x_i}\left(\boldsymbol{x}', -2x_N\right)\right) \\ &\quad + s\left(\frac{\partial u_i}{\partial x_N}\left(\boldsymbol{x}', -x_N\right) + \frac{\partial u_N}{\partial x_i}\left(\boldsymbol{x}', -2x_N\right)\right) \end{aligned}$$

provided we take $-2p = r$ and $-q = s$. Hence,

$$\begin{aligned} p + q = 1, &\quad r + s = 1, \\ -2p = r, &\quad -q = s, \end{aligned}$$

which gives $p = -2, q = 3, r = 4, s = -3$. In conclusion we have shown that for all $i, j = 1, \ldots, N$ and $\boldsymbol{x} \in \mathbb{R}^N_-$ we have

$$\left|\frac{\partial v_i}{\partial x_j}(\boldsymbol{x}) + \frac{\partial v_j}{\partial x_i}(\boldsymbol{x})\right| \leq 4\left|\frac{\partial u_i}{\partial x_j}\left(\boldsymbol{x}', -2x_N\right) + \frac{\partial u_j}{\partial x_i}\left(\boldsymbol{x}', -2x_N\right)\right| + 3\left|\frac{\partial u_i}{\partial x_j}\left(\boldsymbol{x}', -x_N\right) + \frac{\partial u_j}{\partial x_i}\left(\boldsymbol{x}', -x_N\right)\right|,$$

and so

$$\begin{aligned}\int_{\mathbb{R}^N_-} |\boldsymbol{E}(\boldsymbol{v})|^2 \, d\boldsymbol{x} &\leq c \int_{\mathbb{R}^N_-} \left|\boldsymbol{E}(\boldsymbol{u})\left(\boldsymbol{x}', -2x_N\right)\right|^2 d\boldsymbol{x} \\ &\quad + c \int_{\mathbb{R}^N_-} \left|\boldsymbol{E}(\boldsymbol{v})\left(\boldsymbol{x}', -x_N\right)\right|^2 d\boldsymbol{x} \\ &= c \int_{\mathbb{R}^N_+} |\boldsymbol{E}(\boldsymbol{u})|^2 \, d\boldsymbol{x},\end{aligned}$$

where we have used the change of variables $\boldsymbol{y} = \left(\boldsymbol{x}', -2x_N\right)$ and $\boldsymbol{z} = \left(\boldsymbol{x}', -x_N\right)$. It follows by the previous theorem applied to $\boldsymbol{v}$ that

$$\begin{aligned}\int_{\mathbb{R}^N_+} |\nabla \boldsymbol{u}|^2 \, d\boldsymbol{x} \leq \int_{\mathbb{R}^N} |\nabla \boldsymbol{v}|^2 \, d\boldsymbol{x} &\leq 4 \int_{\mathbb{R}^N} |\boldsymbol{E}(\boldsymbol{v})|^2 \, d\boldsymbol{x}, \\ &= 4 \int_{\mathbb{R}^N_+} |\boldsymbol{E}(\boldsymbol{u})|^2 \, d\boldsymbol{x} + 4 \int_{\mathbb{R}^N_-} |\boldsymbol{E}(\boldsymbol{v})|^2 \, d\boldsymbol{x} \\ &\leq c \int_{\mathbb{R}^N_+} |\boldsymbol{E}(\boldsymbol{u})|^2 \, d\boldsymbol{x}.\end{aligned}$$

This concludes the proof. □

Remark 2.4. In view of Remark 2.2, the previous proof continues to work if we assume that $\boldsymbol{u} \in H^1(Q' \times (0, \infty); \mathbb{R}^N)$, where $Q' \subset \mathbb{R}^{N-1}$ is a cube and $\boldsymbol{u}(\cdot, x_N)$ is Q'-periodic. In this case we obtain

$$\|\nabla \boldsymbol{u}\|_{L^2(Q' \times \mathbb{R})} \leq c \, \|\boldsymbol{E}(\boldsymbol{u})\|_{L^2(Q' \times \mathbb{R})} \, .$$

Next we consider the case of a Lipschitz domain, which is almost flat.

Theorem 2.5. *Let*

$$\Omega = \left\{\boldsymbol{x} \in \mathbb{R}^N : x_N > f\left(\boldsymbol{x}'\right)\right\},$$

where $f : \mathbb{R}^{N-1} \to \mathbb{R}$ *is a Lipschitz function, and let* $\boldsymbol{u} \in H^1\left(\Omega; \mathbb{R}^N\right)$. *Then there exists a constant* $\theta \in (0, 1)$ *such that*

$$\|\nabla \boldsymbol{u}\|_{L^2(\Omega)} \leq c \, \|\boldsymbol{E}(\boldsymbol{u})\|_{L^2(\Omega)} \, ,$$

provided

$$\|\nabla_{x'} f\|_{L^\infty(\mathbb{R}^{N-1})} \leq \theta, \tag{2.2}$$

where $c > 0$ is a constant independent of f and $\boldsymbol{u}$.

Proof. Consider the transformation $\boldsymbol{\Psi} : \mathbb{R}^N \to \mathbb{R}^N$ defined by

$$\boldsymbol{\Psi}(\boldsymbol{y}) := \left(\boldsymbol{y}', y_N + f(\boldsymbol{y}')\right) := \boldsymbol{x}.$$

Note that Ψ sends points of the form $(\boldsymbol{y}', 0)$ into $(\boldsymbol{y}', f(\boldsymbol{x}'))$, so it maps $\mathbb{R}^N_+$ into Ω. Moreover,

$$\boldsymbol{\Psi}^{-1}(\boldsymbol{x}) = \left(\boldsymbol{x}', x_N - f(\boldsymbol{x}')\right) = \boldsymbol{y}$$

and

$$\nabla_{\boldsymbol{y}} \boldsymbol{\Psi}(\boldsymbol{y}) = \begin{pmatrix} & 0 \\ I_{N-1} & \vdots \\ & 0 \\ \nabla_{y'} f(\boldsymbol{y}') & 1 \end{pmatrix} \tag{2.3}$$

for $\mathcal{L}^{N-1}$ a.e. $\boldsymbol{y}' \in \mathbb{R}^{N-1}$. Hence,

$$\det \nabla_{\boldsymbol{y}} \boldsymbol{\Psi}(\boldsymbol{y}) = 1$$

for $\mathcal{L}^{N-1}$ a.e. $\boldsymbol{y}' \in \mathbb{R}^{N-1}$ and all $y_N \in \mathbb{R}$. Consider the function

$$\boldsymbol{w}(\boldsymbol{y}) := \boldsymbol{u}(\boldsymbol{\Psi}(\boldsymbol{y})) = \boldsymbol{u}\left(\boldsymbol{y}', y_N + f(\boldsymbol{y}')\right). \tag{2.4}$$

By the previous theorem applied to $\boldsymbol{w}$,

$$\int_{\mathbb{R}^N_+} \left|\nabla_{\boldsymbol{y}} \boldsymbol{w}(\boldsymbol{y})\right|^2 d\boldsymbol{y} \leq c \int_{\mathbb{R}^N_+} |\boldsymbol{E}(\boldsymbol{w})(\boldsymbol{y})|^2 d\boldsymbol{y}.$$

Now $\boldsymbol{u}(\boldsymbol{x}) = \boldsymbol{w}\left(\boldsymbol{\Psi}^{-1}(\boldsymbol{x})\right) = \boldsymbol{w}\left(\boldsymbol{x}', x_N - f(\boldsymbol{x}')\right)$ and so for $i = 1, \ldots, N$, $j = 1, \ldots, N-1$ and $\boldsymbol{x} \in \Omega$ we have

$$\begin{aligned} \frac{\partial u_i}{\partial x_j}(\boldsymbol{x}) &= \frac{\partial w_i}{\partial y_j}\left(\boldsymbol{\Psi}^{-1}(\boldsymbol{x})\right) - \frac{\partial f}{\partial x_j}(\boldsymbol{x}') \frac{\partial w_i}{\partial y_N}\left(\boldsymbol{\Psi}^{-1}(\boldsymbol{x})\right), \\ \frac{\partial u_i}{\partial x_N}(\boldsymbol{x}) &= \frac{\partial w_i}{\partial y_N}\left(\boldsymbol{\Psi}^{-1}(\boldsymbol{x})\right). \end{aligned} \tag{2.5}$$

It follows that

$$\begin{aligned}
\int_{\Omega} |\nabla_x \boldsymbol{u}(\boldsymbol{x})|^2 \, d\boldsymbol{x} &\le 2\left(1+\theta^2\right) \int_{\Omega} \left|\nabla_y \boldsymbol{w}\left(\boldsymbol{\Psi}^{-1}(\boldsymbol{x})\right)\right|^2 d\boldsymbol{x} \\
&= 2\left(1+\theta^2\right) \int_{\Omega} \left|\nabla_y \boldsymbol{w}\left(\boldsymbol{\Psi}^{-1}(\boldsymbol{x})\right)\right|^2 \left|\det \nabla_x \boldsymbol{\Psi}^{-1}(\boldsymbol{x})\right| d\boldsymbol{x} \\
&= 2\left(1+\theta^2\right) \int_{\mathbb{R}^N_+} \left|\nabla_y \boldsymbol{w}(\boldsymbol{y})\right|^2 d\boldsymbol{y} \\
&\le c \int_{\mathbb{R}^N_+} |\boldsymbol{E}(\boldsymbol{w})(\boldsymbol{y})|^2 \, d\boldsymbol{y},
\end{aligned}$$

where we have used (2.2), (2.3), the fact that $\det \nabla_x \boldsymbol{\Psi}^{-1}(\boldsymbol{x}) = 1$, and the change of variables $\boldsymbol{y} := \boldsymbol{\Psi}^{-1}(\boldsymbol{x})$. On the other hand, since for $i = 1, \ldots, N$, $j = 1, \ldots, N-1$ and $\boldsymbol{y} \in \mathbb{R}^N_+$ we have

$$\begin{aligned}
\frac{\partial w_i}{\partial y_j}(\boldsymbol{y}) &= \frac{\partial u_i}{\partial x_j}(\boldsymbol{\Psi}(\boldsymbol{y})) + \frac{\partial f}{\partial y_j}\left(\boldsymbol{y}'\right) \frac{\partial u_i}{\partial x_N}(\boldsymbol{\Psi}(\boldsymbol{y})), \\
\frac{\partial w_i}{\partial y_N}(\boldsymbol{y}) &= \frac{\partial u_i}{\partial x_N}(\boldsymbol{\Psi}(\boldsymbol{y})),
\end{aligned}$$

it follows that

$$|\boldsymbol{E}(\boldsymbol{w})(\boldsymbol{y})|^2 \le 2\,|\boldsymbol{E}(\boldsymbol{u})(\boldsymbol{\Psi}(\boldsymbol{y}))|^2 + 2\theta^2 \left|\frac{\partial \boldsymbol{u}}{\partial x_N}(\boldsymbol{\Psi}(\boldsymbol{y}))\right|^2,$$

and so, reasoning as above,

$$\begin{aligned}
&\int_{\mathbb{R}^N_+} |\boldsymbol{E}(\boldsymbol{w})(\boldsymbol{y})|^2 \, d\boldsymbol{y} \\
&\quad\le 2 \int_{\mathbb{R}^N_+} |\boldsymbol{E}(\boldsymbol{u})(\boldsymbol{\Psi}(\boldsymbol{y}))|^2 \, d\boldsymbol{y} + 2\theta^2 \int_{\mathbb{R}^N_+} \left|\frac{\partial \boldsymbol{u}}{\partial x_N}(\boldsymbol{\Psi}(\boldsymbol{y}))\right|^2 d\boldsymbol{y} \\
&\quad= 2 \int_{\Omega} |\boldsymbol{E}(\boldsymbol{u})(\boldsymbol{x})|^2 \, d\boldsymbol{x} + 2\theta^2 \int_{\Omega} \left|\frac{\partial \boldsymbol{u}}{\partial x_N}(\boldsymbol{x})\right|^2 d\boldsymbol{x}.
\end{aligned}$$

In conclusion, we have shown that

$$\int_{\Omega} |\nabla \boldsymbol{u}|^2 \, d\boldsymbol{x} \le c \int_{\Omega} |\boldsymbol{E}(\boldsymbol{u})|^2 \, d\boldsymbol{x} + c\theta^2 \int_{\Omega} |\nabla \boldsymbol{u}|^2 \, d\boldsymbol{x},$$

and so if $\theta \in (0, 1)$ is so small that $c\theta^2 \le \frac{1}{2}$, we get

$$\frac{1}{2} \int_{\Omega} |\nabla \boldsymbol{u}|^2 \, d\boldsymbol{x} \le c \int_{\Omega} |\boldsymbol{E}(\boldsymbol{u})|^2 \, d\boldsymbol{x},$$

which concludes the proof. □

Remark 2.6. If $\boldsymbol{u} = \boldsymbol{0}$ outside $\Omega \cap B(\boldsymbol{0}, R)$ for some $R > 0$ then we can replace $\|\nabla_{x'} f\|_{L^\infty(\mathbb{R}^{N-1})}$ with $\|\nabla_{x'} f\|_{L^\infty(B_{N-1}(\boldsymbol{0},R))}$ in (2.2).

Theorem 2.7. *Let $\Omega \subset \mathbb{R}^N$ be an open bounded set with C^1 boundary. Then there exists a constant $c = c(N, \Omega) > 0$ such that*

$$\|\nabla \boldsymbol{u}\|_{L^2(\Omega)} \le c\left(\|\boldsymbol{E}(\boldsymbol{u})\|_{L^2(\Omega)} + \|\boldsymbol{u}\|_{L^2(\Omega)}\right) \tag{2.6}$$

for all $\boldsymbol{u} \in H^1\left(\Omega; \mathbb{R}^N\right)$.

Proof. Since $\partial\Omega$ is of class C^1, for every $\boldsymbol{x}_0 \in \partial\Omega$ there exist $r > 0$, local coordinates $\boldsymbol{y} = (\boldsymbol{y}', y_N)$ and a function $f : \mathbb{R}^{N-1} \to \mathbb{R}$ of class C^1 such that $\nabla_{y'} f\left(\boldsymbol{y}_0'\right) = \boldsymbol{0}$ and

$$\Omega \cap B(\boldsymbol{x}_0, r) = \left\{\boldsymbol{y} \in B(\boldsymbol{x}_0, r) : y_N > f\left(\boldsymbol{y}'\right)\right\},$$

where $(\boldsymbol{y}_0', y_{0,N})$ are the local coordinates of $\boldsymbol{x}_0$. Since f is of class C^1, there exists $0 < r_{x_0} \le r$ such that for all $\boldsymbol{y}' \in B_{N-1}\left(\boldsymbol{y}_0', r_{x_0}\right)$ and $i = 1, \ldots, N-1$,

$$\left|\frac{\partial f}{\partial y_i}\left(\boldsymbol{y}'\right)\right| = \left|\frac{\partial f}{\partial y_i}\left(\boldsymbol{y}'\right) - \frac{\partial f}{\partial y_i}\left(\boldsymbol{y}_0'\right)\right| \le \frac{\theta}{\sqrt{N-1}}. \tag{2.7}$$

Hence, $\left\|\nabla_{y'} f\right\|_{C(B_{N-1}(\boldsymbol{y}_0', r_{x_0}))} \le \theta$. Since $\partial\Omega$ is compact and

$$\partial\Omega \subset \bigcup_{\boldsymbol{x}\in\partial\Omega} B\left(\boldsymbol{x}, \tfrac{1}{2} r_{\boldsymbol{x}}\right),$$

there exists a finite number of such balls with centers $\boldsymbol{x}_i$ and radii $r_i := r_{r_{x_i}}$, $i = 1, \ldots, \ell$, such that

$$\partial\Omega \subset \bigcup_{i=1}^{\ell} B\left(\boldsymbol{x}_i, \tfrac{1}{2} r_i\right).$$

Set $r_0 := \min_i r_i > 0$ and define

$$\Omega_0 := \{\boldsymbol{x} \in \Omega : \operatorname{dist}(\boldsymbol{x}, \partial\Omega) > r_0/2\}.$$

Then

$$\Omega \subseteq \Omega_0 \cup \bigcup_{i=1}^{\ell} B(\boldsymbol{x}_i, r_i)$$

Consider a partition of unity $\{\phi_n\}_{n=1}^{M}$ subordinated to the family of open sets $\{\Omega_0\} \cup \{B(\boldsymbol{x}_i, r_i)\}_{i=1}^{\ell}$. For each n, if $\operatorname{supp}\phi_n \subseteq \Omega_0$, then we can apply Theorem 2.1 to the function $\phi_n \boldsymbol{u}$ (extended to be zero whenever $\phi_n = 0$) to conclude that

$$\begin{aligned}\|\nabla(\phi_n \boldsymbol{u})\|_{L^2(\Omega)} &= \|\nabla(\phi_n \boldsymbol{u})\|_{L^2(\mathbb{R}^2)} \le c\,\|\boldsymbol{E}\,(\phi_n \boldsymbol{u})\|_{L^2(\mathbb{R}^2)} \\ &= c\,\|\boldsymbol{E}\,(\phi_n \boldsymbol{u})\|_{L^2(\Omega)}\,.\end{aligned}$$

Otherwise $\operatorname{supp}\phi_n \subseteq B\,(\boldsymbol{x}_i, r_i)$. In this case we consider a rotation R_i such that $R_i \boldsymbol{e}_N$ is the unit outer normal to $\partial\Omega$ at $\boldsymbol{x}_i$ and define

$$\boldsymbol{v}_n(\boldsymbol{y}) := R_i^T\,(\phi_n \boldsymbol{u})(R_i\,\boldsymbol{y}),$$

where once again $\phi_n \boldsymbol{u}$ has been extended to be zero whenever $\phi_n = 0$. Then

$$\begin{aligned}\nabla_{\boldsymbol{y}} \boldsymbol{v}_n(\boldsymbol{y}) &= R_i^T\,\nabla_{\boldsymbol{x}}(\phi_n \boldsymbol{u})(R_i\,\boldsymbol{y})R_i\,,\\ \nabla_{\boldsymbol{y}} \boldsymbol{v}_n(\boldsymbol{y}) + (\nabla_{\boldsymbol{y}} \boldsymbol{v}_n(\boldsymbol{y}))^T &= R_i^T\,\nabla_{\boldsymbol{x}}(\phi_n \boldsymbol{u})(R_i\,\boldsymbol{y})R_i + (R_i^T\,\nabla_{\boldsymbol{x}}(\phi_n \boldsymbol{u})(R_i\,\boldsymbol{y})R_i)^T \\ &= R_i^T\,(\nabla_{\boldsymbol{x}}(\phi_n \boldsymbol{u}) + \nabla_{\boldsymbol{x}}(\phi_n \boldsymbol{u})^T)(R_i\,\boldsymbol{y})R_i\,.\end{aligned}$$

Hence,

$$\begin{aligned}\left\|\boldsymbol{E}_{\boldsymbol{y}}\,(\boldsymbol{v}_n)\right\|_{L^2(R_i^T\Omega)} &= \|\boldsymbol{E}_{\boldsymbol{x}}\,(\phi_n \boldsymbol{u})\|_{L^2(\Omega)}\,,\\ \left\|\nabla_{\boldsymbol{y}} \boldsymbol{v}_n\right\|_{L^2(R_i^T\Omega)} &= \|\nabla_{\boldsymbol{x}}(\phi_n \boldsymbol{u})\|_{L^2(\Omega)}\,.\end{aligned}$$

Since in local coordinates $\left\|\nabla_{\boldsymbol{y}'} f_i\right\|_{C(B_{N-1}(y_i', r_{x_i}))} \le \theta$, the function $\boldsymbol{v}_n$ satisfies the hypotheses of Theorem 2.5 (using also Remark 2.6), and so

$$\begin{aligned}\|\nabla(\phi_n \boldsymbol{u})\|_{L^2(\Omega)} &= \left\|\nabla_{\boldsymbol{y}} \boldsymbol{v}_n\right\|_{L^2(R_i^T\Omega)} \le \left\|\boldsymbol{E}_{\boldsymbol{y}}\,(\boldsymbol{v}_n)\right\|_{L^2(R_i^T\Omega)} \\ &= \|\boldsymbol{E}_{\boldsymbol{x}}\,(\phi_n \boldsymbol{u})\|_{L^2(\Omega)}\,.\end{aligned}$$

Since $\sum_{n=1}^{M} \phi_n = 1$, we have

$$\begin{aligned}\|\nabla \boldsymbol{u}\|_{L^2(\Omega)} &= \left\|\nabla\left(\sum_{n=1}^{M} \phi_n \boldsymbol{u}\right)\right\|_{L^2(\Omega)} = \left\|\sum_{n=1}^{M} \nabla(\phi_n \boldsymbol{u})\right\|_{L^2(\Omega)} \\ &\le \sum_{n=1}^{M} \|\nabla(\phi_n \boldsymbol{u})\|_{L^2(\Omega)} \le c \sum_{n=1}^{M} \|\boldsymbol{E}\,(\phi_n \boldsymbol{u})\|_{L^2(\Omega)}\,.\end{aligned}$$

Now

$$\frac{\partial\,(\phi_n u_i)}{\partial x_j} + \frac{\partial\,(\phi_n u_j)}{\partial x_i} = \phi_n \left(\frac{\partial u_i}{\partial x_j} + \frac{\partial u_j}{\partial x_i}\right) + \frac{\partial \phi_n}{\partial x_j} u_i + \frac{\partial \phi_n}{\partial x_i} u_j.$$

Hence,

$$\|\boldsymbol{E}\,(\phi_n \boldsymbol{u})\|_{L^2(\Omega)} \le \|\phi_n \boldsymbol{E}\,(\boldsymbol{u})\|_{L^2(\Omega)} + \|\nabla\phi_n\|_{L^\infty(\Omega)}\,\|\boldsymbol{u}\|_{L^2(\Omega)}\,,$$

and so

$$\|\nabla \boldsymbol{u}\|_{L^2(\Omega)} \leq c(N) M \|\boldsymbol{E}(\boldsymbol{u})\|_{L^2(\Omega)} + c(N) \max_n \|\nabla \phi_n\|_{L^\infty(\Omega)} \|\boldsymbol{u}\|_{L^2(\Omega)},$$

which concludes the proof. □

Finally, we prove the general C^1 case.

Theorem 2.8. *Let $\Omega \subset \mathbb{R}^N$ be an open bounded connected set with C^1 boundary and let $B \subseteq \Omega$ be a ball. Then there exists a constant $c = c(\Omega, B) > 0$ such that*

$$\|\boldsymbol{u}\|_{H^1(\Omega)} \leq c \|\boldsymbol{E}(\boldsymbol{u})\|_{L^2(\Omega)}$$

for all $\boldsymbol{u} \in H^1(\Omega; \mathbb{R}^N)$ with

$$\int_B \boldsymbol{u}(\boldsymbol{x})\, d\boldsymbol{x} = \boldsymbol{0} \quad \textit{and} \quad \int_B \left(\nabla \boldsymbol{u}(\boldsymbol{x}) - \nabla^T \boldsymbol{u}(\boldsymbol{x})\right) d\boldsymbol{x} = \boldsymbol{0}.$$

Proof. Assume by contradiction that the result is false. Then for every n there exists $\boldsymbol{u}_n \in H^1(\Omega; \mathbb{R}^N)$ such that

$$\int_B \boldsymbol{u}_n(\boldsymbol{x})\, d\boldsymbol{x} = \boldsymbol{0} \quad \text{and} \quad \int_B \left(\nabla \boldsymbol{u}_n(\boldsymbol{x}) - \nabla^T \boldsymbol{u}_n(\boldsymbol{x})\right) d\boldsymbol{x} = \boldsymbol{0}$$

and

$$\|\boldsymbol{u}_n\|_{H^1(\Omega)} > n \|\boldsymbol{E}(\boldsymbol{u}_n)\|_{L^2(\Omega)}.$$

Define $\boldsymbol{v}_n := \boldsymbol{u}_n / \|\boldsymbol{u}_n\|_{H^1(\Omega)}$. Then

$$\|\boldsymbol{v}_n\|_{H^1(\Omega)} = 1, \quad \|\boldsymbol{E}(\boldsymbol{v}_n)\|_{L^2(\Omega)} < \frac{1}{n}. \tag{2.8}$$

Hence, up to a subsequence, there exists $\boldsymbol{v} \in H^1(\Omega; \mathbb{R}^N)$ such that $\boldsymbol{v}_n \rightharpoonup \boldsymbol{v}$ in $H^1(\Omega; \mathbb{R}^N)$. By the Rellich-Kondrachov theorem (see Theorem 11.10 in [59]), up to a further subsequence, $\boldsymbol{v}_n \to \boldsymbol{v}$ in $L^2(\Omega; \mathbb{R}^N)$. On the other hand, by (2.6),

$$\|\nabla \boldsymbol{v}_n - \nabla \boldsymbol{v}_m\|_{L^2(\Omega)} \leq c\left(\|\boldsymbol{E}(\boldsymbol{v}_n - \boldsymbol{v}_m)\|_{L^2(\Omega)} + \|\boldsymbol{v}_n - \boldsymbol{v}_m\|_{L^2(\Omega)}\right) \to 0$$

as $n, m \to \infty$. Hence, $\{\boldsymbol{v}_n\}$ is a Cauchy sequence in $H^1(\Omega; \mathbb{R}^N)$ and so $\boldsymbol{v}_n \to \boldsymbol{v}$ in $H^1(\Omega; \mathbb{R}^N)$. In particular,

$$1 = \|\boldsymbol{v}_n\|_{H^1(\Omega)} \to \|\boldsymbol{v}\|_{H^1(\Omega)},$$

while $\boldsymbol{E}(\boldsymbol{v}_n) \to \boldsymbol{E}(\boldsymbol{v})$ in $L^2(\Omega; \mathbb{R}^{N\times N})$ and so by (2.8), $\|\boldsymbol{E}(\boldsymbol{v}_n)\|_{L^2(\Omega)} = 0$, which implies that $\boldsymbol{v}(\boldsymbol{x}) = \boldsymbol{c} + \boldsymbol{A}\boldsymbol{x}$, where $\boldsymbol{A}$ is skew-symmetric[1]. But since

$$\int_B \boldsymbol{v}(\boldsymbol{x})\, d\boldsymbol{x} = \boldsymbol{0} \quad \text{and} \quad \int_B \left(\nabla \boldsymbol{v}(\boldsymbol{x}) - \nabla^T \boldsymbol{v}(\boldsymbol{x})\right) d\boldsymbol{x} = \boldsymbol{0},$$

we have that $\boldsymbol{c} = \boldsymbol{0}$ and $\boldsymbol{A} = \boldsymbol{0}$. Hence, $\boldsymbol{v} = \boldsymbol{0}$ which contradicts the fact that $\|\boldsymbol{v}\|_{H^1(\Omega)} = 1$. □

To study the case of Lipschitz domains, we need to introduce the notion of regularized distance. The proof of the theorem that we present here is due to Lieberman [60]. We refer to the book of Stein [79] for the original proof.

Theorem 2.9 (regularized distance). *Let $V \subset \mathbb{R}^N$ be an open set with nonempty boundary and let*

$$d(\boldsymbol{x}) := \begin{cases} +\operatorname{dist}(\boldsymbol{x}, \partial V) & \text{if } \boldsymbol{x} \in V, \\ -\operatorname{dist}(\boldsymbol{x}, \partial V) & \text{if } \boldsymbol{x} \notin V. \end{cases}$$

Then there exists a Lipschitz function $d_{\text{reg}} : \mathbb{R}^N \to \mathbb{R}$ such that $d_{\text{reg}} \in C^\infty(\mathbb{R}^N \setminus \partial V)$,

$$\frac{1}{2} \le \frac{d(\boldsymbol{x})}{d_{\text{reg}}(\boldsymbol{x})} \le \frac{3}{2} \tag{2.9}$$

for all $\boldsymbol{x} \in \mathbb{R}^N \setminus \partial V$, and

$$\left| \frac{\partial^{\boldsymbol{\alpha}} d_{\text{reg}}}{\partial \boldsymbol{x}^{\boldsymbol{\alpha}}}(\boldsymbol{x}) \right| \le \frac{C_{\boldsymbol{\alpha}}}{\left| d_{\text{reg}}(\boldsymbol{x}) \right|^{|\boldsymbol{\alpha}|-1}} \tag{2.10}$$

for all $\boldsymbol{x} \in \mathbb{R}^N \setminus \partial V$ and for every multi-index $\boldsymbol{\alpha}$ with length $|\boldsymbol{\alpha}| \le 2$, for some constant $C_{\boldsymbol{\alpha}} > 0$.

Proof. **Step 1:** Let $\varphi \in C_c^\infty(\mathbb{R}^N)$ be a nonnegative function with

$$\operatorname{supp} \varphi \subseteq \overline{B(\boldsymbol{0}, 1)}, \qquad \int_{\mathbb{R}^N} \varphi(\boldsymbol{x})\, d\boldsymbol{x} = 1. \tag{2.11}$$

[1] This follows from the fact that

$$\frac{\partial^2 u_i}{\partial x_j \partial x_k} = \frac{1}{2}\frac{\partial}{\partial x_j}\left(\frac{\partial u_i}{\partial x_k} + \frac{\partial u_k}{\partial x_i}\right) + \frac{1}{2}\frac{\partial}{\partial x_k}\left(\frac{\partial u_i}{\partial x_j} + \frac{\partial u_j}{\partial x_i}\right) - \frac{1}{2}\frac{\partial}{\partial x_i}\left(\frac{\partial u_j}{\partial x_k} + \frac{\partial u_k}{\partial x_j}\right).$$

Let

$$G(\boldsymbol{x}, t) := \int_{\mathbb{R}^N} d\left(\boldsymbol{x} - \tfrac{t}{2}\boldsymbol{y}\right) \varphi(\boldsymbol{y})\, d\boldsymbol{y}, \quad \boldsymbol{x} \in \mathbb{R}^N, \quad t \in \mathbb{R}. \tag{2.12}$$

Then for all $\boldsymbol{x} \in \mathbb{R}^N$ and for all $t_1, t_2 \in \mathbb{R}$, by (2.11) and the fact that d is 1-Lipschitz,

$$\begin{aligned} |G(\boldsymbol{x}, t_1) - G(\boldsymbol{x}, t_2)| &\leq \int_{B(\boldsymbol{0},1)} \left| d\left(\boldsymbol{x} - \tfrac{t_1}{2}\boldsymbol{y}\right) - d\left(\boldsymbol{x} - \tfrac{t_2}{2}\boldsymbol{y}\right) \right| \varphi(\boldsymbol{y})\, d\boldsymbol{y} \\ &\leq \frac{1}{2}|t_1 - t_2| \int_{B(\boldsymbol{0},1)} \varphi(\boldsymbol{y})\, d\boldsymbol{y} = \frac{1}{2}|t_1 - t_2|, \end{aligned} \tag{2.13}$$

while for all $t \in \mathbb{R}$,

$$\begin{aligned} |G(\boldsymbol{x}_1, t) - G(\boldsymbol{x}_2, t)| &\leq \int_{B(\boldsymbol{0},1)} \left| d\left(\boldsymbol{x}_1 - \tfrac{t}{2}\boldsymbol{y}\right) - d\left(\boldsymbol{x}_2 - \tfrac{t}{2}\boldsymbol{y}\right) \right| \varphi(\boldsymbol{y})\, d\boldsymbol{y} \\ &\leq |\boldsymbol{x}_1 - \boldsymbol{x}_2| \end{aligned} \tag{2.14}$$

for all $\boldsymbol{x}_1, \boldsymbol{x}_2 \in \mathbb{R}^N$.

Step 2: Using the change of variables $\boldsymbol{x} - \frac{t}{2}\boldsymbol{y} = \boldsymbol{z}$, for $t \neq 0$ we can rewrite G as

$$G(\boldsymbol{x}, t) = \frac{2^N}{t^N} \int_{\mathbb{R}^N} \varphi\left(\frac{2(\boldsymbol{x} - \boldsymbol{z})}{t}\right) d(\boldsymbol{z})\, d\boldsymbol{z} = \left(d * \varphi_{t/2}\right)(\boldsymbol{x}), \tag{2.15}$$

which implies that G belongs to $C^\infty(\mathbb{R}^N \times (\mathbb{R} \setminus \{0\}))$. In particular, by the properties of mollifiers,

$$\begin{aligned} \frac{\partial G}{\partial x_i}(\boldsymbol{x}, t) &= \frac{2^{N+1}}{t^{N+1}} \int_{\mathbb{R}^N} d(\boldsymbol{z}) \frac{\partial \varphi}{\partial x_i}\left(\frac{\boldsymbol{x} - \boldsymbol{z}}{t/2}\right) d\boldsymbol{z} \\ &= \frac{2}{t} \int_{\mathbb{R}^N} d\left(\boldsymbol{x} - \frac{t}{2}\boldsymbol{y}\right) \frac{\partial \varphi}{\partial x_i}(\boldsymbol{y})\, d\boldsymbol{y} \end{aligned} \tag{2.16}$$

and

$$\begin{aligned} \frac{\partial^2 G}{\partial x_j \partial x_i}(\boldsymbol{x}, t) &= \frac{2^{N+2}}{t^{N+2}} \int_{\mathbb{R}^N} d(\boldsymbol{z}) \frac{\partial^2 \varphi}{\partial x_j \partial x_i}\left(\frac{\boldsymbol{x} - \boldsymbol{z}}{t/2}\right) d\boldsymbol{z} \\ &= \frac{4}{t^2} \int_{\mathbb{R}^N} d\left(\boldsymbol{x} - \frac{t}{2}\boldsymbol{y}\right) \frac{\partial^2 \varphi}{\partial x_j \partial x_i}(\boldsymbol{y})\, d\boldsymbol{y}. \end{aligned} \tag{2.17}$$

Note that since φ has compact support, $\int_{\mathbb{R}^N} \frac{\partial^2 \varphi}{\partial x_j \partial x_i}(\boldsymbol{y})\, d\boldsymbol{y} = 0$, and so

$$\frac{\partial^2 G}{\partial x_j \partial x_i}(\boldsymbol{x}, t) = \frac{4}{t^2} \int_{\mathbb{R}^N} \left(d\left(\boldsymbol{x} - \frac{t}{2}\boldsymbol{y}\right) - d(\boldsymbol{x}) \right) \frac{\partial^2 \varphi}{\partial x_j \partial x_i}(\boldsymbol{y})\, d\boldsymbol{y},$$

which gives

$$\left|\frac{\partial^2 G}{\partial x_j \partial x_i}(\boldsymbol{x}, t)\right| \leq \frac{4}{t^2} \int_{\mathbb{R}^N} \frac{|t|}{2} |\boldsymbol{y}| \left|\frac{\partial^2 \varphi}{\partial x_j \partial x_i}(\boldsymbol{y})\right| d\boldsymbol{y} \leq \frac{c}{|t|}. \tag{2.18}$$

Similar estimates hold for $\frac{\partial^2 G}{\partial t \partial x_i}$ and $\frac{\partial^2 G}{\partial t^2}$. We omit the details.
Step 3: In view of (2.13), it follows by Banach's fixed point theorem that for every $\boldsymbol{x} \in \mathbb{R}^N$ there exists a unique $t_{\boldsymbol{x}} \in \mathbb{R}$ such that

$$t_{\boldsymbol{x}} = G(\boldsymbol{x}, t_{\boldsymbol{x}}).$$

Define

$$d_{\text{reg}}(\boldsymbol{x}) := t_{\boldsymbol{x}}.$$

By (2.13) and (2.14),

$$\left|d_{\text{reg}}(\boldsymbol{x}_1) - d_{\text{reg}}(\boldsymbol{x}_2)\right| \leq 2|\boldsymbol{x}_1 - \boldsymbol{x}_2|$$

for all $\boldsymbol{x}_1, \boldsymbol{x}_2 \in \mathbb{R}^N$. We will prove that d_{reg} is of class $C^\infty(\mathbb{R}^N \setminus \partial V)$. Consider the function

$$H(\boldsymbol{x}, t) = t - G(\boldsymbol{x}, t)$$

for $(\boldsymbol{x}, t) \in \mathbb{R}^N \times (\mathbb{R} \setminus \{0\})$. Then

$$\frac{\partial H}{\partial t}(\boldsymbol{x}, t) = 1 - \frac{\partial G}{\partial t}(\boldsymbol{x}, t) \geq 1 - \frac{1}{2},$$

since $\left|\frac{\partial G}{\partial t}(\boldsymbol{x}, t)\right| \leq \frac{1}{2}$ by (2.13). Hence, we are in a position to apply the implicit function theorem at every point $(\boldsymbol{x}_0, t_{\boldsymbol{x}_0}) \in \mathbb{R}^N \times (\mathbb{R} \setminus \{0\})$ to conclude that in a neighborhood of $(\boldsymbol{x}_0, t_{\boldsymbol{x}_0})$ there is a unique function f of class C^∞ such that $H(\boldsymbol{x}, f(\boldsymbol{x})) = 0$. On the other hand, by the uniqueness of d_{reg} we must have that $f = d_{\text{reg}}$, which shows that d_{reg} is of class C^∞.
Step 4: We will show that

$$\frac{1}{2} \leq \frac{d(\boldsymbol{x})}{d_{\text{reg}}(\boldsymbol{x})} \leq \frac{3}{2}$$

for all $\boldsymbol{x} \in \mathbb{R}^N \setminus \partial V$. By (2.11), we have that

$$G(\boldsymbol{x}, 0) = \int_{\mathbb{R}^N} d(\boldsymbol{x}) \varphi(\boldsymbol{y})\, d\boldsymbol{y} = d(\boldsymbol{x}).$$

Using the fact that

$$d_{\text{reg}}(\boldsymbol{x}) = G(\boldsymbol{x}, d_{\text{reg}}(\boldsymbol{x})),$$

it follows by (2.12) and (2.13) that

$$\left|d_{\text{reg}}(\boldsymbol{x}) - d\,(\boldsymbol{x})\right| = \left|G(\boldsymbol{x}, d_{\text{reg}}(\boldsymbol{x})) - G(\boldsymbol{x}, 0)\right| \le \frac{1}{2}\left|d_{\text{reg}}(\boldsymbol{x}) - 0\right|,$$

which shows that

$$d_{\text{reg}}(\boldsymbol{x}) - \frac{1}{2}\left|d_{\text{reg}}(\boldsymbol{x})\right| \le d(x) \le d_{\text{reg}}(\boldsymbol{x}) + \frac{1}{2}\left|d_{\text{reg}}(\boldsymbol{x})\right|.$$

Step 5: Finally we will prove that for every multi-index $\boldsymbol{\alpha}$ with length $|\boldsymbol{\alpha}| \le 2$,

$$\left|\frac{\partial^{\boldsymbol{\alpha}} d_{\text{reg}}}{\partial \boldsymbol{x}^{\boldsymbol{\alpha}}}(\boldsymbol{x})\right| \le \frac{C_{\boldsymbol{\alpha}}}{\left|d_{\text{reg}}(\boldsymbol{x})\right|^{|\boldsymbol{\alpha}|-1}}$$

for all $\boldsymbol{x} \in \mathbb{R}^N \setminus \partial V$ and for some constant $C_{\boldsymbol{\alpha}} > 0$. Since

$$d_{\text{reg}}(\boldsymbol{x}) = G(\boldsymbol{x}, d_{\text{reg}}(\boldsymbol{x})),$$

by differentiating with respect to x_i, we have

$$\frac{\partial d_{\text{reg}}}{\partial x_i}(\boldsymbol{x}) = \frac{\partial G}{\partial x_i}(\boldsymbol{x}, d_{\text{reg}}(\boldsymbol{x})) + \frac{\partial G}{\partial t}(\boldsymbol{x}, d_{\text{reg}}(\boldsymbol{x}))\frac{\partial d_{\text{reg}}}{\partial x_i}(\boldsymbol{x}). \tag{2.19}$$

Hence,

$$\frac{\partial d_{\text{reg}}}{\partial x_i}(\boldsymbol{x}) = \frac{\frac{\partial G}{\partial x_i}(\boldsymbol{x}, d_{\text{reg}}(\boldsymbol{x}))}{1 - \frac{\partial G}{\partial t}(\boldsymbol{x}, d_{\text{reg}}(\boldsymbol{x}))}.$$

Since $\left|\frac{\partial G}{\partial x_i}(\boldsymbol{x}, t)\right| \le 1$ and $\left|\frac{\partial G}{\partial t}(\boldsymbol{x}, t)\right| \le \frac{1}{2}$ by (2.13), it follows that $\left|\frac{\partial d_{\text{reg}}}{\partial x_i}(\boldsymbol{x})\right| \le 2$. Differentiating (2.19) with respect to x_j, we have

$$\begin{aligned}\frac{\partial^2 d_{\text{reg}}}{\partial x_j \partial x_i}(\boldsymbol{x}) = {} & \frac{\partial^2 G}{\partial x_j \partial x_i}(\boldsymbol{x}, d_{\text{reg}}(\boldsymbol{x})) + \frac{\partial^2 G}{\partial t \partial x_i}(\boldsymbol{x}, d_{\text{reg}}(\boldsymbol{x}))\frac{\partial d_{\text{reg}}}{\partial x_j}(\boldsymbol{x}) \\ & + \frac{\partial^2 G}{\partial x_j \partial t}(\boldsymbol{x}, d_{\text{reg}}(\boldsymbol{x}))\frac{\partial d_{\text{reg}}}{\partial x_i}(\boldsymbol{x}) \\ & + \frac{\partial^2 G}{\partial t^2}(\boldsymbol{x}, d_{\text{reg}}(\boldsymbol{x}))\frac{\partial d_{\text{reg}}}{\partial x_j}(\boldsymbol{x})\frac{\partial d_{\text{reg}}}{\partial x_i}(\boldsymbol{x}) \\ & + \frac{\partial G}{\partial t}(\boldsymbol{x}, d_{\text{reg}}(\boldsymbol{x}))\frac{\partial^2 d_{\text{reg}}}{\partial x_j \partial x_i}(\boldsymbol{x}).\end{aligned}$$

Hence,

$$\frac{\partial^2 d_{\mathrm{reg}}}{\partial x_j \partial x_i}(\boldsymbol{x})$$
$$= \frac{1}{1-\frac{\partial G}{\partial t}(\boldsymbol{x}, d_{\mathrm{reg}}(\boldsymbol{x}))}\left[\frac{\partial^2 G}{\partial x_j \partial x_i}(\boldsymbol{x}, d_{\mathrm{reg}}(\boldsymbol{x})) + \frac{\partial^2 G}{\partial t \partial x_i}(\boldsymbol{x}, d_{\mathrm{reg}}(\boldsymbol{x}))\frac{\partial d_{\mathrm{reg}}}{\partial x_j}(\boldsymbol{x})\right.$$
$$\left. + \frac{\partial^2 G}{\partial x_j \partial t}(\boldsymbol{x}, d_{\mathrm{reg}}(\boldsymbol{x}))\frac{\partial d_{\mathrm{reg}}}{\partial x_i}(\boldsymbol{x}) + \frac{\partial^2 G}{\partial x_j \partial x_i}(\boldsymbol{x}, d_{\mathrm{reg}}(\boldsymbol{x}))\frac{\partial d_{\mathrm{reg}}}{\partial x_j}(\boldsymbol{x})\frac{\partial d_{\mathrm{reg}}}{\partial x_i}\right].$$

Since $\left|\frac{\partial G}{\partial t}(\boldsymbol{x}, t)\right| \leq \frac{1}{2}$, the result follows from (2.18). □

Remark 2.10. The inequality (2.10) actually holds for every multi-index $\boldsymbol{\alpha}$ but this will not be needed here.

Theorem 2.11. *Let*

$$\Omega = \left\{\boldsymbol{x} \in \mathbb{R}^N : x_N > f\left(\boldsymbol{x}'\right)\right\},$$

where $f : \mathbb{R}^{N-1} \to \mathbb{R}$ *is a Lipschitz function with Lipschitz constant* L*, let* $\boldsymbol{u} \in H^1\left(\Omega; \mathbb{R}^N\right)$*. Then there exists a constant* $c = c\,(N) > 0$ *such that*

$$\|\nabla \boldsymbol{u}\|_{L^2(\Omega)} \leq c(1 + L^2)\, \|\boldsymbol{E}\,(\boldsymbol{u})\|_{L^2(\Omega)}\,.$$

Proof. **Step 1:** We will prove that for every $\boldsymbol{x} = (\boldsymbol{x}', x_N) \in \mathbb{R}^N \setminus \overline{\Omega}$,

$$\frac{1}{\sqrt{1+L^2}}\left(f\left(\boldsymbol{x}'\right) - x_N\right) \leq \operatorname{dist}\left(\boldsymbol{x}, \overline{\Omega}\right) \leq \left(f\left(\boldsymbol{x}'\right) - x_N\right). \qquad (2.20)$$

We claim that $\operatorname{dist}\left(\boldsymbol{x}, \overline{\Omega}\right) = \operatorname{dist}\left(\boldsymbol{x}, \partial\Omega\right)$. To see this, consider a point $\boldsymbol{x} \in \mathbb{R}^N \setminus \overline{\Omega}$ and a point $\boldsymbol{y} \in \Omega$. Then the segment of endpoints $\boldsymbol{x}$ and $\boldsymbol{y}$ must intersect $\partial\Omega$ at some point $\boldsymbol{z}$ and so $|\boldsymbol{x} - \boldsymbol{y}| \geq |\boldsymbol{x} - \boldsymbol{z}|$, which proves the claim.

Next, given $\boldsymbol{x} \in \mathbb{R}^N \setminus \overline{\Omega}$ and $\boldsymbol{y} \in \partial\Omega$, note that

$$f\left(\boldsymbol{x}'\right) - x_N = f\left(\boldsymbol{x}'\right) - f\left(\boldsymbol{y}'\right) + f\left(\boldsymbol{y}'\right) - x_N \leq L\left|\boldsymbol{x}' - \boldsymbol{y}'\right| + f\left(\boldsymbol{y}'\right) - x_N.$$

Hence,

$$\begin{aligned}\left(f\left(\boldsymbol{x}'\right) - x_N\right)^2 &\leq L^2\left|\boldsymbol{x}' - \boldsymbol{y}'\right|^2 + \left(f\left(\boldsymbol{y}'\right) - x_N\right)^2 \\ &\quad + 2L\left|\boldsymbol{x}' - \boldsymbol{y}'\right|\left(f\left(\boldsymbol{y}'\right) - x_N\right) \\ &\leq L^2\left|\boldsymbol{x}' - \boldsymbol{y}'\right|^2 + \left(f\left(\boldsymbol{y}'\right) - x_N\right)^2 \\ &\quad + \left|\boldsymbol{x}' - \boldsymbol{y}'\right|^2 + L^2\left(f\left(\boldsymbol{y}'\right) - x_N\right)^2 \\ &= (1 + L^2)\left(\left|\boldsymbol{x}' - \boldsymbol{y}'\right|^2 + \left(f\left(\boldsymbol{y}'\right) - x_N\right)^2\right).\end{aligned}$$

It follows that

$$\frac{\left(f\left(\boldsymbol{x}'\right)-x_N\right)^2}{1+L^2}\le|\boldsymbol{x}-\boldsymbol{y}|^2,$$

and so

$$\frac{1}{\sqrt{1+L^2}}\left(f\left(\boldsymbol{x}'\right)-x_N\right)\le|\boldsymbol{x}-\boldsymbol{y}|$$

for all $\boldsymbol{y}\in\partial\Omega$. Hence, (2.20) holds. In turn, also by (2.9), for all $\boldsymbol{x}\in\mathbb{R}^N\setminus\overline{\Omega}$,

$$0<\frac{2}{3\sqrt{1+L^2}}\left(f\left(\boldsymbol{x}'\right)-x_N\right)\le d_{\text{reg}}(\boldsymbol{x})\le 2\left(f\left(\boldsymbol{x}'\right)-x_N\right). \quad (2.21)$$

Define

$$d_1(\boldsymbol{x}):=3\sqrt{1+L^2}d_{\text{reg}}(\boldsymbol{x}). \quad (2.22)$$

Then

$$0<2\left(f\left(\boldsymbol{x}'\right)-x_N\right)\le d_1(\boldsymbol{x})\le 6\sqrt{1+L^2}\left(f\left(\boldsymbol{x}'\right)-x_N\right). \quad (2.23)$$

Step 2: Let $\phi\in C([1,2])$ be such that

$$\int_1^2\phi(t)\,dt=1,\quad \int_1^2 t\phi(t)\,dt=0,\quad \int_1^2 t^2\phi(t)\,dt=0. \quad (2.24)$$

Given $\boldsymbol{u}\in H^1\left(\Omega;\mathbb{R}^N\right)$, let $\boldsymbol{v}:\mathbb{R}^N\setminus\partial\Omega\to\mathbb{R}$ be the function defined by

$$v_i(\boldsymbol{x})=\int_1^2\phi(t)\left\{u_i(\boldsymbol{x}_t)+t\frac{\partial d_1}{\partial x_i}(\boldsymbol{x})u_N(\boldsymbol{x}_t)\right\}dt, \quad (2.25)$$

$i=1,\ldots,N$, where

$$\boldsymbol{x}_t:=(\boldsymbol{x}',x_N+d_1(\boldsymbol{x})t). \quad (2.26)$$

Note that in view of (2.23), if $\boldsymbol{x}\in\mathbb{R}^N\setminus\overline{\Omega}$ and $t\ge 1$ then

$$x_N+d_1(\boldsymbol{x})t\ge x_N+2\left(f\left(\boldsymbol{x}'\right)-x_N\right)>x_N+f\left(\boldsymbol{x}'\right)-x_N=f(\boldsymbol{x}'),$$

and so $\boldsymbol{x}_t\in\Omega$. Thus, $\boldsymbol{v}$ is well-defined.

Moreover, if $\boldsymbol{u}\in C^1\left(\overline{\Omega};\mathbb{R}^N\right)$, then we can extend $\boldsymbol{u}$ to a function $\boldsymbol{u}\in C^1\left(\mathbb{R}^N;\mathbb{R}^N\right)$. By the fundamental theorem of calculus applied to the function $g(t):=u_N(\boldsymbol{x}_t)$,

$$u_N(\boldsymbol{x}_t)=u_N(\boldsymbol{x}_1)+d_1(\boldsymbol{x})\int_1^t\frac{\partial u_N}{\partial x_N}(\boldsymbol{x}_s)\,ds.$$

Hence, also by (2.24) and Fubini's theorem,

$$\int_1^2 \phi(t) t u_N(\boldsymbol{x}_t)\, dt = d_1(\boldsymbol{x}) \int_1^2 \frac{\partial u_N}{\partial x_N}(\boldsymbol{x}_s) \int_s^2 \phi(t) t\, dt\, ds. \tag{2.27}$$

It follows that

$$\begin{aligned} v_i(\boldsymbol{x}) &= \int_1^2 \phi(t) \left\{ u_i(\boldsymbol{x}_t) + t \frac{\partial d_1}{\partial x_i}(\boldsymbol{x}) u_N(\boldsymbol{x}_t) \right\} dt \\ &= \int_1^2 \phi(t) u_i(\boldsymbol{x}_t)\, dt + \frac{\partial d_1}{\partial x_i}(\boldsymbol{x}) d_1(\boldsymbol{x}) \int_1^2 \frac{\partial u_N}{\partial x_N}(\boldsymbol{x}_s) \int_s^2 \phi(t) t\, dt ds. \end{aligned}$$

Hence, if $\boldsymbol{x} \in \mathbb{R}^N \setminus \overline{\Omega}$ approaches $\boldsymbol{x}_0 := (\boldsymbol{x}_0', f(\boldsymbol{x}_0'))$ by (2.10) and (2.24), $v_i(\boldsymbol{x}) \to u_i(\boldsymbol{x}_0)$, where we used the fact that

$$\int_1^2 \int_s^2 \phi(t) t\, dt ds = \int_1^2 \phi(t) t \int_1^t 1\, ds dt = 0 \tag{2.28}$$

by (2.24).

By differentiating (2.25), from (2.26) and Theorem 2.9 we get

$$\begin{aligned} \frac{\partial v_i}{\partial x_k}(\boldsymbol{x}) = \int_1^2 \phi(t) \Big\{ & \frac{\partial u_i}{\partial x_k}(\boldsymbol{x}_t) + t \frac{\partial d_1}{\partial x_k}(\boldsymbol{x}) \frac{\partial u_i}{\partial x_N}(\boldsymbol{x}_t) + t \frac{\partial d_1}{\partial x_i}(\boldsymbol{x}) \frac{\partial u_N}{\partial x_k}(\boldsymbol{x}_t) \\ & + t \frac{\partial d_1}{\partial x_i}(\boldsymbol{x}) \frac{\partial d_1}{\partial x_k}(\boldsymbol{x}) \frac{\partial u_N}{\partial x_N} + t \frac{\partial^2 d_1}{\partial x_k \partial x_i}(\boldsymbol{x}) u_N(\boldsymbol{x}_t) \Big\}\, dt, \end{aligned}$$

It follows from (2.27) that

$$\begin{aligned} \frac{\partial v_i}{\partial x_k}(\boldsymbol{x}) = \int_1^2 \phi(t) \Big\{ & \frac{\partial u_i}{\partial x_k}(\boldsymbol{x}_t) + t \frac{\partial d_1}{\partial x_k}(\boldsymbol{x}) \frac{\partial u_i}{\partial x_N}(\boldsymbol{x}_t) + t \frac{\partial d_1}{\partial x_i}(\boldsymbol{x}) \frac{\partial u_N}{\partial x_k}(\boldsymbol{x}_t) \\ & + t \frac{\partial d_1}{\partial x_i}(\boldsymbol{x}) \frac{\partial d_1}{\partial x_k}(\boldsymbol{x}) \frac{\partial u_N}{\partial x_N}(\boldsymbol{x}_t) \Big\}\, dt \\ & + \frac{\partial^2 d_1}{\partial x_k \partial x_i}(\boldsymbol{x}) d_1(\boldsymbol{x}) \int_1^2 \frac{\partial u_N}{\partial x_N}(\boldsymbol{x}_s) \int_s^2 \phi(t) t\, dt ds. \end{aligned}$$

Hence, if $\boldsymbol{x} \in \mathbb{R}^N \setminus \overline{\Omega}$ approaches $\boldsymbol{x}_0 := (\boldsymbol{x}_0', f(\boldsymbol{x}_0'))$ by (2.10) and (2.24),

$$\frac{\partial v_i}{\partial x_k}(\boldsymbol{x}) \to \frac{\partial u_i}{\partial x_k}(\boldsymbol{x}_0),$$

where we used (2.28). This shows that $\boldsymbol{v} \in H^1_{\mathrm{loc}}\left(\mathbb{R}^N; \mathbb{R}^N\right)$.

Moreover,

$$(\boldsymbol{E}(\boldsymbol{v}))_{i,k}(\boldsymbol{x}) = \int_1^2 \phi(t)\left\{E^t_{i,k} + t\frac{\partial d_1}{\partial x_k}(\boldsymbol{x})E^t_{i,N} + t\frac{\partial d_1}{\partial x_i}(\boldsymbol{x})E^t_{k,N} + t\frac{\partial d_1}{\partial x_i}(\boldsymbol{x})\frac{\partial d_1}{\partial x_k}(\boldsymbol{x})E^t_{N,N}\right\} dt, + \frac{\partial^2 d_1}{\partial x_k \partial x_i}(\boldsymbol{x})d_1(\boldsymbol{x})\int_1^2 E^s_{N,N}\int_s^2 \phi(t)t\,dt\,ds, \tag{2.29}$$

where

$$E^t_{i,k} := \frac{1}{2}\left(\frac{\partial u_i}{\partial x_k}(\boldsymbol{x}_t) + \frac{\partial u_k}{\partial x_i}(\boldsymbol{x}_t)\right).$$

In turn, by (2.10), (2.27), and (2.29), and the fact that $1 < t < 2$

$$\begin{aligned}|\boldsymbol{E}(\boldsymbol{v})(\boldsymbol{x})| &\le c(1+L^2)\Big(1 + |\nabla d_{\mathrm{reg}}(\boldsymbol{x})| + |\nabla d_{\mathrm{reg}}(\boldsymbol{x})|^2 \\ &\quad + d_{\mathrm{reg}}(\boldsymbol{x})|\nabla^2 d_{\mathrm{reg}}(\boldsymbol{x})|\Big)\int_1^2 |\boldsymbol{E}(\boldsymbol{u})(\boldsymbol{x}_t)|\,dt \\ &\le c(1+L^2)\int_1^2 |\boldsymbol{E}(\boldsymbol{u})(\boldsymbol{x}_t)|t^{-1}\,dt,\end{aligned}$$

where $c > 0$ depends only on ϕ and N. Fix $\boldsymbol{x}' \in \mathbb{R}^{N-1}$ and assume, after a translation, that $f(\boldsymbol{x}') = 0$. Then for $x_N < 0$, we have

$$\begin{aligned}|\boldsymbol{E}(\boldsymbol{v})(\boldsymbol{x}', x_N)| &\le c(1+L^2)\int_1^2 |\boldsymbol{E}(\boldsymbol{u})(\boldsymbol{x}', x_N + d_1(\boldsymbol{x})t)|t^{-1}\,dt \\ &= c(1+L^2)\int_{x_N+d_1(\boldsymbol{x})}^{x_N+2d_1(\boldsymbol{x})} |\boldsymbol{E}(\boldsymbol{u})(\boldsymbol{x}', s)|(s-x_N)^{-1}\,ds \\ &\le c(1+L^2)\int_{|x_N|}^{12(1+L)|x_N|} |\boldsymbol{E}(\boldsymbol{u})(\boldsymbol{x}', s)|s^{-1}\,ds,\end{aligned}$$

where we used (2.23). Squaring both sides we get

$$|\boldsymbol{E}(\boldsymbol{v})(\boldsymbol{x}', x_N)|^2 \le c(1+L^4)\left(\int_{|x_N|}^{12(1+L)|x_N|} s^{-1}|\boldsymbol{E}(\boldsymbol{u})(\boldsymbol{x}', s)|ds\right)^2.$$

Integrating both sides in x_N by Fubini's theorem and Hardy's inequality we obtain

$$\begin{aligned}\int_{-\infty}^{0} &|\boldsymbol{E}(\boldsymbol{v})(\boldsymbol{x}', x_N)|^2 dx_N \\ &\leq c(1+L^4)\int_{-\infty}^{0}\left(\int_{|x_N|}^{12(1+L)|x_N|} s^{-1}|\boldsymbol{E}(\boldsymbol{u})(\boldsymbol{x}', s)|ds\right)^2 dx_N \\ &\leq c(1+L^4)\int_{0}^{\infty}|\boldsymbol{E}(\boldsymbol{u})(\boldsymbol{x}', s)|^2 ds.\end{aligned}$$

Without the translation, we have that

$$\int_{-\infty}^{f(\boldsymbol{x}')} |\boldsymbol{E}(\boldsymbol{v})(\boldsymbol{x}', x_N)|^2 dx_N \leq c(1+L^4)\int_{f(\boldsymbol{x}')}^{\infty}|\boldsymbol{E}(\boldsymbol{u})(\boldsymbol{x}', s)|^2 ds.$$

Integrating both sides with respect to $\boldsymbol{x}'$ gives

$$\int_{\mathbb{R}^N\setminus\Omega} |\boldsymbol{E}(\boldsymbol{v})(\boldsymbol{x})|^2 d\boldsymbol{x} \leq c(1+L^4)\int_{\Omega}|\boldsymbol{E}(\boldsymbol{u})(\boldsymbol{x})|^2 d\boldsymbol{x}.$$

We are now in a position to apply Theorem 2.1. The general case follows by density. □

Problem 2.12. It would be interesting to find the sharp dependence on L in the previous inequality. I am not aware of any such result.

Remark 2.13. Combining the proofs of Theorems 2.1 and 2.11, one can show that Korn's inequality continues to hold in domains Ω of the form

$$\Omega = \left\{\boldsymbol{x}\in\mathbb{R}^N : M > x_N > f\left(\boldsymbol{x}'\right), \quad \boldsymbol{x}'\in R'\right\},$$

where $f:\mathbb{R}^{N-1}\to\mathbb{R}$ is a Lipschitz function with Lipschitz constant L, and R' is a rectangle in $\mathbb{R}^{N-1}$. Moreover, it can be shown that the constant in Korn's inequality depends only on L, M and R'. We omit the details.

Remark 2.14. In view of Remarks 2.2 and 2.4, the previous proof continues to work for domains of the form

$$\Omega = \left\{\boldsymbol{x}\in Q'\times\mathbb{R} : x_N > f\left(\boldsymbol{x}'\right)\right\},$$

where Q' is a cube in $\mathbb{R}^{N-1}$, $f:\mathbb{R}^{N-1}\to\mathbb{R}$ is a Q'-periodic Lipschitz function with Lipschitz constant L, and $\boldsymbol{u}\in H^1\left(\Omega;\mathbb{R}^N\right)$ is such that $\boldsymbol{u}(\cdot, x_N)$ is Q'-periodic for every fixed x_N.

The proof of the following theorem is similar to the one of Theorems 2.7 and 2.8, and therefore we omit it.

Theorem 2.15. *Let $\Omega \subset \mathbb{R}^N$ be an open bounded connected set with Lipschitz boundary and let $B \subseteq \Omega$ be a ball. Then there exists a constant $c = c(\Omega, B) > 0$ such that*

$$\|\boldsymbol{u}\|_{H^1(\Omega)} \leq c \, \|\boldsymbol{E}(\boldsymbol{u})\|_{L^2(\Omega)}$$

for all $\boldsymbol{u} \in H^1(\Omega; \mathbb{R}^N)$ with

$$\int_B \boldsymbol{u}(\boldsymbol{x}) \, d\boldsymbol{x} = \boldsymbol{0} \quad \text{and} \quad \int_B \left(\nabla \boldsymbol{u}(\boldsymbol{x}) - \nabla^T \boldsymbol{u}(\boldsymbol{x})\right) d\boldsymbol{x} = \boldsymbol{0}.$$

Remark 2.16. Although this will not be needed in what follows, it is important to note that the previous theorems continue to hold if the exponent 2 are replaced by $p > 1$, so that $H^1(\Omega)$ and $L^2(\Omega)$ should be replaced by $W^{1,p}(\Omega)$ and $L^p(\Omega)$, respectively, (see, *e.g.*, [84]), while they fail for $p = 1$ (see the classical paper of Ornstein [68] as well as the more recent paper of Conti, Faraco, and Maggi [24]).

3 The Lamé system in a polygonal domain

Let Ω be a bounded open set in $\mathbb{R}^2$ whose boundary can be decomposed in three curves

$$\partial\Omega = \Gamma_1 \cup \Gamma_2 \cup \Gamma_3,$$

where Γ_1 and Γ_2 are two segments meeting at the origin with an (internal) angle $\omega \in (0, 2\pi)$ and Γ_3 is a regular curve joining the two remaining endpoints of Γ_1 and Γ_2 in a smooth way. Denote by $\omega_0 \in (\pi, 2\pi)$ the solution of the equation

$$\omega_0 = \tan \omega_0.$$

In this section we assume that W in (1.4) is the bulk energy density of a linearly isotropic material, *i.e.*,

$$W(\boldsymbol{E}) = \frac{1}{2}\lambda \left[\operatorname{tr}(\boldsymbol{E})\right]^2 + \mu \operatorname{tr}\left(\boldsymbol{E}^2\right), \tag{3.1}$$

where λ and μ are the (constant) Lamé moduli with

$$\mu > 0, \qquad \mu + \lambda > 0. \tag{3.2}$$

Note that in this range, the quadratic form W is coercive. Also, the Euler-Lagrange system of equations associated to W is called the *Lamé system* and is given by

$$\mu \Delta \boldsymbol{u} + (\lambda + \mu) \nabla (\operatorname{div} \boldsymbol{u}) = \boldsymbol{0} \quad \text{in } \Omega.$$

The following result is a simplified version of a general theorem due to Grisvard [46], to which we refer for the general case. The sketch of the proof we present here can be found in [65] and in [73].

Theorem 3.1. *Let* $\Omega \subset \mathbb{R}^2$ *be as above and let* $\boldsymbol{u} \in W^{1,2}\left(\Omega; \mathbb{R}^2\right)$ *be a weak solution of the Neumann problem*

$$\begin{cases} \mu \Delta \boldsymbol{u} + (\lambda + \mu) \nabla (\operatorname{div} \boldsymbol{u}) = \boldsymbol{f} & \text{in } \Omega, \\ \left[\mu \left(\nabla \boldsymbol{u} + \nabla \boldsymbol{u}^T\right) + \lambda (\operatorname{div} \boldsymbol{u}) I\right] \boldsymbol{\nu} = \boldsymbol{0} & \text{on } \partial\Omega, \end{cases}$$

where $\boldsymbol{f} \in L^2\left(\Omega; \mathbb{R}^2\right)$. *Then*

(i) *if* $\omega \in (0, \pi]$ *then* $\boldsymbol{u} \in W^{2,2}\left(\Omega; \mathbb{R}^2\right)$ *and*

$$\|\boldsymbol{u}\|_{W^{2,2}(\Omega)} \leq c(\Omega)\left(\|\boldsymbol{f}\|_{L^2(\Omega)} + \|\boldsymbol{u}\|_{L^2(\Omega)}\right);$$

(ii) *if* $\omega \in (\pi, 2\pi)$, $\omega \neq \omega_0$, *then* $\boldsymbol{u}$ *may be decomposed as*

$$\boldsymbol{u} = \boldsymbol{u}_{\text{reg}} + \sum_{\alpha} c_\alpha \mathcal{S}_\alpha, \tag{3.3}$$

where α *ranges among all complex numbers with* $\operatorname{Re} \alpha \in (0, 1)$ *which are solutions of the equation*

$$\sin^2(\alpha\omega) = \alpha^2 \sin^2 \omega, \tag{3.4}$$

the functions $\mathcal{S}_\alpha$ *are independent of* $\boldsymbol{f}$ *and in polar coordinates*

$$\mathcal{S}_\alpha(r, \theta) = r^\alpha \boldsymbol{g}_\alpha(r, \theta),$$

with $\boldsymbol{g}_\alpha \in W^{2,2}\left(\Omega; \mathbb{R}^2\right)$. *Moreover*

$$\left\|\boldsymbol{u}_{\text{reg}}\right\|_{W^{2,2}(\Omega)} + \sum_{\alpha} |c_\alpha| \leq c(\Omega)\left(\|\boldsymbol{f}\|_{L^2(\Omega)} + \|\boldsymbol{u}\|_{L^2(\Omega)}\right); \tag{3.5}$$

(iii) *if* $\omega = \omega_0$ *then* $\boldsymbol{u}$ *may be decomposed as in* (3.3) *with the only difference that* α *ranges among all complex numbers with* $\operatorname{Re} \alpha \in (0, 1]$ *which are solutions of* (3.4) *and the estimate* (3.5) *should be replaced by*

$$\left\|\boldsymbol{u}_{\text{reg}}\right\|_{W^{s,2}(\Omega)} + \sum_{\alpha} |c_\alpha| \leq c(s, \Omega)\, \|\boldsymbol{f}\|_{L^2(\Omega)}$$

for every $1 < s < 2$.

Sketch of the proof. We only show how to obtain condition (3.4). Consider the infinite sector defined in polar coordinates by

$$S := \left\{ \boldsymbol{x} \in \mathbb{R}^2 : \ \boldsymbol{x} = (r\cos\theta, r\sin\theta), \ r \geq 0, \ -\tfrac{\omega}{2} < \theta < \tfrac{\omega}{2} \right\}$$

and the boundary value problem

$$\begin{cases} \mu\Delta\boldsymbol{u} + (\lambda+\mu)\,\nabla\,(\operatorname{div}\boldsymbol{u}) = \boldsymbol{0} & \text{in } S, \\ \left[\mu\left(\nabla\boldsymbol{u} + \nabla\boldsymbol{u}^T\right) + \lambda\,(\operatorname{div}\boldsymbol{u})\,I\right]\boldsymbol{\nu} = \boldsymbol{0} & \text{on } \partial S. \end{cases} \tag{3.6}$$

Introducing the polar base vectors

$$\boldsymbol{e}_r := (\cos\theta, \sin\theta), \qquad \boldsymbol{e}_r := (-\sin\theta, \cos\theta)$$

and writing

$$\boldsymbol{u} = u_r\boldsymbol{e}_r + u_\theta\boldsymbol{e}_\theta,$$

we have that the system of pdes in (3.6) becomes

$$\begin{aligned} &(\lambda+2\mu)\left(\frac{\partial^2 u_r}{\partial r^2} + \frac{1}{r}\frac{\partial u_r}{\partial r} - \frac{1}{r^2}u_r\right) \\ &\quad + \frac{\mu}{r^2}\frac{\partial^2 u_r}{\partial\theta^2} + \frac{\lambda+\mu}{r}\frac{\partial^2 u_\theta}{\partial\theta\partial r} - \frac{\lambda+3\mu}{r^2}\frac{\partial u_\theta}{\partial\theta} = 0, \\ &\mu\left(\frac{\partial^2 u_\theta}{\partial r^2} + \frac{1}{r}\frac{\partial u_\theta}{\partial r} - \frac{1}{r^2}u_\theta\right) \\ &\quad + \frac{\lambda+2\mu}{r^2}\frac{\partial^2 u_\theta}{\partial\theta^2} + \frac{\lambda+\mu}{r}\frac{\partial^2 u_\theta}{\partial\theta\partial r} + \frac{\lambda+3\mu}{r^2}\frac{\partial u_r}{\partial\theta} \qquad = 0, \end{aligned}$$

in S, while the Neuman boundary conditions transform into

$$\begin{aligned} \frac{\lambda+2\mu}{r}\left(\frac{\partial u_\theta}{\partial\theta} + u_r\right) + \lambda\frac{\partial u_r}{\partial r} = 0, \\ \frac{1}{r}\frac{\partial u_r}{\partial\theta} + \frac{\partial u_\theta}{\partial r} - \frac{1}{r}u_\theta = 0, \end{aligned}$$

on ∂S.

Now we use the Mellin transform. Given a function $v = v(r,\theta)$, for $\alpha \in \mathbb{C}$ we define

$$\hat{v}(\alpha,\theta) := \int_0^\infty r^{-\alpha-1}v(r,\theta)\,dr.$$

Observe that the Mellin transform consists in taking an ansatz with special functions of the form $v(r,\theta) = r^\alpha g(\theta)$.

With the Mellin transformation, the partial differential system reduces to a system of ordinary differential equations in the variable θ. To be precise,

$$\begin{aligned}&\mu\frac{\partial^2\hat{u}_r}{\partial\theta^2}+(\lambda+2\mu)(\alpha^2-1)\hat{u}_r+[(\lambda+\mu)\alpha-(\lambda+3\mu)]\frac{\partial\hat{u}_\theta}{\partial\theta}=0,\\&(\lambda+2\mu)\frac{\partial^2\hat{u}_\theta}{\partial\theta^2}+\mu(\alpha^2-1)\hat{u}_\theta+[(\lambda+\mu)\alpha+(\lambda+3\mu)]\frac{\partial\hat{u}_r}{\partial\theta}=0,\end{aligned}\tag{3.7}$$

for $-\frac{\omega}{2}<\theta<\frac{\omega}{2}$ and $\alpha\in\mathbb{C}$. On the other hand, the Neumann boundary conditions change into

$$\begin{aligned}\mu\left(\frac{\partial\hat{u}_r}{\partial\theta}+(\alpha-1)\hat{u}_\theta\right)&=0,\\(\lambda+2\mu)\left(\frac{\partial\hat{u}_\theta}{\partial\theta}+\hat{u}_r\right)+\lambda\alpha\hat{u}_r&=0,\end{aligned}\tag{3.8}$$

at $\theta=\pm\frac{\omega}{2}$.

For $\alpha\neq0$ the general solution $\boldsymbol{v}=(\hat{u}_r,\hat{u}_\theta)$ of the system (3.7) is given by

$$\begin{aligned}\boldsymbol{v}=\,&c_1\big(\cos((1+\alpha)\theta),-\sin((1+\alpha)\theta)\big)\\&+c_2\big(\sin((1+\alpha)\theta),\cos((1+\alpha)\theta)\big)\\&+c_3\big((\lambda+3\mu-\alpha(\lambda+\mu))\cos((1-\alpha)\theta),\\&\qquad-(\lambda+3\mu+\alpha(\lambda+\mu))\sin((1-\alpha)\theta)\big)\\&+c_4\big((\lambda+3\mu-\alpha(\lambda+\mu))\sin((1-\alpha)\theta),\\&\qquad(\lambda+3\mu+\alpha(\lambda+\mu))\cos((1-\alpha)\theta)\big),\end{aligned}$$

where $c_1,c_2,c_3,c_4\in\mathbb{R}$ depend on α, while for $\alpha=0$ we have

$$\begin{aligned}\boldsymbol{v}=\,&c_1(\cos\theta,-\sin\theta)+c_2(\sin\theta,\cos\theta)\\&+c_3(-(\lambda+3\mu)\theta\cos\theta+\mu\sin\theta,(\lambda+3\mu)\theta\sin\theta+(\lambda+2\mu)\cos\theta)\\&+c_4((\lambda+3\mu)\theta\sin\theta+\mu\cos\theta,(\lambda+3\mu)\theta\cos\theta-(\lambda+2\mu)\sin\theta),\end{aligned}$$

where $c_1,c_2,c_3,c_4\in\mathbb{R}$.

In turn, (3.8) become

$$\begin{aligned}(0,0)=\,&c_1\big(-2\alpha\mu\sin((1+\alpha)\theta),-2\alpha\mu\cos((1+\alpha)\theta)\big)\\&+c_2\big(2\alpha\mu\cos((1+\alpha)\theta),-2\alpha\mu\sin((1+\alpha)\theta)\big)\\&+c_3\big(-2\alpha\mu\,(\alpha-1)\,(\lambda+\mu)\sin((1-\alpha)\theta),\\&\qquad2\alpha\mu\,(\alpha+1)\,(\lambda+\mu)\cos((1-\alpha)\theta)\big)\\&+c_4\big(2\mu\alpha\,(\alpha-1)\,(\lambda+\mu)\cos((1-\alpha)\theta),\\&\qquad2\alpha\mu\,(\alpha+1)\,(\lambda+\mu)\sin((1-\alpha)\theta)\big)\end{aligned}$$

for $\theta = \pm\frac{\omega}{2}$ and $\alpha \neq 0$, and

$$\begin{aligned}(0,0) = {} & c_1(0,0) + c_2(0,0) + c_3(2\mu(\lambda+\mu)\cos\theta, -2\mu(\lambda+\mu)\sin\theta) \\ & + c_4(2\mu(\lambda+\mu)\sin\theta, 2\mu(\lambda+\mu)\cos\theta)\end{aligned}$$

for $\theta = \pm\frac{\omega}{2}$ and $\alpha = 0$. Thus in both cases we get a system of four equations in the four unknowns c_1, c_2, c_3, c_4. These systems have a nontrivial solution if and only if the determinant of the coefficients is zero. For $\alpha \neq 0$ the determinant of the matrix

$$\begin{pmatrix} -\sin x & \cos x & (\lambda+\mu)(1-\alpha)\sin y & -(\lambda+\mu)(1-\alpha)\cos y \\ -\cos x & -\sin x & (\lambda+\mu)(1+\alpha)\cos y & (\lambda+\mu)(1+\alpha)\sin y \\ \sin x & \cos x & -(\lambda+\mu)(1-\alpha)\sin y & -(\lambda+\mu)(1-\alpha)\cos y \\ -\cos x & \sin x & (\lambda+\mu)(1+\alpha)\cos y & -(\lambda+\mu)(1+\alpha)\sin y \end{pmatrix},$$

where $x = (1+\alpha)\frac{\omega}{2}$ and $y = (1-\alpha)\frac{\omega}{2}$, is given by

$$\begin{aligned} & -2(\lambda+\mu)^2\left(\cos(2x-2y) + \alpha^2 - \alpha^2\cos(2x+2y) \quad 1\right) \\ & = -4(\lambda+\mu)(-\sin^2\alpha\omega + \alpha^2\sin^2\omega). \end{aligned}$$

This shows (3.4). □

Remark 3.2. It can be seen that all the solutions to (3.4) with $\mathrm{Re}\,\alpha \in (0,1)$ lie in a bounded region of the complex plane. Hence, by analyticity, their cardinality is finite.

The following theorem is due to Nicaise [64].

Theorem 3.3. *If* $\omega \in (0, 2\pi)$ *then the equation* (3.4) *has no solutions with* $\mathrm{Re}\,\alpha \in (0, \frac{1}{2}]$.

Proof. A complex number α solves (3.4) if and only if either

$$\sin(\alpha\omega) = \alpha\sin\omega, \tag{3.9}$$

or

$$\sin(\alpha\omega) = -\alpha\sin\omega. \tag{3.10}$$

Assume that $\alpha = \xi + i\eta$ solves (3.9), where $\xi, \eta \in \mathbb{R}$. Using the identity

$$\sin(x+iy) = \sin x\cosh y + i\cos x\sinh y, \qquad x, y \in \mathbb{R},$$

and taking the real and imaginary part of (3.9) we arrive at the system

$$\sin(\xi\omega)\cosh(\eta\omega) = \xi \sin\omega,$$
$$\cos(\omega\xi)\sinh(\eta\omega) = \eta \sin\omega.$$

Consider the functions

$$f(x) := \sin(x\omega)\cosh(\eta\omega), \quad g(x) := x\sin\omega, \quad x \in \mathbb{R}.$$

We claim that $f(\frac{1}{2}) > g(\frac{1}{2})$. Indeed, if $\omega \in [\pi, 2\pi)$, then $g(\frac{1}{2}) = \frac{1}{2}\sin\omega \le 0$, while $f(\frac{1}{2}) = \sin\frac{\omega}{2}\cosh(\eta\omega) > 0$, while if $\omega \in (0,\pi)$, then $g(\frac{1}{2}) = \sin\frac{\omega}{2}\cos\frac{\omega}{2} < \sin\frac{\omega}{2}\cosh(\eta\omega)$.

Since $f''(x) = -\omega^2\sin(x\omega)\cosh(\eta\omega) < 0$ for $x \in (0, \frac{\pi}{\omega})$, we have that f is concave in $[0, \frac{\pi}{\omega}]$. Observing that $\frac{1}{2} < \frac{\pi}{\omega}$, for all $x \in (0, \frac{1}{2}]$, writing $x = 2x\frac{1}{2} + (1-2x)0$, by the concavity of f and the fact that $f(0) = 0$, we have

$$f(x) \ge 2xf(\tfrac{1}{2}) > 2xg(\tfrac{1}{2}) = g(x).$$

Hence, (3.9) has no solution α with $\operatorname{Re}\alpha \in (0, \frac{1}{2}]$. A similar result holds for (3.10). We omit the proof. □

From the previous two theorems we derive the following decay estimate for solutions of the Lamé system at a corner point (see [33]).

Theorem 3.4 (decay estimate). *Let Ω be as in Theorem 3.1 with Γ_3 piecewise smooth and r_0 such that $\overline{B}(\mathbf{0}, r_0) \cap \Gamma_3 = \emptyset$. Then there exist a constant $c > 0$ and an exponent $\beta \in \left(\frac{1}{2}, 1\right)$ such that for every weak solution $\boldsymbol{u} \in W^{1,2}\left(\Omega; \mathbb{R}^2\right)$ of the Lamé problem*

$$\mu\Delta\boldsymbol{u} + (\lambda+\mu)\nabla(\operatorname{div}\boldsymbol{u}) = \mathbf{0} \quad \text{in } \Omega,$$
$$\left[\mu\left(\nabla\boldsymbol{u} + \nabla\boldsymbol{u}^T\right) + \lambda(\operatorname{div}\boldsymbol{u})I\right]\boldsymbol{\nu} = \mathbf{0} \quad \text{on } \Gamma_1 \cup \Gamma_2,$$

we have the following decay estimate

$$\int_{B(\mathbf{0},r)\cap\Omega} |\nabla\boldsymbol{u}|^2\,d\boldsymbol{x} \le cr^{2\beta}\int_\Omega \left(|\boldsymbol{u}|^2 + |\nabla\boldsymbol{u}|^2\right)d\boldsymbol{x} \tag{3.11}$$

for all $0 < r < r_0$.

Before proving the theorem we need the following auxiliary lemma.

Lemma 3.5. *Let $\Omega \subset \mathbb{R}^2$ be as in Theorem 3.1 and let $\boldsymbol{g} \in W^{\frac{1}{2},2}\left(\partial\Omega;\mathbb{R}^2\right)$ be a function vanishing in a neighborhood of the origin. Then there exists a function $\boldsymbol{v} \in W^{2,2}\left(\Omega;\mathbb{R}^2\right)$ such that*

$$\left[\mu\left(\nabla \boldsymbol{v} + \nabla \boldsymbol{v}^T\right) + \lambda\,(\operatorname{div} \boldsymbol{v})\, I\right] \boldsymbol{\nu} = \boldsymbol{g} \quad on\ \partial\Omega$$

and

$$\|\boldsymbol{v}\|_{W^{2,2}(\Omega)} \le c\,(\Omega)\, \|\boldsymbol{g}\|_{W^{\frac{1}{2},2}(\partial\Omega)}.$$

Proof. Writing $\boldsymbol{v} = (v_1, v_2)$, $\boldsymbol{\nu} = (\nu_1, \nu_2)$ and $\boldsymbol{g} = (g_1, g_2)$, a straightforward calculation shows that the equality

$$\left[\mu\left(\nabla \boldsymbol{v} + \nabla \boldsymbol{v}^T\right) + \lambda\,(\operatorname{div} \boldsymbol{v})\, I\right] \boldsymbol{\nu} = \boldsymbol{g}$$

is equivalent to

$$\begin{aligned}
&\frac{\partial v_1}{\partial x}(2\mu+\lambda)\,\nu_1 + \frac{\partial v_1}{\partial y}\mu\nu_2 + \frac{\partial v_2}{\partial x}\mu\nu_2 + \frac{\partial v_2}{\partial y}\lambda\nu_1 = g_1,\\
&\frac{\partial v_1}{\partial x}\lambda\nu_2 + \frac{\partial v_1}{\partial y}\mu\nu_1 + \frac{\partial v_2}{\partial x}\mu\nu_1 + \frac{\partial v_2}{\partial y}(2\mu+\lambda)\,\nu_2 = g_2.
\end{aligned}$$

Since

$$\frac{\partial v}{\partial x} = \frac{\partial v}{\partial \nu}\nu_1 + \frac{\partial v}{\partial \tau}\nu_2, \qquad \frac{\partial v}{\partial y} = \frac{\partial v}{\partial \nu}\nu_2 - \frac{\partial v}{\partial \tau}\nu_1$$

the previous equations can be rewritten as

$$\begin{aligned}
&\frac{\partial v_1}{\partial \nu}\left[\mu + (\mu+\lambda)\,\nu_1^2\right] + \frac{\partial v_1}{\partial \tau}(\mu+\lambda)\,\nu_1\nu_2\\
&\quad + \frac{\partial v_2}{\partial \nu}(\mu+\lambda)\,\nu_1\nu_2 + \frac{\partial v_2}{\partial \tau}\left(\mu\nu_2^2 - \lambda\nu_1^2\right) = g_1,\\
&\frac{\partial v_1}{\partial \nu}(\mu+\lambda)\,\nu_1\nu_2 + \frac{\partial v_1}{\partial \tau}\left(\lambda\nu_2^2 - \mu\nu_1^2\right)\\
&\quad + \frac{\partial v_2}{\partial \nu}\left[\mu + (\mu+\lambda)\,\nu_2^2\right] - \frac{\partial v_2}{\partial \tau}(\mu+\lambda)\,\nu_1\nu_2 = g_2.
\end{aligned}$$

Choosing $\frac{\partial v_1}{\partial \tau} = \frac{\partial v_2}{\partial \tau} = 0$ the previous system becomes

$$\begin{aligned}
&\frac{\partial v_1}{\partial \nu}\left[\mu + (\mu+\lambda)\,\nu_1^2\right] + \frac{\partial v_2}{\partial \nu}(\mu+\lambda)\,\nu_1\nu_2 = g_1,\\
&\frac{\partial v_1}{\partial \nu}(\mu+\lambda)\,\nu_1\nu_2 + \frac{\partial v_2}{\partial \nu}\left[\mu + (\mu+\lambda)\,\nu_2^2\right] = g_2,
\end{aligned}$$

which yields

$$\frac{\partial v_1}{\partial \nu} = \frac{g_1\left[\mu + (\mu+\lambda)\,\nu_2^2\right] - g_2\,(\mu+\lambda)\,\nu_1\nu_2}{(2\mu+\lambda)\,\mu}, \tag{3.12}$$

$$\frac{\partial v_2}{\partial \nu} = \frac{g_2\left[\mu + (\mu+\lambda)\,\nu_1^2\right] - g_1\,(\mu+\lambda)\,\nu_1\nu_2}{(2\mu+\lambda)\,\mu}. \tag{3.13}$$

Note that even if ν is discontinuous at the origin by the assumption on $\boldsymbol{g}$ the right-hand sides of the previous equations are zero in a neighborhood of the origin, hence they are both in the space $W^{\frac{1}{2},2}(\partial\Omega)$. We can now apply [47, Theorem 1.5.2.8] to get a function $\boldsymbol{v}\in W^{2,2}\left(\Omega;\mathbb{R}^2\right)$ such that its trace $\boldsymbol{v} = \boldsymbol{0}$ on $\partial\Omega$, the components of its normal derivative satisfy (3.12) and (3.13), and its $W^{2,2}$ norm is bounded by the $W^{\frac{1}{2},2}$ norm of the right-hand side of (3.12) and (3.13)

$$\|\boldsymbol{v}\|_{W^{2,2}(\Omega)} \le c\left(\Omega'\right)\|\boldsymbol{g}\|_{W^{\frac{1}{2},2}(\partial\Omega)},$$

which concludes the proof. □

We are now ready to prove the theorem.

Proof. Let ω be the angle of Ω at the origin. We only give the proof in the case where ω satisfies condition (ii) of Theorem 3.1, that is when $\omega\in(\pi,2\pi)$, $\omega\neq\omega_0$, since case (i) is significantly simpler, while case (iii) is completely analogous to case (ii).

Let $\Omega'\subset\Omega$ be as in Theorem 3.1 and such that $\Omega'\supset\Omega\cap B(\boldsymbol{0},r_0)$ and the distance between Γ_3' and Γ_3 is strictly positive, where $\Gamma_3' := \Omega\cap\partial\Omega'$. By standard elliptic estimates on flat domains we have that $\boldsymbol{u}$ is C^∞ outside a neighborhood U of the origin, with

$$\int_{\Omega'\setminus U}\left|\nabla^2\boldsymbol{u}\right|^2 d\boldsymbol{x} \le c\int_\Omega |\nabla\boldsymbol{u}|^2\, d\boldsymbol{x}. \tag{3.14}$$

Set

$$\sigma(\boldsymbol{u}) := \left[\mu\left(\nabla\boldsymbol{u}+\nabla\boldsymbol{u}^T\right)+\lambda\,(\operatorname{div}\boldsymbol{u})\,I\right].$$

Since $\sigma(\boldsymbol{u})\,\boldsymbol{\nu} = 0$ on $U\cap\partial\Omega'$ and is smooth we are in position to apply the previous lemma with $\boldsymbol{g} := \sigma(\boldsymbol{u})\,\boldsymbol{\nu}$ to find a function $\boldsymbol{v}\in W^{2,2}\left(\Omega';\mathbb{R}^2\right)$ such that

$$\left[\mu\left(\nabla\boldsymbol{v}+\nabla\boldsymbol{v}^T\right)+\lambda\,(\operatorname{div}\boldsymbol{v})\,I\right]\boldsymbol{\nu} = \sigma(\boldsymbol{u})\,\nu \quad \text{on } \partial\Omega',$$

and

$$\|\boldsymbol{v}\|_{W^{2,2}(\Omega')} \le c\left(\Omega'\right)\|\sigma\left(\boldsymbol{u}\right)\boldsymbol{\nu}\|_{W^{\frac{1}{2},2}(\partial\Omega')}$$
$$\le c\left(\Omega'\right)\left(\int_{\Omega'\setminus U}\left(|\nabla \boldsymbol{u}|^2 + \left|\nabla^2 \boldsymbol{u}\right|^2\right)dx\right)^{\frac{1}{2}}.$$

Therefore from the estimate (3.14) we conclude that

$$\|\boldsymbol{v}\|_{W^{2,2}(\Omega')} \le c\left(\Omega\right)\|\nabla \boldsymbol{u}\|_{L^2(\Omega)}. \tag{3.15}$$

Defining $\boldsymbol{w} := \boldsymbol{u} - \boldsymbol{v}$ we get that $\boldsymbol{w}$ is a weak solution of

$$\mu\Delta\boldsymbol{w} + (\lambda+\mu)\nabla\left(\operatorname{div}\boldsymbol{w}\right) = \boldsymbol{f} \quad \text{in } \Omega',$$
$$\left[\mu\left(\nabla\boldsymbol{w} + \nabla\boldsymbol{w}^T\right) + \lambda\left(\operatorname{div}\boldsymbol{w}\right)I\right]\boldsymbol{\nu} = \boldsymbol{0} \quad \text{on } \partial\Omega',$$

where

$$\boldsymbol{f} := \mu\Delta\boldsymbol{v} + (\lambda+\mu)\nabla\left(\operatorname{div}\boldsymbol{v}\right).$$

By (3.15) we have that

$$\|\boldsymbol{f}\|_{L^2(\Omega')} \le c\left(\Omega\right)\|\boldsymbol{u}\|_{W^{1,2}(\Omega)}.$$

By Theorem 3.1 we may write

$$\boldsymbol{w} = \boldsymbol{w}_{\mathrm{reg}} + \sum_{\alpha} c_\alpha \mathcal{S}_\alpha$$

with

$$\left\|\boldsymbol{w}_{\mathrm{reg}}\right\|_{W^{2,2}(\Omega')} + \sum_{\alpha}|c_\alpha| \le c\left(\Omega'\right)\left(\|\boldsymbol{f}\|_{L^2(\Omega)} + \|\boldsymbol{u}\|_{L^2(\Omega)}\right)$$
$$\le c\left(\Omega\right)\|\boldsymbol{u}\|_{W^{1,2}(\Omega)}.$$

Here and in the remaining part of the proof α ranges among all complex numbers with $\operatorname{Re}\alpha \in (0, 1)$ that are solutions of the equation (3.4). We recall that such numbers are finitely many, as observed in Remark 3.2. Using Hölder's inequality and the Sobolev–Gagliardo–Nirenberg embedding theorem (see Theorem 11.2 and Exercise 11.6 in [59]) and the previous estimate we have for any $p > 2$ and $0 < r \le r_0$,

$$\begin{aligned}\int_{B(\mathbf{0},r)\cap\Omega} |\nabla \boldsymbol{u}|^2\,d\boldsymbol{x} &\le 2\left(\int_{B(\mathbf{0},r)\cap\Omega} |\nabla \boldsymbol{w}|^2\,d\boldsymbol{x} + \int_{B(\mathbf{0},r)\cap\Omega} |\nabla \boldsymbol{v}|^2\,d\boldsymbol{x}\right)\\ &\le c\left(\int_{B(\mathbf{0},r)\cap\Omega} \left|\nabla \boldsymbol{w}_{\mathrm{reg}}\right|^2 d\boldsymbol{x} + \int_{B(\mathbf{0},r)\cap\Omega} |\nabla \boldsymbol{v}|^2\,d\boldsymbol{x}\right)\\ &\quad + c\sum_{\alpha} |c_\alpha|^2 \int_{B(\mathbf{0},r)\cap\Omega} |\nabla \mathcal{S}_\alpha|^2\,d\boldsymbol{x}\\ &\le c\left(\int_{B(\mathbf{0},r)\cap\Omega} \left|\nabla \boldsymbol{w}_{\mathrm{reg}}\right|^p d\boldsymbol{x} + \int_{B(\mathbf{0},r)\cap\Omega} |\nabla \boldsymbol{v}|^p\,d\boldsymbol{x}\right)^{\frac{2}{p}} r^{2-\frac{4}{p}}\\ &\quad + c\sum_{\alpha} |c_\alpha|^2 r^{2\,\mathrm{Re}\,\alpha}\\ &\le c\left(\left\|\boldsymbol{w}_{\mathrm{reg}}\right\|^2_{W^{2,2}(\Omega')} + \|\boldsymbol{v}\|^2_{W^{2,2}(\Omega')}\right) r^{2-\frac{4}{p}}\\ &\quad + c\,\|\boldsymbol{u}\|^2_{W^{1,2}(\Omega)} \sum_{\alpha} r^{2\,\mathrm{Re}\,\alpha}\\ &\le c\,\|\boldsymbol{u}\|^2_{W^{1,2}(\Omega)} \left(r^{2-\frac{4}{p}} + \sum_{\alpha} r^{2\,\mathrm{Re}\,\alpha}\right).\end{aligned}$$

Choosing p so large that

$$1 - \frac{2}{p} \ge \beta := \min_{\alpha} \{\mathrm{Re}\,\alpha\} \in \left(\tfrac{1}{2}, 1\right)$$

by Theorem 3.3, we obtain (3.11) for all $0 < r \le r_0$. □

4 Existence of minimizers

Next we discuss the existence of minimizers for the functional $\mathcal{F}$ defined in (1.7). We will see that $\mathcal{F}$ is not lower semicontinuous, and thus in general one cannot expect the existence of minimizers. In what follows we assume that the surface energy density of the film is less than or equal to the one of the substrate, that is,

$$\gamma_{\mathrm{sub}} \ge \gamma_{\mathrm{film}}. \tag{4.1}$$

We will see that this condition favors the SK growth mode over the VW mode (see Section 1).

Definition 4.1. Given a function $\boldsymbol{w} : [0, b] \to \mathbb{R}^N$, we define the *pointwise variation* of $\boldsymbol{w}$ as

$$\mathrm{var}_{[0,b]}\,\boldsymbol{w} := \sup\left\{\sum_{i=i}^{n} |\boldsymbol{w}(x_i) - \boldsymbol{w}(x_{i-1})|\right\},$$

where the supremum is taken over all partitions $x_0 = 0 < x_1 < \cdots < x_n = b, n \in \mathbb{N}$.

Remark 4.2. It can be shown that if $\boldsymbol{w}$ is Lipschitz, then

$$\operatorname{var}_{[0,b]} \boldsymbol{w} = \int_0^b |\boldsymbol{w}'(x)|\, dx.$$

Remark 4.3. If $\{\boldsymbol{w}_n\}$ converges pointwise to $\boldsymbol{w}$, then

$$\liminf_{n\to\infty} \operatorname{var}_{[0,b]} \boldsymbol{w}_n \geq \operatorname{var}_{[0,b]} \boldsymbol{w}.$$

Definition 4.4. Given a function $h : [0, b] \to \mathbb{R}$ we define the *length* of the (possibly discontinuous) curve Γ_h as

$$\operatorname{length} \Gamma_h := \operatorname{var}_{[0,b]} \boldsymbol{w},$$

where $\boldsymbol{w}(x) := (x, h(x))$.

Remark 4.5. Hence,

$$\operatorname{var}_{[0,b]} \boldsymbol{w} = \sup \left\{ \sum_{i=i}^{n} \sqrt{(x_i - x_{i-1})^2 + (h(x_i) - h(x_{i-1}))^2} \right\},$$

where the supremum is taken over all partitions $x_0 = 0 < x_1 < \cdots < x_n = b, n \in \mathbb{N}$. It follows that

$$\operatorname{var}_{[0,b]} h \leq \operatorname{length} \Gamma_h \leq b - 0 + \operatorname{var}_{[0,b]} h.$$

Thus, the length of Γ_h is finite if and only if h has finite pointwise variation. Moreover, if h is Lipschitz continuous, then by Remark 4.2,

$$\operatorname{length} \Gamma_h = \int_0^b \sqrt{1 + (h'(x))^2}\, dx.$$

Let

$$X := \Big\{ (\boldsymbol{u}, h) : h : [0, b] \to [0, \infty) \text{ Lipschitz}, \int_0^b h\, dx = d, \quad \boldsymbol{u} \in H^1_{\mathrm{loc}}(\Omega_h; \mathbb{R}^2) \Big\}$$

and

$$X_0 = \Big\{ (\boldsymbol{u}, h) : h : [0, b] \to [0, \infty) \text{ lower semicontinuous}, \operatorname{var}_{[0,b]} h < \infty, \quad \int_0^b h\, dx = d, \quad \boldsymbol{u} \in H^1_{\mathrm{loc}}(\Omega_h; \mathbb{R}^2) \Big\},$$

where we recall that (see (1.1))

$$\Omega_h := \{\boldsymbol{x} = (x, y) : 0 < x < b,\ y < h(x)\}.$$

For $(\boldsymbol{u}, h) \in X_0$ define

$$\mathcal{G}(\boldsymbol{u}, h) := \int_{\Omega_h} W(\boldsymbol{E}(\boldsymbol{u})(\boldsymbol{x}) - \boldsymbol{E}_0(y))\, d\boldsymbol{x} + \gamma_{\text{film}} \operatorname{length} \Gamma_h. \tag{4.2}$$

Theorem 4.6 (liminf inequality). *There exists $(\boldsymbol{u}_0, h_0) \in X_0$ such that*

$$\inf_X \mathcal{F} \geq \mathcal{G}(\boldsymbol{u}_0, h_0).$$

Proof. Consider a minimizing sequence $\{(\boldsymbol{u}_n, h_n)\} \subset X$, that is,

$$\lim_{n\to\infty} \mathcal{F}(\boldsymbol{u}_n, h_n) = \inf \mathcal{F} < \infty. \tag{4.3}$$

By extracting a subsequence, not relabeled, we can assume that there exist the limits

$$\lim_{n\to\infty} \int_{\Omega_{h_n}} W(\boldsymbol{E}(\boldsymbol{u}_n) - \boldsymbol{E}_0)\, d\boldsymbol{x} \quad \text{and} \quad \lim_{n\to\infty} \int_{\Gamma_{h_n}} \psi(y)\, ds, \tag{4.4}$$

where we recall that ψ is the function given in (1.6). This will allow us to take further subsequences without further mention.

Since $\psi \geq \gamma_{\text{film}} > 0$, it follows that

$$\sup_n \operatorname{length} \Gamma_{h_n} < \infty.$$

In particular, by Remark 4.5,

$$\sup_n \operatorname{var}_{[0,b]} h_n < \infty.$$

Since $\int_0^b h_n\, dx = d$, by the mean value theorem there exists $x_n \in [0, b]$ such that $h_n(x_n) = d/b$. Together with the fact that $\sup_n \operatorname{var}_{[0,b]} h_n < \infty$, it follows that $\sup_n \|h_n\|_\infty < \infty$. Hence, by the Helly theorem[2] there

[2] For a proof of the following theorem we refer to [59].

Theorem 4.7 (Helly's selection theorem). *Let $I \subset \mathbb{R}$ be an interval and let $\mathcal{H}$ be an infinite family of functions with finite pointwise variation. Assume that there exist $x_0 \in I$ and a constant $c > 0$ such that*

$$|f(x_0)| + \operatorname{var} f \leq c \tag{4.5}$$

for all $f \in \mathcal{H}$. Then there exists a sequence $\{f_n\} \subset \mathcal{H}$ and a function g with finite pointwise variation such that

$$f_n(x) \to g(x)$$

for all $x \in I$.

exist a subsequence of $\{h_n\}$ not relabeled and a nonnegative function f_0 of finite pointwise variation such that $\{h_n\}$ converges pointwise to f_0. By Remark 4.3, it follows that

$$\infty > \liminf_{n\to\infty} \operatorname{length} \Gamma_{h_n} \geq \operatorname{length} \Gamma_{f_0}.$$

Since f_0 has finite pointwise variation, f_0 has a left and a right limit $f_0^-(x)$ and $f_0^+(x)$ at every point $x \in [0, b]$ (only the left at 0 and the right at b). Note that if

$$f_0(x_0) > \limsup_{x\to x_0} f_0(x) = \max\left\{f_0^-(x_0), f_0^+(x_0)\right\},$$

then the closed subgraph of f_0 contains a vertical segment

$$\{x_0\} \times \left[\limsup_{x\to x_0} f_0(x), f_0(x_0)\right].$$

This segment will have no contribution in the elastic energy but it will count needlessly in the surface energy. Since we are interesting in minimizers, there is no advantage in keeping this segment. Thus, we define

$$h_0(x) := \min\left\{f_0(x), \liminf_{z\to x} f_0(z)\right\},$$

that is, the lower semicontinuous envelope of f_0. Note that $\operatorname{length} \Gamma_{f_0} \geq \operatorname{length} \Gamma_{h_0}$ and h_n converges to h_0 for all but countably many points. Thus, since $\gamma_{\text{sub}} \geq \gamma_{\text{film}}$,

$$\liminf_{n\to\infty} \int_{\Gamma_{h_n}} \psi(y)\, ds(\mathbf{x}) \geq \gamma_{\text{film}} \liminf_{n\to\infty} \operatorname{length} \Gamma_{h_n} \geq \gamma_{\text{film}} \operatorname{length} \Gamma_{h_0}.$$

Define

$$\Omega_{h_0} := \{\mathbf{x} = (x, y) : 0 < x < b,\ y < h_0(x)\}.$$

Let $L := \sup_n \|h_n\|_\infty < \infty$ and define the compact set

$$K_n := \{(x, y) : 0 \leq x \leq b,\ h_n(x) \leq y \leq L + 1\}.$$

By the Blaschke compactness theorem, there exist a subsequence of $\{K_n\}$, not relabeled, and a compact set K such that $K_n \to K$ in the Hausdorff distance $d_{\mathcal{H}}$.[3] It can be shown that (Exercise)

$$K \subseteq \{(x, y) : 0 \leq x \leq b,\ h_0(x) \leq y \leq L + 1\}.$$

[3] We recall that given two sets $A, B \subseteq \mathbb{R}^N$, the *Hausdorff distance* between A and B is defined by

$$d_{\mathcal{H}}(A, B) := \inf\{\varepsilon > 0 : A \subseteq N_\varepsilon(B) \text{ and } B \subseteq N_\varepsilon(A)\},$$

In particular, if U_1 is an open bounded set compactly contained in Ω_{h_0}, then, since $\operatorname{dist}(K, U_1) > 0$, we have that $U_1 \cap K_n = \emptyset$ for all n sufficiently large, so that $U_1 \subseteq \Omega_{h_n}$ for all n sufficiently large, say, $n \geq n_1$. Hence, by (1.3) and (4.3) we have that

$$\sup_{n \geq n_1} \int_{U_1} |\boldsymbol{E}(\boldsymbol{u}_n)|^2 \, dxdy \leq c.$$

Take U_1 to be of class C^1 and connected and let $B \subseteq U_1$ be a fixed ball. By Theorem 2.8,

$$\|\boldsymbol{u}_n - \boldsymbol{v}_n\|^2_{H^1(U_1)} \leq c(U_1) \int_{U_1} |\boldsymbol{E}(\boldsymbol{u}_n)|^2 \, dxdy \leq c,$$

where $\boldsymbol{v}_n$ is a rigid motion such that

$$\int_B (\boldsymbol{u}_n - \boldsymbol{v}_n) \, d\boldsymbol{x} = \boldsymbol{0} \quad \text{and} \quad \int_B \left(\nabla(\boldsymbol{u}_n - \boldsymbol{v}_n) - \nabla^T(\boldsymbol{u}_n - \boldsymbol{v}_n)\right) \, d\boldsymbol{x} = \boldsymbol{0}.$$

It follows that there exist a subsequence $\{\boldsymbol{u}_{n^{(1)}} - \boldsymbol{v}_{n^{(1)}}\}$ of $\{\boldsymbol{u}_n - \boldsymbol{v}_n\}$ and a function $\boldsymbol{w}_1 \in H^1(U_1)$ such that $\boldsymbol{u}_{n^{(1)}} - \boldsymbol{v}_{n^{(1)}} \rightharpoonup \boldsymbol{w}_1$ in $H^1(U_1)$. By the lower semicontinuity of the convex functional $\boldsymbol{u} \mapsto \int_{U_1} W(\boldsymbol{E}(\boldsymbol{u}) - \boldsymbol{E}_0) \, d\boldsymbol{x}$ with respect to weak convergence, we get that

$$\begin{aligned}
\int_{U_1} W(\boldsymbol{E}(\boldsymbol{w}_1) - \boldsymbol{E}_0) \, d\boldsymbol{x} &\leq \liminf_{n^{(1)} \to \infty} \int_{U_1} W(\boldsymbol{E}(\boldsymbol{u}_{n^{(1)}} - \boldsymbol{v}_{n^{(1)}}) - \boldsymbol{E}_0) \, d\boldsymbol{x} \\
&= \liminf_{n^{(1)} \to \infty} \int_{U_1} W(\boldsymbol{E}(\boldsymbol{u}_{n^{(1)}}) - \boldsymbol{E}_0) \, d\boldsymbol{x} \\
&\leq \liminf_{n \to \infty} \int_{\Omega_{h_n}} W(\boldsymbol{E}(\boldsymbol{u}_n) - \boldsymbol{E}_0) \, d\boldsymbol{x},
\end{aligned}$$

where we used the fact that $\boldsymbol{v}_n$ is a rigid motion (so that $\boldsymbol{E}(\boldsymbol{v}_n) = \boldsymbol{0}$) and (4.4).

---

where $N_\varepsilon(E)$ denotes the ε–neighborhood of a set $E \subseteq \mathbb{R}^N$, *i.e.*,

$$N_\varepsilon(E) := \left\{ \boldsymbol{x} \in \mathbb{R}^N : \operatorname{dist}(\boldsymbol{x}, E) < \varepsilon \right\}.$$

Theorem 4.8 (Blaschke compactness). *Let $K_n \subseteq K \subset \mathbb{R}^N$, $n \in \mathbb{N}$, be compact sets. Then there exist a subsequence $\{K_{n_j}\}$ of $\{K_n\}$ and a compact set $K_0 \subseteq K$ such that $K_{n_j} \to K_0$ in the Hausdorff distance $d_{\mathcal{H}}$.*

Taking $U_1 \Subset U_2 \Subset \Omega_{h_0}$, where U_2 is open, bounded, connected with C^1 boundary, and repeating the previous argument, we can find a subsequence $\{\boldsymbol{u}_{n^{(2)}} - \boldsymbol{v}_{n^{(2)}}\}$ of $\{\boldsymbol{u}_{n^{(1)}} - \boldsymbol{v}_{n^{(1)}}\}$ and a function $\boldsymbol{w}_2 \in H^1(U_2; \mathbb{R}^2)$ such that $\boldsymbol{u}_{n^{(2)}} - \boldsymbol{v}_{n^{(2)}} \rightharpoonup \boldsymbol{w}_2$ in $H^1(U_2; \mathbb{R}^2)$ and

$$\int_{U_2} W\left(\boldsymbol{E}\left(\boldsymbol{w}_2\right) - \boldsymbol{E}_0\right) d\boldsymbol{x} \leq \liminf_{n\to\infty} \int_{\Omega_{h_n}} W\left(\boldsymbol{E}\left(\boldsymbol{u}_n\right) - \boldsymbol{E}_0\right) d\boldsymbol{x}.$$

By the uniqueness of weak limits, we have that $\boldsymbol{w}_2 = \boldsymbol{w}_1$ in U_1. Continuing in this fashion, by approximating Ω_{h_0} with an increasing sequence of regular sets U_k we can find a function $\boldsymbol{u}_0 \in H^1_{\text{loc}}(\Omega_{h_0}; \mathbb{R}^2)$ such that

$$\int_{U_k} W\left(\boldsymbol{E}\left(\boldsymbol{u}_0\right) - \boldsymbol{E}_0\right) d\boldsymbol{x} \leq \liminf_{n\to\infty} \int_{\Omega_{h_n}} W\left(\boldsymbol{E}\left(\boldsymbol{u}_n\right) - \boldsymbol{E}_0\right) d\boldsymbol{x}$$

for every k. Letting $k \to \infty$ and using Fatou's lemma, we conclude that

$$\int_{\Omega_{h_0}} W\left(\boldsymbol{E}\left(\boldsymbol{u}_0\right) - \boldsymbol{E}_0\right) d\boldsymbol{x} \leq \liminf_{n\to\infty} \int_{\Omega_{h_n}} W\left(\boldsymbol{E}\left(\boldsymbol{u}_n\right) - \boldsymbol{E}_0\right) d\boldsymbol{x}.$$

This concludes the proof. □

Remark 4.9. Since $\gamma_{\text{sub}} \geq \gamma_{\text{film}}$, we have that $\mathcal{F}(\boldsymbol{u}, h) \geq \mathcal{G}(\boldsymbol{u}, h)$ for every $(\boldsymbol{u}, h) \in X$. Hence,

$$\inf_{(\boldsymbol{u},h)\in X} \mathcal{F}(\boldsymbol{u}, h) \geq \inf_{(\boldsymbol{u},h)\in X} \mathcal{G}(\boldsymbol{u}, h).$$

On the other hand, reasoning exactly as in the previous proof (with $\mathcal{F}$ in place of $\mathcal{G}$), we have that

$$\inf_{(\boldsymbol{u},h)\in X} \mathcal{F}(\boldsymbol{u}, h) \geq \inf_{(\boldsymbol{u},h)\in X} \mathcal{G}(\boldsymbol{u}, h) \geq \inf_{(\boldsymbol{u},h)\in X_0} \mathcal{G}(\boldsymbol{u}, h).$$

The following theorem is due to Bonnetier and Chambolle [21].

Theorem 4.10 (Yosida transform). *Let $h : [0, b] \to \mathbb{R}$ be a lower semicontinuous function. For every $n \in \mathbb{N}$ and $x \in [0, b]$ define*

$$h_n(x) := \inf\left\{h(z) + n|x - z| : z \in [0, b]\right\}.$$

Then $h_n \to h$ pointwise, $h_n \leq h_{n+1} \leq h$, h_n is Lipschitz continuous, with $\operatorname{Lip} h_n \leq n$*, and*

$$\text{length}\,\Gamma_{h_n} \leq \text{length}\,\Gamma_h + h(0) - h_n(0) + h(b) - h_n(b) \tag{4.6}$$

for all n.

Proof. The fact that $h_n \le h_{n+1} \le h$ follows from the definition of h_n. In particular, there exists

$$\lim_{n\to\infty} h_n(x) \le h(x)$$

for every $x \in [0, b]$. Let $z_n \in [0, b]$ be such that

$$h_n(x) = h(z_n) + n|x - z_n| \le h(x).$$

Note that z_n exists since h is lower semicontinuous and $[0, b]$ compact. Since h is bounded from below, it follows that $z_n \to x$ and thus, again by the lower semicontinuity of h,

$$h\,(x) \le \liminf_{n\to\infty} h(z_n) \le \liminf_{n\to\infty} [h(z_n) + n|x_n - z|] = \lim_{n\to\infty} h_n(x) \le h(x),$$

which proves pointwise convergence.

The fact that $\operatorname{Lip} h_n \le n$ is left as an exercise.

It remains to prove (4.6). Consider a partition $0 = x_0 < \cdots < x_\ell = b$. Since h is lower semicontinuous and h_n continuous, the set

$$A_n := \{x \in [0, b] : \; h(x) > h_n(x)\}$$

is relatively open. Hence, if $h(x_i) > h_n(x_i)$ for some $i \in \{0, \ldots, \ell\}$, there exists a largest relatively open interval I_i of $[0, b]$ that contains x_i. Note that if one the endpoints of I_i is not 0 or b, then by the maximality of I_i at that endpoint we have that $h = h_n$. Moreover, the same interval I_i could contain more than one point of the partition. Let $\mathcal{J}_1$ be the (possibly empty) set of indices corresponding to distinct intervals I_i. Add the endpoints of all such I_i to the partition, to get a new partition $0 = t_0 < \cdots < t_m = b$. Let

$$\begin{aligned}\mathcal{K}_1 &:= \{k \in \{0, \ldots, m\} : \; [t_{k-1}, t_k] \subseteq \overline{I_i} \text{ for some } i \in J_1\},\\ \mathcal{K}_2 &:= \{0, \ldots, m\} \setminus \mathcal{K}_1.\end{aligned}$$

Note that if $h(0) > h_n(0)$, then $t_1 \in \mathcal{K}_1$ by construction. Similarly, if $h(b) > h_n(b)$, then $b \in \mathcal{K}_1$. By the additivity property of the the pointwise variation (see Proposition 2.6 in [59]), we have that

$$\begin{aligned}\operatorname{length}\Gamma_h &= \sum_{k\in\mathcal{K}_1} \operatorname{length}\Gamma_{h|_{[t_{k-1},t_k]}} + \sum_{k\in K_2} \operatorname{length}\Gamma_{h|_{[t_{k-1},t_k]}}\\ &= \sum_{i\in\mathcal{J}_1} \operatorname{length}\Gamma_{h|_{\overline{I_i}}} + \sum_{k\in\mathcal{K}_2} \operatorname{length}\Gamma_{h|_{[t_{k-1},t_k]}}.\end{aligned} \tag{4.7}$$

We divide the proof in a few steps.

Step 1: We claim that if $k \in \mathcal{K}_2$, then $h(t_{k-1}) = h_n(t_{k-1})$ and $h(t_k) = h_n(t_k)$. Indeed, if $h(t_{k-1}) > h_n(t_{k-1})$ and $0 < t_{k-1} < b$, then t_{k-1} cannot be an endpoint of one of the I_i's and so it must be a point of the original partition. But then t_{k-1} would be in the interior of some I_i, which would imply that $[t_{k-1}, t_k] \subseteq \overline{I_i}$ for some $i \in J_1$, and contradict the fact that $k \in \mathcal{K}_2$. If $t_{k-1} = 0$, so that $k = 1$, and $h(0) > h_n(0)$, then we already observed that $t_1 \in \mathcal{K}_1$, which is a contradiction. Thus, $h(t_{k-1}) = h_n(t_{k-1})$. Similarly, $h(t_k) = h_n(t_k)$.

Thus the claim holds and so

$$\begin{aligned}\sum_{k\in\mathcal{K}_2} \text{length}\,\Gamma_{h|_{[t_{k-1},t_k]}} &\geq \sum_{k\in\mathcal{K}_2} \sqrt{(t_k - t_{k-1})^2 + (h(t_k) - h(t_{k-1}))^2} \\ &= \sum_{k\in\mathcal{K}_2} \sqrt{(t_k - t_{k-1})^2 + (h_n(t_k) - h_n(t_{k-1}))^2}.\end{aligned} \tag{4.8}$$

Step 2: Let I be one of the intervals I_i, $i \in J_1$, and let $\alpha < \beta$ be the endpoints of I.

Substep 2a: We claim that for all $x \in (\alpha, \beta)$,

$$h_n(x) = \min\{h(\alpha) + n|x - \alpha|, h(\beta) + n|x - \beta|\}. \tag{4.9}$$

Indeed, using the facts that $\operatorname{Lip} h_n \leq n$ and $h_n \leq h$ we have that

$$h_n(x) \leq h_n(\alpha) + n|x - \alpha| \leq h(\alpha) + n|x - \alpha|.$$

Similarly, $h_n(x) \leq h(\beta) + n|x - \beta|$. To prove the other inequality, assume that there exists $x_0 \in (\alpha, \beta)$ such that

$$h_n(x_0) < \min\{h(\alpha) + n|x_0 - \alpha|, h(\beta) + n|x_0 - \beta|\}.$$

Let $z_0 \in [0, b]$ be such that $h_n(x_0) = h(z_0) + n|x_0 - z_0|$. Then $z_0 \neq \alpha, \beta$. If $0 \leq z_0 < \alpha$, then

$$h_n(x_0) = h(z_0) + n(x_0 - z_0) < h(\alpha) + n(x_0 - \alpha).$$

It follows that $h_n(\alpha) \leq h(z_0) + n(\alpha - z_0) < h(\alpha)$, which contradicts the fact that $I = (\alpha, \beta)$ was a maximal interval in which $h > h_n$. Hence, $\alpha < z_0$ and, similarly, $z_0 < \beta$. Let $w_0 \in [0, b]$ be such that $h_n(z_0) = h(w_0) + n|w_0 - z_0|$. Then, using again the facts that $\operatorname{Lip} h_n \leq n$ and $h_n \leq h$, we have

$$\begin{aligned}h_n(x_0) &\leq h_n(w_0) + n|w_0 - x_0| \leq h(w_0) + n|w_0 - z_0| + n|z_0 - x_0| \\ &= h_n(z_0) + n|z_0 - x_0| \leq h(z_0) + n|z_0 - x_0| = h_n(x_0).\end{aligned}$$

In turn, all the inequalities must be equalities, and so $h_n(z_0) = h(z_0)$, which is a contradiction, since $z_0 \in I$ and $h_n < h$. This proves (4.9).

Substep 2b: Next we claim that

$$\text{length}\,\Gamma_{h_n|_{[\alpha,\beta]}} \leq \text{length}\,\Gamma_{h|_{[\alpha,\beta]}} \tag{4.10}$$

if $[\alpha, \beta] \subset (0, b)$, that

$$\text{length}\,\Gamma_{h_n|_{[0,\beta]}} \leq \text{length}\,\Gamma_{h|_{[0,\beta]}} + h(0) - h_n(0) \tag{4.11}$$

if $\alpha = 0$ and $\beta < b$, that

$$\text{length}\,\Gamma_{h_n|_{[\alpha,b]}} \leq \text{length}\,\Gamma_{h|_{[\alpha,b]}} + h(b) - h_n(b) \tag{4.12}$$

if $0 < \alpha$ and $\beta = b$, and that

$$\text{length}\,\Gamma_{h_n|_{[0,b]}} \leq \text{length}\,\Gamma_{h|_{[0,b]}} + h(0) - h_n(0) + h(b) - h_n(b) \tag{4.13}$$

if $\alpha = 0$ and $\beta = b$.

Assume first that for all $x \in [\alpha, \beta]$,

$$h_n(x) = h(\alpha) + n(x - \alpha). \tag{4.14}$$

If $[\alpha, \beta] \subset (0, b)$, then $h_n(\alpha) = h(\alpha)$ and $h_n(\beta) = h(\beta)$. Then $\Gamma_{h_n|_{[\alpha,\beta]}}$ and $\Gamma_{h|_{[\alpha,\beta]}}$ are two curves joining the points $(\alpha, h_n(\alpha))$ and $(\beta, h_n(\beta))$. The curve of shortest length joining two points is a line, hence (4.10) holds when (4.14) is satisfied.

If $\alpha = 0 < \beta < b$, then $h_n(0) \leq h(0)$ and $h_n(\beta) = h(\beta)$. Consider the curve $\tilde{\Gamma}_n$ obtained by adding to $\Gamma_{h|_{[\alpha,\beta]}}$ the vertical segment $\{0\} \times [h_n(0), h(0)]$. Then $\Gamma_{h_n|_{[0,\beta]}}$ and $\tilde{\Gamma}_n$ are two curves joining the points $(0, h_n(0))$ and $(\beta, h_n(\beta))$ and so as before

$$\text{length}\,\Gamma_{h_n|_{[0,\beta]}} \leq \text{length}\,\tilde{\Gamma}_n = \text{length}\,\Gamma_{h|_{[0,\beta]}} + h(0) - h_n(0),$$

which proves (4.11) when (4.14) is satisfied. The case $0 < \alpha < \beta = b$ is similar. Finally, if $0 = \alpha < \beta = b$ consider the curve $\tilde{\Gamma}_n$ obtained by adding to $\Gamma_{h|_{[\alpha,\beta]}}$ the vertical segments $\{0\} \times [h_n(0), h(0)]$ and $\{b\} \times [h_n(b), h(b)]$. Then $\Gamma_{h_n|_{[0,\beta]}}$ and $\tilde{\Gamma}_n$ are two curves joining the points $(0, h_n(0))$ and $(b, h_n(b))$ and so as before

$$\begin{aligned}\text{length}\,\Gamma_{h_n|_{[\alpha,\beta]}} &\leq \text{length}\,\tilde{\Gamma}_n \\ &= \text{length}\,\Gamma_{h|_{[\alpha,\beta]}} + h(0) - h_n(0) + h(b) - h_n(b).\end{aligned}$$

This shows (4.13) when (4.14) is satisfied.

Next assume that (4.14) does not hold. Then there exists $c \in (\alpha, \beta)$ such that

$$h_n(x) = \begin{cases} h(\alpha) + n(x-\alpha) & \text{if } \alpha \le x \le c, \\ h(\beta) + n(\beta - x) & \text{if } c \le x \le \beta. \end{cases}$$

It follows that $\max\{h(\alpha), h(\beta)\} \le h_n(c) \le h(c)$. If $0 < \alpha$, then $h_n(\alpha) = h(\alpha)$. In this case the length of the line joining the points $(\alpha, h_n(\alpha))$ and $(c, h_n(c))$ is less than the length of the line joining the points $(\alpha, h_n(\alpha))$ and $(c, h(c))$, which in turn is less than length $\Gamma_{h|_{[\alpha,c]}}$, so that

$$\text{length}\, \Gamma_{h_n|_{[\alpha,c]}} \le \text{length}\, \Gamma_{h|_{[\alpha,c]}}.$$

If $\alpha = 0$, then consider the curve $\tilde{\Gamma}_n$ obtained by adding to $\Gamma_{h|_{[0,\beta]}}$ the vertical segment $\{0\} \times [h_n(0), h(0)]$. Then

$$\text{length}\, \Gamma_{h_n|_{[0,c]}} \le \text{length}\, \tilde{\Gamma}_n = \text{length}\, \Gamma_{h|_{[0,c]}} + h(0) - h_n(0).$$

Similarly, if $\beta < b$,

$$\text{length}\, \Gamma_{h_n|_{[c,\beta]}} \le \text{length}\, \Gamma_{h|_{[c,\beta]}},$$

while if $\beta = b$,

$$\text{length}\, \Gamma_{h_n|_{[c,b]}} \le \text{length}\, \Gamma_{h|_{[c,b]}} + (h(b) - h_n(b)).$$

This concludes the proof of the claim.

Step 3: In view of Step 2 we have that

$$\begin{aligned} &\sum_{i \in \mathcal{J}_1} \text{length}\, \Gamma_{h|_{\overline{I_i}}} + (h(0) - h_n(0)) + (h(b) - h_n(b)) \\ &\qquad \ge \sum_{i \in \mathcal{J}_1} \text{length}\, \Gamma_{h_n|_{\overline{I_i}}} \ge \sum_{k \in \mathcal{K}_1} \sqrt{(t_k - t_{k-1})^2 + (h_n(t_k) - h_n(t_{k-1}))^2}, \end{aligned}$$

which, together with (4.7) and (4.8), implies that

$$\begin{aligned} &\text{length}\, \Gamma_h + (h(0) - h_n(0)) + (h(b) - h_n(b)) \\ &\qquad \ge \sum_{k=1}^{m} \sqrt{(t_k - t_{k-1})^2 + (h_n(t_k) - h_n(t_{k-1}))^2} \\ &\qquad \ge \sum_{i=1}^{\ell} \sqrt{(x_i - x_{i-1})^2 + (h_n(x_i) - h_n(x_{i-1}))^2}. \end{aligned}$$

the supremum over all partitions of $[0, b]$ concludes the proof. ☐

Using the previous theorem we can prove the following result, which, together with Theorem 4.6 and Remark 4.9, shows that

$$\inf_X \mathcal{F} = \min_{X_0} \mathcal{G}.$$

Theorem 4.11 (limsup inequality). *For every $(\boldsymbol{u}, h) \in X_0$ there exists a sequence $(\boldsymbol{u}_n, h_n) \in X$ such that*

$$\limsup_{n\to\infty} \mathcal{F}(\boldsymbol{u}_n, h_n) \leq \mathcal{G}(\boldsymbol{u}, h). \tag{4.15}$$

Proof. It suffices to consider the case $\mathcal{G}(\boldsymbol{u}, h) < \infty$. Let h_n be the Yosida transform of h given in Theorem 4.10 and define

$$\varepsilon_n := \frac{d}{b} - \frac{1}{b}\int_0^b h_n \, dx \geq 0.$$

Since $0 \leq h_n \leq h$ and $h_n \to h$ pointwise, it follows by the Lebesgue dominated convergence theorem that $\varepsilon_n \to 0$. Note that since $h_n \leq h_{n+1}$, if $\varepsilon_n = 0$ for some n, then $\varepsilon_m = 0$ for all $m \geq n$.

Assume first that $\varepsilon_n > 0$ for all n. By taking n sufficiently large we may assume that $\varepsilon_n \leq 1$ for all n. Since $\mathcal{G}(\boldsymbol{u}, h) < \infty$, we have that $\boldsymbol{E}(\boldsymbol{u}) \in L^2((0,b) \times (0,-k); \mathbb{R}^{2\times 2})$ for every $k > 0$ and thus by Korn's inequality in Lipschitz domains (see Theorem 2.15), $\boldsymbol{u} \in H^1((0,b) \times (0,-k); \mathbb{R}^2)$ for every $k > 0$. By a slicing argument (see, *e.g.*, Theorem 10.35 in [59]) for $\mathcal{L}^1$ a.e. $y < 0$, $\boldsymbol{u}(\cdot, y) \in H^1((0,b); \mathbb{R}^2)$. Let $y_0 < 0$ be any such y and define[4]

$$\boldsymbol{u}_n(x,y) := \begin{cases} \boldsymbol{u}(x,y) & \text{if } y < y_0, \\ \boldsymbol{u}(x,y_0) & \text{if } y_0 \leq y \leq y_0 + \varepsilon_n, \\ \boldsymbol{u}(x, y-\varepsilon_n) & \text{if } y_0 + \varepsilon_n \leq y \leq h_n(x) + \varepsilon_n =: f_n(x). \end{cases} \tag{4.16}$$

Then $\int_0^b f_n \, dx = d$ by the definition of ε_n. Since $f_n \geq \varepsilon_n > 0$, we have that $\varphi(f_n(x)) = \gamma_{\text{film}}$. Hence, by (1.3),

$$\mathcal{F}(\boldsymbol{u}_n, f_n) = \mathcal{G}(\boldsymbol{u}_n, f_n) = \int_0^b \int_{-\infty}^{y_0} W(\boldsymbol{E}(\boldsymbol{u}))\, dydx$$
$$+ \varepsilon_n \int_{(0,b)} W(\boldsymbol{E}(\boldsymbol{v})(x, y_0))\, dx$$
$$+ \int_0^b \int_{y_0+\varepsilon_n}^{h_n(x)+\varepsilon_n} W(\boldsymbol{E}(\boldsymbol{u})(x, y-\varepsilon_n) - \boldsymbol{E}_0(y))\, dydy + \gamma_{\text{film}} \operatorname{length} \Gamma_{f_n},$$

[4] Alternatively, one could use the simpler sequence of functions

$$u_n(x,y) := u(x, y-\varepsilon_n) \quad \text{for} \quad y \leq f_n(x).$$

where $\boldsymbol{v}(\boldsymbol{x}) := \boldsymbol{u}(x, y_0)$. By a change of variables, (1.3), and Theorem 4.10 we have

$$\begin{aligned}\mathcal{F}(\boldsymbol{u}_n, f_n) \leq & \int_0^b \int_{-\infty}^{y_0} W(\boldsymbol{E}(\boldsymbol{u}))\, dydx \\ & + \int_0^b \int_{y_0}^{h(x)} W(\boldsymbol{E}(\boldsymbol{u})(x,s) - \boldsymbol{E}_0(s+\varepsilon_n))\, dsdx \\ & + \gamma_{\text{film}} \operatorname{length} \Gamma_h + c\varepsilon_n \int_{(0,b)} \left| \frac{\partial \boldsymbol{u}}{\partial x}(x, y_0) \right|^2 dx \\ & + \gamma_{\text{film}}(h(0) - h_n(0)) + \gamma_{\text{film}}(h(b) - h_n(b)).\end{aligned}$$

Letting $n \to \infty$, we get (4.15) in the case $\varepsilon_n > 0$ for all n.

Assume next that $\varepsilon_n = 0$ for all n sufficiently large. Since $h_n \leq h$, it follows that $h_n = h$ for all n sufficiently large, and so h is Lipschitz continuous. However, in this case $\mathcal{F}(\boldsymbol{u}, h)$ could be strictly above $\mathcal{G}(\boldsymbol{u}, h)$ if $\gamma_{\text{film}} < \gamma_{\text{sub}}$. Thus we first raise h and then truncate it to preserve the volume constraint. Define

$$f_k := \min\{h + 1/k, t_k\},$$

where $t_k > 0$ has been chosen so that $\int_0^b f_k\, dx = d$. Note that length $\Gamma_{f_k} \leq$ length Γ_h. Indeed, for every partition $x_0 = 0 < \cdots < x_n = b$, we have that

$$(f_k(x_i) - f_k(x_{i-1}))^2 \leq (h(x_i) - h(x_{i-1}))^2$$

for all $i = 1, \ldots, n$. Since $f_k \geq \min\{1/k, t_k\} > 0$, we have that that $\varphi(f_n(x)) = \gamma_{\text{film}}$. As before let $y_0 < 0$ be $\boldsymbol{u}(\cdot, y) \in H^1((0,b); \mathbb{R}^2)$ and define[5]

$$\boldsymbol{u}_k(x,y) := \begin{cases} \boldsymbol{u}(x,y) & \text{if } y < y_0, \\ \boldsymbol{u}(x, y_0) & \text{if } y_0 \leq y \leq y_0 + 1/k, \\ \boldsymbol{u}(x, y - 1/k) & \text{if } y_0 + 1/k \leq y \leq f_k(x). \end{cases}$$

[5] Alternatively, one could use the simpler sequence of functions

$$u_k(x,y) := u(x, y - 1/k) \quad \text{for} \quad y \leq f_k(x).$$

Then as before

$$\mathcal{F}(\boldsymbol{u}, f_k) = \mathcal{G}(\boldsymbol{u}, f_k) \leq \int_0^b \int_{-\infty}^{y_0} W(\boldsymbol{E}(\boldsymbol{u}))\, dydx$$
$$+ \int_0^b \int_{y_0}^{h(x)} W\left(\boldsymbol{E}(\boldsymbol{u})(x,s) - \boldsymbol{E}_0\left(s + \tfrac{1}{k}\right)\right) dsdx$$
$$+ \gamma_{\text{film}} \operatorname{length} \Gamma_h + \tfrac{c}{k} \int_{(0,b)} \left| \frac{\partial \boldsymbol{u}}{\partial x}(x, y_0) \right|^2 dx$$

and the proof is complete. □

Remark 4.12. We observe that in the construction of (4.16), in view of (4.1), it is energetically more favorable to have a thin wetting layer of the film over the substrate rather than leaving the substrate exposed. Thus, (4.1) favors the SK growth mode over the VW mode (see Section 1).

5 Regularity of minimizers

Before discussing the regularity of minimizers of $\mathcal{G}$, we prove that the volume constraint $\int_0^b h(x)\, dx = d$ can be replaced by a volume penalization. This will give significantly more freedom in the choice of test functions in the regularity results.

Theorem 5.1 (volume penalization). *Let* $(\boldsymbol{u}_0, h_0) \in X_0$ *be a minimizer of the functional* $\mathcal{G}$ *defined in* (4.2) *with* $\int_0^b h_0(x)\, dx = d$. *Then there exists* $k_0 > 0$ *such that for all* $k \geq k_0$, $(\boldsymbol{u}_0, h_0)$ *is a minimizer of the penalized functional*

$$\mathcal{G}_k(\boldsymbol{u}, h) := \int_{\Omega_h} W(\boldsymbol{E}(\boldsymbol{u}) - \boldsymbol{E}_0)\, d\boldsymbol{x} + \gamma_{\text{film}} \operatorname{length} \Gamma_h + k \left| \int_0^b h\, dx - d \right| \tag{5.1}$$

over all $(\boldsymbol{u}, h) \in X_0$.

Proof. Reasoning as in the proof of Theorem 4.6, we can prove that for every k there exists a minimimizer $(\boldsymbol{v}_k, f_k)$ of $\mathcal{G}_k$. In turn,

$$\mathcal{G}_k(\boldsymbol{v}_k, f_k) \leq \mathcal{G}_k(\boldsymbol{u}_0, h_0) = \mathcal{G}(\boldsymbol{u}_0, h_0) < \infty. \tag{5.2}$$

Then by (5.1) and (5.2) we have that $\int_0^b f_k\, dx \to d$ as $k \to \infty$ and $\sup_k \operatorname{length} \Gamma_{f_k} < \infty$. Let k_1 be so large that $\int_0^b f_k\, dx > \frac{d}{2}$ and $\|f_k\|_\infty \leq c$ for some constant c independent of k and set $k_2 := c\mathcal{G}(\boldsymbol{u}_0, h_0) + k_1$. We claim that for all $k \geq k_2$ sufficiently large

$$\int_0^b f_k\, dx \geq d. \tag{5.3}$$

To see this, assume by contradiction that there is $k \geq k_2$ large such that

$$d > \int_0^b f_k \, dx.$$

Define

$$t_k := \frac{d}{\int_0^b f_k \, dx}, \qquad h_k(x) = t_k f_k(x)$$

for all $x \in (a, b)$. Then

$$\int_0^b h_k \, dx = d, \qquad 1 < t_k < 2.$$

Consider a partition $0 = x_0 < \cdots < x_\ell = b$. Then

$$\begin{aligned}
&\sum_{i=1}^{\ell} \sqrt{(x_i - x_{i-1})^2 + (h_k(x_i) - h_k(x_{i-1}))^2} \\
&\quad = \sum_{i=1}^{\ell} \sqrt{(x_i - x_{i-1})^2 + t_k^2 (f_k(x_i) - f_k(x_{i-1}))^2} \\
&\quad \leq t_k \sum_{i=1}^{\ell} \sqrt{(x_i - x_{i-1})^2 + (f_k(x_i) - f_k(x_{i-1}))^2} \\
&\quad \leq t_k \, \text{length} \, \Gamma_{f_k}.
\end{aligned}$$

Hence,

$$\text{length} \, \Gamma_{h_k} \leq t_k \, \text{length} \, \Gamma_{f_k},$$

and so, by (5.2),

$$\begin{aligned}
\gamma_{\text{film}} \, \text{length} \, \Gamma_{h_k} - \gamma_{\text{film}} \, \text{length} \, \Gamma_{f_k} &\leq (t_k - 1) \gamma_{\text{film}} \, \text{length} \, \Gamma_{f_k} \\
&\leq (t_k - 1) \mathcal{G}(\boldsymbol{u}_0, h_0).
\end{aligned} \tag{5.4}$$

For $(x, y') \in \Omega_{h_k}$ define

$$\boldsymbol{w}_k(x, y') := \left((\boldsymbol{v}_k)_1 \left(x, \frac{y'}{t_k} \right), \frac{1}{t_k} (\boldsymbol{v}_k)_2 \left(x, \frac{y'}{t_k} \right) \right).$$

By a change of variables and (1.3), we have

$$\begin{aligned}
&\int_{\Omega_{h_k}} W(\boldsymbol{E}(\boldsymbol{w}_k)(x, y') - \boldsymbol{E}_0(y')) \, dx dy' \\
&\quad = \frac{1}{t_k} \int_{\Omega_{f_k}} W(\widetilde{\boldsymbol{E}}(\boldsymbol{v}_k)(\boldsymbol{x}) - \boldsymbol{E}_0(y)) \, d\boldsymbol{x},
\end{aligned}$$

where $\widetilde{\boldsymbol{E}}(\boldsymbol{v}_k)(\boldsymbol{x})$ is the 2×2 matrix whose entries are

$$\begin{aligned} \widetilde{\boldsymbol{E}}_{11}(\boldsymbol{v}_k)(\boldsymbol{x}) &= \boldsymbol{E}_{11}(\boldsymbol{v}_k)(\boldsymbol{x}), \\ \widetilde{\boldsymbol{E}}_{12}(\boldsymbol{v}_k)(\boldsymbol{x}) &= \frac{1}{t_k}\boldsymbol{E}_{12}(\boldsymbol{v}_k)(\boldsymbol{x}), \\ \widetilde{\boldsymbol{E}}_{22}(\boldsymbol{v}_k)(\boldsymbol{x}) &= \frac{1}{t_k^2}\boldsymbol{E}_{22}(\boldsymbol{v}_k)(\boldsymbol{x}). \end{aligned} \tag{5.5}$$

Since $W(\boldsymbol{E})$ is a positive definite quadratic form over the 2×2 symmetric matrices (see (1.4)), we have that

$$|W(\boldsymbol{E}) - W(\boldsymbol{E}_1)| \leq c\,(|\boldsymbol{E}| + |\boldsymbol{E}_1|)\,|\boldsymbol{E} - \boldsymbol{E}_1|$$

for all 2×2 symmetric matrices $\boldsymbol{E}$ and $\boldsymbol{E}_1$. Hence by (1.3) and (5.5),

$$\begin{aligned} &\int_{\Omega_{h_k}} W(\boldsymbol{E}(\boldsymbol{w}_k)(x, y') - \boldsymbol{E}_0(y'))\,d\boldsymbol{x}' - \int_{\Omega_{f_k}} W(\boldsymbol{E}(\boldsymbol{v}_k)(\boldsymbol{x}) - \boldsymbol{E}_0\,(y))\,d\boldsymbol{x} \\ &\leq \frac{1}{t_k}\int_{\Omega_{f_k}} \left[W(\widetilde{\mathbf{E}}(\boldsymbol{v}_k)(\boldsymbol{x}) - \boldsymbol{E}_0\,(y)) - W(\boldsymbol{E}(\boldsymbol{v}_k)(\boldsymbol{x}) - \boldsymbol{E}_0(y))\right]\,d\boldsymbol{x} \\ &\leq \frac{c}{t_k}\int_{\Omega_{f_k}} \left(|\widetilde{\mathbf{E}}(\boldsymbol{v}_k) - \boldsymbol{E}_0| + |\boldsymbol{E}(\boldsymbol{v}_k) - \boldsymbol{E}_0|\right)\left(|\widetilde{\mathbf{E}}(\boldsymbol{v}_k) - \boldsymbol{E}(\boldsymbol{v}_k)|\right)\,d\boldsymbol{x} \\ &\leq c(t_k - 1)\int_{\Omega_{f_k}} |\boldsymbol{E}(\boldsymbol{v}_k) - \boldsymbol{E}_0||\boldsymbol{E}(\boldsymbol{v}_k)|\,d\boldsymbol{x} \\ &\leq c_0(t_k - 1)(\mathcal{G}_k\,(\boldsymbol{v}_k, f_k) + |\hat{\boldsymbol{E}}_0|^2) \leq c_0(t_k - 1)(\mathcal{G}\,(\boldsymbol{u}_0, h_0) + |\hat{\boldsymbol{E}}_0|^2), \end{aligned} \tag{5.6}$$

where c_0 depends only on the ellipticity constants of W and $\sup_k \|f_k\|_\infty$. By (5.2), (5.4), and (5.6), we have that

$$\begin{aligned} \mathcal{G}\,(\boldsymbol{u}_0, h_0) &\leq \mathcal{G}\,(\boldsymbol{u}_k, h_k) \\ &\leq \mathcal{G}\,(\boldsymbol{v}_k, f_k) + (t_k - 1)\left[(c_0 + 1)\mathcal{G}\,(\boldsymbol{u}_0, h_0) + c_0|\hat{\boldsymbol{E}}_0|^2\right] \\ &\leq \mathcal{G}_k\,(\boldsymbol{v}_k, f_k) + (t_k - 1)\left[(c_0 + 1)\mathcal{G}\,(\boldsymbol{u}_0, h_0) + c_0|\hat{\boldsymbol{E}}_0|^2\right] \\ &\quad - k\left(d - \int_0^b f_k\,dx\right) \\ &= \mathcal{G}_k\,(\boldsymbol{v}_k, f_k) + (t_k - 1)\left[(c_0 + 1)\mathcal{G}\,(\boldsymbol{u}_0, h_0) + c_0|\hat{\boldsymbol{E}}_0|^2\right] \\ &\quad - (t_k - 1)k\int_0^b f_k\,dx \\ &\leq \mathcal{G}\,(\boldsymbol{u}_0, h_0) + (t_k - 1)\left[(c_0 + 1)\mathcal{G}\,(\boldsymbol{u}_0, h_0) + c_0|\hat{\boldsymbol{E}}_0|^2 - k\frac{d}{2}\right]. \end{aligned}$$

Thus, if

$$k \geq k_3 := \frac{2}{d}\left[(c_0+1)\mathcal{G}(\boldsymbol{u}_0, h_0) + c_0|\hat{\boldsymbol{E}}_0|^2\right] + 1,$$

we get a contradiction. This contradiction proves the claim (5.3).

It remains to treat the case in which

$$\int_0^b f_k\, dx > d \tag{5.7}$$

for some $k > k_1$. Define

$$h_k := \min\{f_k, t_k\},$$

where $t_k > 0$ has been chosen so that $\int_0^b h_k\, dx = d$. Note that $\operatorname{length}\Gamma_{h_k} \leq \operatorname{length}\Gamma_{f_k}$. Indeed, for every partition $x_0 = 0 < \cdots < x_n = b$, we have that

$$(h_k(x_i) - h_k(x_{i-1}))^2 \leq (f_k(x_i) - f_k(x_{i-1}))^2$$

for all $i = 1, \ldots, n$. Hence,

$$\mathcal{G}(\boldsymbol{v}_k, h_k) = \mathcal{G}_k(\boldsymbol{v}_k, h_k) < \mathcal{G}_k(\boldsymbol{v}_k, f_k),$$

which is a contradiction. This completes the proof. □

The following theorem was first proved in a slightly different context by Chambolle and Larsen [22]. The present proof is taken from the paper of Fusco and Morini [17] (see also [33]).

Theorem 5.2 (internal sphere's condition). *Let* $(\boldsymbol{u}_0, h_0) \in X_0$ *be a minimizer of the functional* $\mathcal{G}$ *defined in* (4.2). *Then there exists* $r_0 > 0$ *with the property that for every* $z_0 \in \overline{\Gamma}_{h_0}$ *there exists an open ball* $B(\boldsymbol{x}_0, r_0)$, *with* $B(\boldsymbol{x}_0, r_0) \cap ((0, b) \times \mathbb{R}) \subseteq \Omega_{h_0}$, *such that*

$$\partial B(\boldsymbol{x}_0, r_0) \cap \overline{\Gamma}_{h_0} = \{z_0\}.$$

Proof. **Step 1:** We will prove that if $r > 0$ is sufficiently small there cannot be a ball $B(\boldsymbol{x}_0, r) \cap ((0, b) \times \mathbb{R}) \subseteq \Omega_{h_0}$ whose boundary touches $\overline{\Gamma}_{h_0}$ in more than one point. Thus, assume by contradiction that there exists a ball $B(\boldsymbol{x}_0, r) \cap ((0, b) \times \mathbb{R}) \subseteq \Omega_{h_0}$ such that $\partial B(\boldsymbol{x}_0, r)$ intersects $\overline{\Gamma}_{h_0}$ in at least two points $\boldsymbol{P} = (x_1, y_1)$ and $\boldsymbol{Q} = (x_2, y_2)$, with $x_1 < x_2$. Let $\boldsymbol{\gamma}$ be the shortest arc on $B(\boldsymbol{x}_0, r)$ connecting $\boldsymbol{P}$ and $\boldsymbol{Q}$ (if $\boldsymbol{P}$ and $\boldsymbol{Q}$

are antipodal, take $\boldsymbol{\gamma}$ to be one of the two arcs) and let $g : [x_1, x_2] \to \mathbb{R}$ be a regular function with graph $\boldsymbol{\gamma}$. Define

$$D := \{\boldsymbol{x} = (x, y) : x_1 < x < x_2,\ g(x) < y < h_0(x)\}$$

and let

$$\bar{h}(x) := \begin{cases} g(x) & \text{if } x_1 < x < x_2, \\ h_0(x) & \text{otherwise in } (0, b). \end{cases}$$

By the minimality of $(\boldsymbol{u}_0, h_0)$ and Theorem 5.1 we have $\mathcal{G}_k(\boldsymbol{u}_0, h_0) \leq \mathcal{G}_k(\boldsymbol{u}_0|_{\Omega_{h_0}}, \bar{h})$, that is,

$$\int_{\Omega_{h_0}} W(\boldsymbol{E}(\boldsymbol{u}_0) - \boldsymbol{E}_0)\, d\boldsymbol{x} + \gamma_{\text{film}} \operatorname{length} \Gamma_{h_0} \leq \int_{\Omega_{\bar{h}}} W(\boldsymbol{E}(\boldsymbol{u}_0) - \boldsymbol{E}_0)\, d\boldsymbol{x} + \gamma_{\text{film}} \operatorname{length} \Gamma_{\bar{h}} + k \left| \int_0^b \bar{h}\, dx - d \right|.$$

It follows that

$$\gamma_{\text{film}}(L - \ell) \leq k\mathcal{L}^2(D), \tag{5.8}$$

where L is the length of the graph of h_0 in the interval $[x_1, x_2]$ and ℓ is the length of $\boldsymbol{\gamma}$.

Assume that h_0 is Lipschitz. By the convexity of the function $f(t) = \sqrt{1 + t^2}$, we have the inequality $f(t) \geq f(s) + f'(s)(t - s)$. Hence,

$$\begin{aligned} L - \ell &= \int_{x_1}^{x_2} \sqrt{1 + (h_0'(x))^2}\, dx - \int_{x_1}^{x_2} \sqrt{1 + (g'(x))^2}\, dx \\ &\geq \int_{x_1}^{x_2} \frac{(h_0'(x) - g'(x))g'(x)}{\sqrt{1 + (g'(x))^2}}\, dx. \end{aligned}$$

Integrating by parts the right-hand side of the previous inequality we obtain

$$\begin{aligned} L - \ell &\geq -\int_{x_1}^{x_2} (h_0(x) - g(x)) \left(\frac{g'(x)}{\sqrt{1 + (g'(x))^2}} \right)' dx \\ &= \frac{1}{r} \int_{x_1}^{x_2} (h_0(x) - g(x))\, dx = \frac{1}{r}\mathcal{L}^2(D), \end{aligned}$$

where we have used , and the facts that $h_0(x_i) = g(x_i)$, $i = 1, 2$, and that $-\left(\frac{g'(x)}{\sqrt{1+(g'(x))^2}} \right)' = \frac{1}{r}$. If h_0 is not Lipschitz, the same inequality can be obtained by approximating h_0 with smooth functions.

Combining the last inequality with (5.8) gives

$$\frac{\gamma_{\text{film}}}{k}(L-\ell) \leq \mathcal{L}^2(D) \leq r(L-\ell).$$

Hence, if $\ell < L$, it follows that $\frac{\gamma_{\text{film}}}{k} \leq r$. In particular, if $r_1 < \frac{\gamma_{\text{film}}}{k}$, then every ball $B(\boldsymbol{x}_0, r_1) \cap ((0,b)\times\mathbb{R}) \subseteq \Omega_{h_0}$ either intersects $\overline{\Gamma}_{h_0}$ at most at one point or one of its subarcs coincides with $\overline{\Gamma}_{h_0}$. Taking $r < r_1$, we have proved the claim.

Step 2: We now deduce from Step 1 the uniform internal sphere condition in the statement of the theorem. Consider the union U of all balls of radius r_0 that are contained in

$$\widetilde{\Omega} := \Omega \cup [(\mathbb{R}\setminus(0,b))\times\mathbb{R}].$$

Then the thesis is equivalent to showing that $\Omega \subseteq U \cap ((0,b)\times\mathbb{R})$. Assume by contradiction that such an inclusion doesn't hold. Then there exist $\boldsymbol{x_0} \in \Omega\cap\partial U$, a sequence of balls $B(\boldsymbol{y}_n, r_0) \subseteq \widetilde{\Omega}$, and $\boldsymbol{x}_n \in \partial B(\boldsymbol{y}_n, r_0)$ such that $\boldsymbol{x}_n \to \boldsymbol{x}_0$. Up to extracting a subsequence (not relabeled) we may assume that $B(\boldsymbol{y}_n, r_0) \to B(\boldsymbol{y}, r_0)$ in the Hausdorff metric, for some ball $B(\boldsymbol{y}, r_0) \subseteq \widetilde{\Omega}$ having $\boldsymbol{x}_0$ at its boundary. Note that the intersection of $\partial B(\boldsymbol{y}, r_0)$ with $\overline{\Gamma}$ must be nonempty, since if it were we could translate the ball slightly still remaining in $\widetilde{\Omega}$ and this would violate the fact that $\boldsymbol{x_0} \in \partial U$. Hence, by the previous step, $\partial B(\boldsymbol{y}, r_0) \cap \overline{\Gamma} = \{z\}$. If $\boldsymbol{x}_0$ and z are antipodal, then we can find $\delta > 0$ such that $B(\boldsymbol{y}+\delta(\boldsymbol{x}_0 - z), r_0) \subseteq \widetilde{\Omega}$, which would imply that $\boldsymbol{x}_0 \in U$, a contradiction. If $\boldsymbol{x}_0$ and z are not antipodal, then we can rotate $B(\boldsymbol{y}, r_0)$ around $\boldsymbol{x}_0$, slightly away from z, to get a ball B' of radius r_0 such that $\overline{B}' \subset \widetilde{\Omega}$ and $\boldsymbol{x}_0 \in \partial B'$. Translating now B' towards $\boldsymbol{x}_0$ we find a ball of the same radius containing $\boldsymbol{x}_0$ and contained in $\widetilde{\Omega}$, which gives again $\boldsymbol{x}_0 \in U$. This concludes the proof of the proposition. □

Remark 5.3. Theorem 5.2 can be restated in the following way: There is $r_0 > 0$ such that for every $z_0 \in \partial\widetilde{\Omega}$ there exists an open ball $B(\boldsymbol{x}_0, r_0)$, with $B(\boldsymbol{x}_0, r_0) \subseteq \widetilde{\Omega}$, such that

$$\partial B(\boldsymbol{x}_0, r_0) \cap \partial\widetilde{\Omega} = \{z_0\}.$$

Remark 5.4. By Theorem 5.2 there exists $r_0 > 0$ with the property that for every $z_0 \in \overline{\Gamma}_{h_0}$ there exists an open ball $B(\boldsymbol{x}_0, r_0)$, with $B(\boldsymbol{x}_0, r_0) \cap ((0,b)\times\mathbb{R}) \subseteq \Omega_{h_0}$, such that

$$\partial B(\boldsymbol{x}_0, r_0) \cap \overline{\Gamma}_{h_0} = \{z_0\}.$$

Note that if $\boldsymbol{\nu}_0 \in \partial B(\mathbf{0}, 1)$ is the outward unit normal to $B(\boldsymbol{x}_0, r_0)$ at z_0, then $\boldsymbol{x}_0 = z_0 - r_0\boldsymbol{\nu}_0$. Thus, the set

$$N_{z_0} := \{\boldsymbol{\nu} \in \partial B(\mathbf{0}, 1) : \; B(z_0 - r_0\boldsymbol{\nu}, r_0) \cap ((0, b) \times \mathbb{R}) \subseteq \Omega_{h_0}\} \quad (5.9)$$

is nonempty.

Remark 5.5. In what follows we will repeatedly use the fact that given a ball $B(\boldsymbol{x}_0, r_0)$, a point $z \in \partial B(\boldsymbol{x}_0, r_0)$, and a unit vector $\boldsymbol{\tau}$, if $\varepsilon \in (0, 1)$ and $\boldsymbol{\nu} \cdot \boldsymbol{\tau} > \varepsilon$, where $\boldsymbol{\nu} := \frac{z - \boldsymbol{x}_0}{\|z - \boldsymbol{x}_0\|}$, then $z - \delta\boldsymbol{\tau} \in B(\boldsymbol{x}_0, r_0)$ for all $0 < \delta < 2r_0\varepsilon$. To see this, by a rotation and a translation, we may assume that $\boldsymbol{x}_0 = \mathbf{0}$ and $z = (r_0, 0)$, so that $\boldsymbol{\nu} = \boldsymbol{e}_1$. Then $z - \delta\boldsymbol{\tau} \in B(\mathbf{0}, r_0)$ if and only if $(r_0 - \delta\tau_1)^2 + \delta^2\tau_2^2 < r_0^2$, that is, $-2r_0\delta\tau_1 + \delta^2 < 0$. Since $\tau_1 > \varepsilon$, the previous inequality holds for all $\delta < 2r_0\varepsilon$.

Theorems 5.6, 5.7, and 5.10 below were first proved by Bonnetier and Chambolle [21] in a more general situation, that is, when Γ_{h_0} is not assumed to be the graph of a function. The proofs we present here are adapted from [32].

In the next theorem we prove that h_0 admits a left and right derivative at all but countably many points.

Theorem 5.6 (left and right derivatives of h). *Let $(\boldsymbol{u}_0, h_0) \in X_0$ be a minimizer of the functional $\mathcal{G}$ defined in (4.2). Then $\overline{\Gamma}_{h_0}$ admits a left and a right tangent at every point z not of the form $z = (x, h_0(x))$ with $x \in S$, where*

$$S := \left\{ x \in (0, b) : \; h_0(x) < \liminf_{t \to x} h_0(t) \right\}. \quad (5.10)$$

Proof. Fix $z_0 = (x_0, y_0) \in \overline{\Gamma}_{h_0}$.
Step 1: We claim that there exists

$$\lim_{x \to x_0^+} \frac{h_0(x) - h_0^+(x_0)}{x - x_0} = \ell_+ \in \overline{\mathbb{R}}, \quad (5.11)$$

where we recall that

$$h_0^+(x_0) := \lim_{x \to x_0^+} h_0(x). \quad (5.12)$$

Indeed, if not, then we can find two half-lines L and M starting from through $z_+ := (x_0, h_0^+(x_0))$ such that the graph of h_0 intersects L and M infinitely many times. Let $s_n \to x_0^+$ and $t_n \to x_0^+$ be such that $(s_n, h_0(s_n)) \in L$ and $(t_n, h_0(t_n)) \in M$. Let $\boldsymbol{\tau}_L$ be the tangential direction of L, so that,

$$L = \{z_+ + r\boldsymbol{\tau}_L : \; r \geq 0\}.$$

We claim that $-\boldsymbol{\tau}_L^\perp \in N_{z_+}$. To see this, let $z_n := (s_n, h_0(s_n))$ and $\boldsymbol{\nu}_n \in N_{z_n}$. Let $\varepsilon \in (0,1)$. If $\boldsymbol{\nu}_n \cdot \boldsymbol{\tau}_L > \varepsilon$ by Remark 5.5 there exists $\delta_\varepsilon > 0$ such that $z_n - \delta \boldsymbol{\tau}_L \in B(z_n - r_0\boldsymbol{\nu}_n, r_0)$ for all $0 < \delta \leq \delta_\varepsilon$. Since $z_n \in L$, it follows that

$$z_+ = z_n - \|z_n - z_+\|\, \boldsymbol{\tau}_L \in B(z_n - r_0\boldsymbol{\nu}_n, r_0)$$

provided $\|z_n - z_+\| \leq \delta_\varepsilon$. In view of (5.12), this gives a contradiction because $z_+ \in \Gamma_{h_0}$. Thus, $\boldsymbol{\nu}_n \cdot \boldsymbol{\tau}_L \leq \varepsilon$ for all n sufficiently small.

On the other hand, if $\boldsymbol{\nu}_n \cdot \boldsymbol{\tau}_L < -\varepsilon$ then

$$z_m = z_n + \|z_m - z_n\|\, \boldsymbol{\tau}_L \in B(z_n - r_0\boldsymbol{\nu}_n, r_0)$$

provided $\|z_n - z_+\| < \|z_m - z_+\| \leq \delta_\varepsilon$, which is again a contradiction. Thus, we have shown that $|\boldsymbol{\nu}_n \cdot \boldsymbol{\tau}_L| \leq \varepsilon$ for all n sufficiently small.

Up to a subsequence, not relabeled, we can assume that $\boldsymbol{\nu}_n \to \boldsymbol{\nu}_0 \in \partial B(\mathbf{0}, 1)$. It follows that the closed sets $\overline{B(z_n - r_0\boldsymbol{\nu}_n, r_0) \cap ([0,b] \times \mathbb{R})}$, which are contained in $\overline{\Omega_{h_0}}$, converge to $\overline{B(z_+ - r_0\boldsymbol{\nu}_0, r_0) \cap ([0,b] \times \mathbb{R})}$ in the Hausdorff distance $d_{\mathcal{H}}$, and so $\overline{B(z_+ - r_0\boldsymbol{\nu}_0, r_0) \cap ([0,b] \times \mathbb{R})}$ is also contained in $\overline{\Omega_{h_0}}$. In turn, $B(z_+ - r_0\boldsymbol{\nu}_0, r_0) \cap ((0,b) \times \mathbb{R}) \subseteq \Omega_{h_0}$, which shows that $\boldsymbol{\nu}_0$ belongs to N_{z_+}. Since $|\boldsymbol{\nu}_n \cdot \boldsymbol{\tau}_L| \leq \varepsilon$ for all n sufficiently small, letting $n \to \infty$, it follows that $|\boldsymbol{\nu}_0 \cdot \boldsymbol{\tau}_L| \leq \varepsilon$ for all $\varepsilon \in (0,1)$. Thus $\boldsymbol{\nu}_0 \cdot \boldsymbol{\tau}_L = 0$, which implies that $\boldsymbol{\nu}_0 = -\boldsymbol{\tau}_L^\perp$ and proves the claim.

On the other hand, by changing L and M, if necessary, we can assume that $\boldsymbol{\nu}_0 \cdot \boldsymbol{\tau}_M < -\varepsilon$ for some $\varepsilon \in (0,1)$, and so, setting $\boldsymbol{w}_n := (t_n, h_0(t_n))$,

$$\boldsymbol{w}_n = z_+ + \|\boldsymbol{w}_n - z_+\|\, \boldsymbol{\tau}_M \in B(z_+ - r_0\boldsymbol{\nu}_0, r_0)$$

provided $\|\boldsymbol{w}_n - z_+\| \leq \delta_\varepsilon$, which is again a contradiction. This shows that L and M must coincide, and, in turn, that the limit (5.11) exists. With a similar proof we can show that there exists

$$\lim_{x \to x_0^-} \frac{h_0(x) - h_0^-(x_0)}{x - x_0} = \ell_- \in \overline{\mathbb{R}}.$$

Step 2: We have now two cases. If z_0 belongs to the relative interior of a vertical segment of $\overline{\Gamma}_{h_0}$, then x_0 is a jump point of h_0 and either $x_0 \in S$ with

$$h_0(x_0) < y_0 < \liminf_{x \to x_0} h_0(x) \tag{5.13}$$

or $x_0 \notin S$ and

$$h_0(x_0) = \liminf_{x \to x_0} h_0(x) < y_0 < \limsup_{x \to x_0} h_0(x). \tag{5.14}$$

In both cases $\overline{\Gamma}_{h_0}$ admits a vertical tangent at z_0. It remains to consider the cases in which

$$y_0 = h_0^-(x_0) \quad \text{or} \quad y_0 = h_0^+(x_0). \tag{5.15}$$

In view of the previous step, $\overline{\Gamma}_{h_0}$ admits a left and a right tangent at z_0 when (5.15) holds. □

Define

$$\Gamma_{\text{cusps}} := \left\{ z \in \overline{\Gamma}_{h_0} : \pm \boldsymbol{e}_1 \in N_z \right\} \tag{5.16}$$

and

$$\Gamma_{\text{cuts}} := \left\{ (x, y) : x \in (0, b) \cap S,\ h_0(x) \le y \le \liminf_{t \to x} h_0(t) \right\}, \tag{5.17}$$

where N_z is the set defined in (5.9) and S is the set defined in 5.10.

Theorem 5.7 (cusps and cuts). *Let $(\boldsymbol{u}_0, h_0) \in X_0$ be a minimizer of the functional $\mathcal{G}$ defined in* (4.2). *Then the sets Γ_{cusps} and Γ_{cuts} contain finitely many (possibly degenerate) vertical segments.*

Proof. **Step 1:** Let $z_0 \in \overline{\Gamma}_{h_0}$ and let $z_n \in \overline{\Gamma}_{h_0}$ be such that $z_n \to z_0$. Then there exists $\boldsymbol{\nu}_n \in N_{z_n}$. Up to a subsequence, not relabeled, we can assume that $\boldsymbol{\nu}_n \to \boldsymbol{\nu}_0 \in \partial B(\boldsymbol{0}, 1)$. It follows that the closed sets $\overline{B(z_n - r_0\boldsymbol{\nu}_n, r_0) \cap ([0, b] \times \mathbb{R})}$, which are contained in $\overline{\Omega_{h_0}}$, converge to $\overline{B(z_0 - r_0\boldsymbol{\nu}_0, r_0) \cap ([0, b] \times \mathbb{R})}$ in the Hausdorff distance $d_{\mathcal{H}}$ and so $\overline{B(z_0 - r_0\boldsymbol{\nu}_0, r_0) \cap ([0, b] \times \mathbb{R})}$ is also contained in $\overline{\Omega_{h_0}}$. In turn, $B(z_0 - r_0\boldsymbol{\nu}_0, r_0) \cap ((0, b) \times \mathbb{R}) \subseteq \Omega_{h_0}$, which shows that $\boldsymbol{\nu}_0$ belongs to N_{z_0}.

The previous argument with $z_n = z_0$ for all n shows in particular that the set N_{z_0} is closed.
Step 2: Let $z_0 = (x_0, y_0) \in \Gamma_{\text{cuts}}$. Then

$$h_0(x_0) \le y_0 \le \liminf_{x \to x_0} h_0(x).$$

Let $z_n = (x_n, h_0(x_n)) \in \overline{\Gamma}_{h_0}$ be such that $z_n \to z_0^+ := (x_0, h_0^+(x_0))$, with $x_n \to x_0^+$. Then by the previous step, we have that $\boldsymbol{\nu}_n \to \boldsymbol{\nu}_0 \in \partial B(\boldsymbol{0}, 1)$ with $\boldsymbol{\nu}_0 \in N_{z_0^+}$. Then

$$B(z_0^+ - r_0\boldsymbol{\nu}_0, r_0) \cap ((0, b) \times \mathbb{R}) \subseteq \Omega_{h_0}.$$

Since the vertical segment $\{x_0\} \times [h_0(x_0), \liminf_{x \to x_0} h_0(x)]$ belongs to $\overline{\Gamma}_{h_0}$, it follows that $\boldsymbol{\nu}_0 = -\boldsymbol{e}_1$, since otherwise the ball would intersect it. Similarly, if we take $z_n \to z_0^- := (x_0, h_0^-(x_0))$, with $x_n \to x_0^-$, then

$\boldsymbol{v}_n \to \boldsymbol{e}_1$ with $\boldsymbol{e}_1 \in N_{z_0^-}$. Note that by lowering these two balls we remain inside Ω_{h_0}, which shows that $\pm\boldsymbol{e}_1 \in N_{z_0}$.

Step 3: It remains to show that there are finitely many jump points and cusps. Assume by contradiction that there are infinitely many $z_n = (x_n, y_n) \in \Gamma_{\text{cusps}}$ with x_n distinct. Then

$$B(z_n \pm r_0\boldsymbol{e}_1, r_0) \cap ((0,b) \times \mathbb{R}) \subseteq \Omega_{h_0}. \tag{5.18}$$

By extracting a subsequence, if necessary, we may assume that $z_n \to z_0 = (x_0, y_0) \in \overline{\Gamma}_{h_0}$, with, say, $x_n \to x_0^+$. Then by (5.18),

$$B(z_0 \pm r_0\boldsymbol{e}_1, r_0) \cap ((0,b) \times \mathbb{R}) \subseteq \Omega_{h_0}.$$

But then for all n sufficiently small, $z_0 \in B(z_n - r_0\boldsymbol{e}_1, r_0)$, which contradicts (5.18) since $z_0 \in \overline{\Gamma}_{h_0}$. Since points in Γ_{cuts} are cusp points, this concludes the proof. □

Remark 5.8. Reasoning as in the previous proof, if $x_0 \in (0,b) \setminus S$ is a jump point of h_0 and if

$$h_0^+(x_0) = \limsup_{x \to x_0} h_0(x),$$

then $-\boldsymbol{e}_1 \in N_{(x_0, h_0^+(x_0))}$, while if

$$h_0^-(x_0) = \limsup_{x \to x_0} h_0(x),$$

then $\boldsymbol{e}_1 \in N_{(x_0, h_0^-(x_0))}$. In the first case, by lowering the ball, we have that $-\boldsymbol{e}_1 \in N_{(x_0,y)}$ for all $h_0^-(x_0) \le y \le h_0^+(x_0)$, while in the second, we have that $\boldsymbol{e}_1 \in N_{(x_0,y)}$ for all $h_0^+(x_0) \le y \le h_0^-(x_0)$.

Remark 5.9. If $-\boldsymbol{e}_1 \in N_{z_0}$, then since $B((x_0+r_0, y_0), r_0) \cap ((0,b) \times \mathbb{R}) \subseteq \Omega_{h_0}$ and h_0 is lower semicontinuous, for all $x > x_0$ sufficiently close to x_0, we have that

$$h_0(x) \ge y_0 + \sqrt{r_0^2 - (x - (x_0 + r_0))^2},$$

and so

$$\frac{h_0(x) - y_0}{x - x_0} \ge \frac{\sqrt{2r_0 - (x - x_0)}}{\sqrt{x - x_0}} \to \infty$$

as $x \to x_0^+$. By Theorem 5.6 it follows that $\overline{\Gamma}_{h_0}$ admits a right vertical tangent at z_0. Similarly, if $\boldsymbol{e}_1 \in N_{z_0}$ then for $x < x_0$, then $\overline{\Gamma}_{h_0}$ admits a left vertical tangent at z_0. In particular, if $\pm\boldsymbol{e}_1 \in N_{z_0}$ and h_0 is continuous at x_0, then

$$(h_0)'_-(x_0) = -\infty, \quad (h_0)'_+(x_0) = \infty. \tag{5.19}$$

Next we prove that except for cut and cusp points, $\overline{\Gamma}_{h_0}$ is locally Lipschitz.

Theorem 5.10. *Let* $(\boldsymbol{u}_0, h_0) \in X_0$ *be a minimizer of the functional* $\mathcal{G}$ *defined in* (4.2). *If* $z_0 \in \overline{\Gamma}_{h_0} \setminus (\Gamma_{\text{cuts}} \cup \Gamma_{\text{cusps}})$, *then* $\overline{\Gamma}_{h_0}$ *is Lipschitz in a neighborhood of* z_0.

Proof. **Step 1:** Let $z_0 \in \overline{\Gamma}_{h_0} \setminus (\Gamma_{\text{cuts}} \cup \Gamma_{\text{cusps}})$ and consider the set defined in (5.9). It follows from the definition of N_{z_0} that $\boldsymbol{\nu} \cdot \boldsymbol{e}_2 \geq 0$ for all $\boldsymbol{\nu} \in N_{z_0}$, since otherwise $B(z_0 - r_0\boldsymbol{\nu}, r_0)$ would contain a vertical segment starting at z_0 and pointing upwards. Since $z_0 \in \overline{\Gamma}_{h_0}$, this would contradict the fact that Ω_{h_0} is the subgraph of h_0. In particular, if N_{z_0} contains two opposite vectors $\boldsymbol{\nu}_0$ and $-\boldsymbol{\nu}_0$, then necessarily $\boldsymbol{\nu}_0 = \pm\boldsymbol{e}_1$, which is impossible, since $z_0 \notin \Gamma_{\text{cusps}}$.

We claim that N_{z_0} is a closed connected subarc of $\partial B(\boldsymbol{0}, 1)$ of length less than π. Indeed, assume that N_{z_0} contains two unit vectors $\boldsymbol{\nu}_1$ and $\boldsymbol{\nu}_2$, then by what we just observed we can assume that $\boldsymbol{\nu}_1 + \boldsymbol{\nu}_2 \neq \boldsymbol{0}$. Let $\langle \boldsymbol{\nu}_1, \boldsymbol{\nu}_2 \rangle$ be the shortest closed arc of $\partial B(\boldsymbol{0}, 1)$ of endpoints $\boldsymbol{\nu}_1$ and $\boldsymbol{\nu}_2$ and let $\boldsymbol{\nu} \in \langle \boldsymbol{\nu}_1, \boldsymbol{\nu}_2 \rangle$. Then there exists $r > 0$ sufficiently small such that

$$B(z_0 - r\boldsymbol{\nu}, r) \subseteq B(z_0 - r\boldsymbol{\nu}_1, r_0) \cup B(z_0 - r\boldsymbol{\nu}_2, r_0).$$

But then

$$B(z_0 - r_0\boldsymbol{\nu}, r_0) \cap ((0, b) \times \mathbb{R}) \subseteq \Omega_{h_0}$$

since otherwise there would exist $r_1 \in [r, r_0]$ such that $B(z_0 - r_1\boldsymbol{\nu}, r_1)$ meets Γ_{h_0} twice, which contradicts the proof of Theorem 5.2. Hence $\boldsymbol{\nu}$ belongs to N_{z_0}, which shows that $\langle \boldsymbol{\nu}_1, \boldsymbol{\nu}_2 \rangle$ is contained in N_{z_0}. By Step 1 in the proof of Theorem 5.7, N_{z_0} is closed. Since $\boldsymbol{\nu} \cdot \boldsymbol{e}_2 \geq 0$ for all $\boldsymbol{\nu} \in N_{z_0}$, necessarily $\mathcal{H}^1(N_{z_0}) \leq \pi$. If $\mathcal{H}^1(N_{z_0}) = \pi$, since N_{z_0} is closed, necessarily, N_{z_0} must contain $\pm\boldsymbol{e}_1$, which is not possible, since $z_0 \in \overline{\Gamma}_{h_0} \setminus (\Gamma_{\text{cuts}} \cup \Gamma_{\text{cusps}})$. Hence, $\mathcal{H}^1(N_{z_0}) < \pi$.

Step 2: In view of the previous step, N_{z_0} is a closed arc of length less than π. Let I be an open arc of $\partial B(\boldsymbol{0}, 1)$ of endpoints $\boldsymbol{\nu}_1, \boldsymbol{\nu}_2$, containing N_{z_0} and with $\mathcal{H}^1(I) < \pi$. We observe that there exists $\delta_1 > 0$ such that if $\|z - z_0\| < \delta_1, z \in \overline{\Gamma}_{h_0}$, then for all $\boldsymbol{\nu} \in N_z$ we have that $\boldsymbol{\nu} \in I$. Indeed, if not then there would exist $\{z_n\} \subset \overline{\Gamma}_{h_0}$ converging to z_0 and $\boldsymbol{\nu}_n \in N_{z_n} \setminus I$. But then, up to a sequence, $\{\boldsymbol{\nu}_n\}$ would converge to some $\boldsymbol{\nu} \in N_{z_0} \setminus I$, which is impossible.

Set $\overline{\boldsymbol{\nu}} := \frac{\boldsymbol{\nu}_1 + \boldsymbol{\nu}_2}{\|\boldsymbol{\nu}_1 + \boldsymbol{\nu}_2\|}$. Then the angle $\alpha := \widehat{\boldsymbol{\nu}_1 \overline{\boldsymbol{\nu}}} = \widehat{\overline{\boldsymbol{\nu}} \boldsymbol{\nu}_2}$ is strictly smaller than $\frac{\pi}{2}$. Let $\eta := 1 - \frac{1}{8}\cos^2\alpha \in (0, 1)$ and $\delta := \frac{1}{2}\min\{\delta_1, r_0\cos\alpha, x_0 - 0, b - x_0\}$. We claim that if $\boldsymbol{\nu} \in \partial B(\boldsymbol{0}, 1)$ is such that $\boldsymbol{\nu} \cdot \overline{\boldsymbol{\nu}} \geq \eta$, then $\{z - t\boldsymbol{\nu} : 0 < t <$

$2\delta\} \subset \Omega_{h_0}$ for all $z \in \overline{\Gamma}_{h_0} \cap B(z_0, \delta)$. To see this, let $z \in \overline{\Gamma}_{h_0} \cap B(z_0, \delta)$ and let $\boldsymbol{\nu}_z \in N_z$. Then $\boldsymbol{\nu}_z \in I$ by the choice of δ, and so

$$\begin{aligned} \boldsymbol{\nu} \cdot \boldsymbol{\nu}_z &= \overline{\boldsymbol{\nu}} \cdot \boldsymbol{\nu}_z + (\boldsymbol{\nu} - \overline{\boldsymbol{\nu}}) \cdot \boldsymbol{\nu}_z \\ &> \cos\alpha - \|\boldsymbol{\nu} - \overline{\boldsymbol{\nu}}\| . \end{aligned}$$

In turn, $\|\boldsymbol{\nu} - \overline{\boldsymbol{\nu}}\|^2 = 2(1 - \boldsymbol{\nu} \cdot \overline{\boldsymbol{\nu}}) \leq 2(1 - \eta) = \frac{1}{4}\cos^2\alpha$. Therefore, $\boldsymbol{\nu} \cdot \boldsymbol{\nu}_z > \frac{1}{2}\cos\alpha$. By Remark 5.5, $z - t\boldsymbol{\nu} \in B(z - r_0\boldsymbol{\nu}_z, r_0)$ for all $0 < t < r_0 \cos\alpha$.

Step 3: Let L_1 be the line through z_0 orthogonal to $\overline{\boldsymbol{\nu}}$ oriented in the direction $-\overline{\boldsymbol{\nu}}^{\perp}$, and let L_2 be the line through z_0 oriented in the direction $\overline{\boldsymbol{\nu}}$. We claim that the set $\overline{\Gamma}_{h_0} \cap B(z_0, \delta)$ is contained in the graph of a Lipschitz function defined on L_1 in an open neighborhood of z_0. Let Π and $\Pi^{\perp}$ be the projection of $\mathbb{R}^2$ onto L_1 and L_2, respectively.

Let $z_1, z_2 \in \overline{\Gamma}_{h_0} \cap B(z_0, \delta)$ and, without loss of generality, assume that $\Pi^{\perp}(z_2) \leq \Pi^{\perp}(z_1)$. Let $S := z_1 - \{r\overline{\boldsymbol{\nu}} : r \geq 0\}$, and consider the two half-lines S_1 and S_2 with endpoint z_1 and forming on both sides of S an angle of $\arccos\eta$. By hypothesis, the open sector of radius 2δ with center at z_1, bounded by the half-lines S_1 and S_2, and intersecting S, is contained in Ω_{h_0}. Hence, since $z_2 \in \overline{\Gamma}_{h_0}$, we have that z_2 does not belong to this sector, and so

$$\left|\Pi^{\perp}(z_2) - \Pi^{\perp}(z_1)\right| < m\left|\Pi(z_2) - \Pi(z_1)\right|$$

where $m := \tan\left(\frac{\pi}{2} - \arccos\eta\right)$.

Note that this inequality implies that if $z_1, z_2 \in \overline{\Gamma}_{h_0} \cap B(z_0, \delta)$ and $\Pi(z_1) = \Pi(z_2)$, then $z_1 = z_2$. Therefore, setting $P := \Pi\left(\overline{\Gamma}_{h_0} \cap B(z_0, \delta)\right)$, it follows that $\Pi_{|\overline{\Gamma}_{h_0} \cap B_\delta(z_0)}$ is one-to-one and the function $f : P \to L_2$, defined by $f(\boldsymbol{w}) := \Pi^{\perp}\left(\left(\Pi_{|\overline{\Gamma}_{h_0} \cap B(z_0,\delta)}\right)^{-1}(\boldsymbol{w})\right)$, is Lipschitz with Lipschitz constant less than or equal to m.

To complete the proof it suffices to show that P contains an open neighborhood of z_0 in L_1. By taking I sufficiently close to N_{z_0} we can assume that $\overline{\boldsymbol{\nu}} \in N_{z_0}$, so that $B(z_0 - r_0\overline{\boldsymbol{\nu}}, r_0) \cap ((0, b) \times \mathbb{R}) \subseteq \Omega_{h_0}$ with $\overline{\Gamma}_{h_0} \cap \partial B(z_0 - r_0\overline{\boldsymbol{\nu}}, r_0) = \{z_0\}$. Write $z_0 = (x_0, y_0)$. If x_0 is a continuity point of h_0, then by Theorem 5.6, h_0 admits left and right derivatives at x_0. Since the corresponding left and right slopes are not parallel to $\overline{\boldsymbol{\nu}}$, it follows that the projection onto L_1 is nonempty both on the left and right of x_0, and so P contains an open neighborhood of z_0 in L_1. On the other hand, if x_0 is a jump point of h_0, then, since x_0 does not belong to the set S defined in (5.10), we have that

$$h_0(x_0) = \liminf_{x \to x_0} h_0(x) < \limsup_{x \to x_0} h_0(x).$$

If $\liminf_{x\to x_0} h_0(x) < y_0 < \limsup_{x\to x_0} h_0(x)$, then in a neighborhood of z_0, $\overline{\Gamma}_{h_0}$ is a vertical segment, which up to a rotation, is the graph of a constant function (hence Lipschitz). Thus, assume that

$$y_0 = \liminf_{x\to x_0} h_0(x),$$

with, say, $y_0 = h_0^-(x_0)$ (the case $y_0 = h_0^+(x_0)$ is similar). Then by Remark 5.8, we have that $-\boldsymbol{e}_1 \in N_{(x_0,y)}$ for all $h_0^-(x_0) \le y \le h_0^+(x_0)$. Hence, $-\boldsymbol{e}_1$ must be one of the endpoints of N_{z_0}. Since $z_0 \notin \Gamma_{\text{cusps}}$, the other endpoint cannot be $\boldsymbol{e}_1$ but a vector strictly contained in the first and second quadrant. By taking I sufficiently close to N_{z_0}, we can also assume that $\overline{\boldsymbol{\nu}} = (\cos\theta, \sin\theta)$ for some $0 < \theta < \pi$. It follows that $-\overline{\boldsymbol{\nu}}^{\perp}$ is not vertical, and so even in this case P contains an open neighborhood of z_0 in L_1.

Finally, if

$$y_0 = \limsup_{x\to x_0} h_0(x),$$

with, say, $y_0 = h_0^+(x_0)$ (the case $y_0 = h_0^-(x_0)$ is similar) then by Remarks 5.8 and 5.9, we have have that $N_{z_0} = \{-\boldsymbol{e}_1\}$. By taking I symmetric, we can assume that $\overline{\boldsymbol{\nu}} = -\boldsymbol{e}_1$ and so $-\overline{\boldsymbol{\nu}}^{\perp}$ is vertical. Thus once again, P contains an open neighborhood of z_0 in L_1. This concludes the proof. □

In what follows we assume that

$$W(\boldsymbol{E}) = \frac{1}{2}\lambda\,[\operatorname{tr}(\boldsymbol{E})]^2 + \mu \operatorname{tr}\left(\boldsymbol{E}^2\right),$$

and that the matrix $\hat{\boldsymbol{E}}_0$ in (1.3) takes the form

$$\hat{\boldsymbol{E}}_0 = \begin{pmatrix} e_0 & 0 \\ 0 & 0 \end{pmatrix} \tag{5.20}$$

for some $e_0 > 0$.

We now use a blow-up argument to show that for a minimizer $(\boldsymbol{u}_0, h_0) \in X$ for the functional $\mathcal{G}$ defined in (4.2) the domain Ω_{h_0} cannot have corners, *i.e.*, that at every point $z_0 \in \overline{\Gamma}_{h_0} \setminus \left(\Gamma_{\text{cusps}} \cup \Gamma_{\text{cuts}}\right)$ the left and right derivatives of the function h_0 given in Theorem 5.6 coincide.

Theorem 5.11 (blow-up). *Assume* (3.1) *and* (5.20). *Let* $(\boldsymbol{u}_0, h_0) \in X_0$ *be a minimizer of the functional* $\mathcal{G}$ *defined in* (4.2). *Assume that* $\overline{\Gamma}_{h_0}$ *has a corner at some point* $z_0 \in \overline{\Gamma}_{h_0} \setminus \left(\Gamma_{\text{cusps}} \cup \Gamma_{\text{cuts}}\right)$. *Then there exist a constant* $c > 0$, *a radius* r_0, *and an exponent* $\frac{1}{2} < \alpha < 1$ *such that*

$$\int_{B(z_0,r)\cap\Omega_{h_0}} |\nabla \boldsymbol{u}_0|^2 \, d\boldsymbol{x} \le c r^{2\alpha} \tag{5.21}$$

for all $0 < r < r_0$.

Proof. **Step 1:** We claim that there exist an orthonormal basis $\{\boldsymbol{b}_1, \boldsymbol{b}_2\}$ of $\mathbb{R}^2$, three constants $c_1, L > 0$, $\tau_0 \in (0, 1)$, and an exponent $\frac{1}{2} < \beta < 1$ such that for all $\tau \in (0, \tau_0)$ there exists a radius $0 < r_\tau < 1$ such that

$$\int_{A(z_0,\tau r)} |\nabla \boldsymbol{u}_0|^2 \, d\boldsymbol{x} \leq c_1 \tau^{2\beta} \int_{A(z_0,r)} \left(1 + |\nabla \boldsymbol{u}_0|^2\right) d\boldsymbol{x} \tag{5.22}$$

for all $0 < r < r_\tau$, where

$$A(z_0, r) := \Omega_{h_0} \cap \{z_0 + s\boldsymbol{b}_1 + t\boldsymbol{b}_2 : -r < s < r, -4Lr < t < 4Lr\}. \tag{5.23}$$

By Theorems 5.6 and 5.10 there exist an orthonormal basis $\{\boldsymbol{b}_1, \boldsymbol{b}_2\}$ of $\mathbb{R}^2$ and a Lipschitz function $g : (-a', a') \to (-b', b')$, with $\operatorname{Lip} g \leq L$ for some $L > 1$, such that $g(0) = 0$ and

$$\Omega_{h_0} \cap Q = \left\{z_0 + s\boldsymbol{b}_1 + t\boldsymbol{b}_2 : -a' < s < a', \ -b' < t < g(s)\right\}$$

for some a', b', where

$$Q := \left\{z_0 + s\boldsymbol{b}_1 + t\boldsymbol{b}_2 : -a' < s < a', \ -b' < t < b'\right\}.$$

By taking a' small enough we can assume the function g admits left and right derivatives at every point. Since $\overline{\Gamma}_{h_0}$ has a corner at z_0 we have $g'_-(0) \neq g'_+(0)$. By Korn's inequality in Lipschitz domains (see Theorem 2.15) we have that $\boldsymbol{u}_0 \in H^1\left(\Omega_{h_0} \cap Q; \mathbb{R}^2\right)$.

Note that for all $0 < r \leq \min\left\{a', \frac{b'}{4L}\right\}$,

$$A(z_0, r) = \{z_0 + s\boldsymbol{b}_1 + t\boldsymbol{b}_2 : -r < s < r, \ -4Lr < t < g(s)\}.$$

Fix $c_1 > 0$, $\tau_0 \in (0, 1)$ and $\beta > \frac{1}{2}$ to be determined later and assume by contradiction that the corresponding estimate (5.22) is false for some $\tau \in (0, \tau_0)$. Then we can find a sequence $\{r_n\}$ of radii converging to zero such that

$$\int_{A(z_0,\tau r_n)} |\nabla \boldsymbol{u}_0|^2 \, d\boldsymbol{x} > c_1 \tau^{2\beta} \int_{A(z_0,r_n)} \left(1 + |\nabla \boldsymbol{u}_0|^2\right) d\boldsymbol{x}. \tag{5.24}$$

Define the rescaled sets

$$\begin{aligned} A_n &:= \frac{1}{r_n}\left(-z_0 + A(z_0, r_n)\right) \\ &= \left\{s\boldsymbol{b}_1 + t\boldsymbol{b}_2 : -1 < s < 1, \ -4L < t < \frac{g(r_n s) - g(0)}{r_n}\right\}. \end{aligned}$$

Since g is Lipschitz and has a left and right derivative at 0, we have that χ_{A_n} converges a.e. to the function χ_{A_∞}, where

$$A_\infty := \{s\boldsymbol{b}_1 + t\boldsymbol{b}_2 : -1 < s < 1, \ -4L < t < g_\infty(s)\},$$

with $g_\infty(s) := g'_-(0)\,s$ for $s < 0$ and $g_\infty(s) := g'_+(0)\,s$ for $s > 0$. We rescale accordingly the function $\boldsymbol{u}_0$ by setting

$$\boldsymbol{u}_n(z) := \frac{\boldsymbol{u}_0(z_0 + r_n z) - \boldsymbol{a}_n}{l_n r_n},$$

where $z: = \frac{x - z_0}{r_n}$,

$$\boldsymbol{a}_n := \frac{1}{\mathcal{L}^2(A(z_0, r_n))}\int_{A(z_0,r_n)} \boldsymbol{u}_0\, d\boldsymbol{x},$$
$$l_n^2 := \frac{1}{\mathcal{L}^2(A(z_0, r_n))}\int_{A(z_0,r_n)} |\nabla \boldsymbol{u}_0|^2\, d\boldsymbol{x}.$$

Then

$$\frac{1}{\mathcal{L}^2(A_n)}\int_{A_n} |\nabla \boldsymbol{u}_n|^2\, dz = 1.$$

Since by construction $\int_{A_n} \boldsymbol{u}_n\, dz = \boldsymbol{0}$, by Poincaré inequality and a standard extension argument (see Theorem 12.15 in [59]) we may extend each function $\boldsymbol{u}_n$ to the rectangle

$$R := \{s\boldsymbol{b}_1 + t\boldsymbol{b}_2 : -1 < s < 1, \ -4L < t < 4L\}$$

in such a way that the resulting function (still denoted $\boldsymbol{u}_n$) belongs to $W^{1,2}(R; \mathbb{R}^2)$ and satisfies

$$\|\boldsymbol{u}_n\|_{W^{1,2}(R)} \le c(L)\, \|\nabla \boldsymbol{u}_n\|_{L^2(A_n)} \le c.$$

Without loss of generality we may assume that the sequence $\{\boldsymbol{u}_n\}$ weakly converges to some function $\boldsymbol{u}_\infty \in W^{1,2}(R; \mathbb{R}^2)$ and that

$$l_n \to l_\infty \in [0, \infty].$$

Note that by (5.24) necessarily $l_\infty > 0$. Moreover, denoting by x_0 the point in $(0, b)$ such that $z_0 = (x_0, h_0(x_0))$, we have that the functions $\boldsymbol{u}_n$ satisfy the equation

$$\int_{A_n} \boldsymbol{E}(\varphi)\cdot C\,[\boldsymbol{E}(\boldsymbol{u}_n)]\, dz = \frac{1}{l_n}\int_{A_n} \boldsymbol{E}(\varphi)\cdot C\,[\boldsymbol{E}_0(h_0(x_0) + r_n z_2)]\, dz \quad (5.25)$$

for every $\varphi \in C_c^1(R; \mathbb{R}^2)$.

We observe that the sequence $\{\boldsymbol{E}_0(h_0(x_0)+r_n\cdot)\}$ converges to $\boldsymbol{E}_\infty$ in $L^p\left(R;\mathbb{R}^{2\times 2}\right)$ for any $1 \le p < \infty$, where

$$\boldsymbol{E}_\infty(z) := e_\infty(z_2)\,\boldsymbol{e}_1 \otimes \boldsymbol{e}_1, \quad e_\infty(z_2) := \begin{cases} e_0 & \text{if } h_0(x_0) > 0, \\ \chi_{\{z_2>0\}} e_0 & \text{if } h_0(x_0) = 0. \end{cases}$$

Step 2: We now fix a ball B such that

$$B \subset\subset \{s\boldsymbol{b}_1 + t\boldsymbol{b}_2 : -1 < s < 1,\ -4L < t < -3L\}.$$

We claim that for all functions $\phi \in C_c^1(R)$ which vanish in B we have

$$\lim_{n\to\infty} \int_{A_n} \phi^2 \left|\nabla \boldsymbol{u}_n - \nabla \boldsymbol{u}_\infty\right|^2 dz = 0.$$

From (5.25), and the fact that $\chi_{A_n} \to \chi_{A_\infty}$ in $L^2(R)$, $\boldsymbol{u}_n \rightharpoonup \boldsymbol{u}_\infty$ in $W^{1,2}\left(R;\mathbb{R}^2\right)$, and

$$\boldsymbol{E}_0(h_0(x_0)+r_n\cdot) \to \boldsymbol{E}_\infty \text{ in } L^p\left(R;\mathbb{R}^{2\times 2}\right)$$

we get that

$$\int_{A_\infty} \boldsymbol{E}(\varphi)\cdot \mathbb{C}\left[\boldsymbol{E}(\boldsymbol{u}_\infty)\right] dz = \frac{1}{l_\infty}\int_{A_\infty} \boldsymbol{E}(\varphi)\cdot \mathbb{C}\left[\boldsymbol{E}_\infty\right] dz, \tag{5.26}$$

where the right-hand side is understood to be zero when $l_\infty = \infty$.

Fix $\phi \in C_c^1(R)$ and choose $\varphi := \phi^2 \boldsymbol{u}_n$ in (5.25) ($\varphi := \phi^2\boldsymbol{u}_\infty$ in (5.26) respectively) thus getting

$$\begin{aligned} &\int_{A_n} \phi^2 \boldsymbol{E}(\boldsymbol{u}_n)\cdot \mathbb{C}\left[\boldsymbol{E}(\boldsymbol{u}_n)\right] dz \\ &\qquad = \frac{1}{l_n}\int_{A_n} \boldsymbol{E}\left(\phi^2\boldsymbol{u}_n\right)\cdot \mathbb{C}\left[\boldsymbol{E}_0(h_0(x_0)+r_n z_2)\right] dz \\ &\qquad\quad - \int_{A_n} \phi\left(\boldsymbol{u}_n\otimes\nabla\phi + (\boldsymbol{u}_n\otimes\nabla\phi)^T\right)\cdot \mathbb{C}[\boldsymbol{E}(\boldsymbol{u}_n)]dz \end{aligned} \tag{5.27}$$

and

$$\begin{aligned} &\int_{A_\infty} \phi^2 \boldsymbol{E}(\boldsymbol{u}_\infty)\cdot \mathbb{C}\left[\boldsymbol{E}(\boldsymbol{u}_\infty)\right] dz \\ &\quad = \frac{1}{l_\infty}\int_{A_\infty} \boldsymbol{E}\left(\phi^2\boldsymbol{u}_\infty\right)\cdot \mathbb{C}\left[\boldsymbol{E}_\infty\right] dz \\ &\quad\quad - \int_{A_\infty} \phi\left(\boldsymbol{u}_\infty\otimes\nabla\phi + (\boldsymbol{u}_\infty\otimes\nabla\phi)^T\right)\cdot \mathbb{C}\left[\boldsymbol{E}(\boldsymbol{u}_\infty)\right] dz. \end{aligned} \tag{5.28}$$

Letting $n \to \infty$ in (5.27), and using the fact that the right-hand side converges to the right-hand side of (5.28), we get that

$$\lim_{n\to\infty} \int_{A_n} \phi^2 \boldsymbol{E}(\boldsymbol{u}_n) \cdot C[\boldsymbol{E}(\boldsymbol{u}_n)]\, dz = \int_{A_\infty} \phi^2 \boldsymbol{E}(\boldsymbol{u}_\infty) \cdot C[\boldsymbol{E}(\boldsymbol{u}_\infty)]\, dz,$$

or, equivalently,

$$\lim_{n\to\infty} \int_{A_n} \phi^2 \{\boldsymbol{E}(\boldsymbol{u}_n) \cdot C[\boldsymbol{E}(\boldsymbol{u}_n)] - \boldsymbol{E}(\boldsymbol{u}_\infty) \cdot C[\boldsymbol{E}(\boldsymbol{u}_\infty)]\}\, dz = 0,$$

from which we get

$$\lim_{n\to\infty} \int_{A_n} \boldsymbol{E}(\phi(\boldsymbol{u}_n - \boldsymbol{u}_\infty)) \cdot C[\boldsymbol{E}(\phi(\boldsymbol{u}_n - \boldsymbol{u}_\infty))]\, dz = 0.$$

Hence the claim follows from Remark 2.13.

Step 3: We now divide the proof according to the three cases $l_\infty = \infty$, $l_\infty < \infty$ and $h_0(x_0) > 0$, $l_\infty < \infty$ and $h_0(x_0) = 0$. We begin by assuming that $l_\infty = \infty$. In this case it follows from (5.26) that $\boldsymbol{u}_\infty$ is a weak solution of the problem

$$\begin{aligned} \mu \Delta \boldsymbol{u}_\infty + (\lambda + \mu) \nabla (\operatorname{div} \boldsymbol{u}_\infty) &= \boldsymbol{0} \quad \text{in } A_\infty, \\ \left[\mu\left(\nabla \boldsymbol{u}_\infty + \nabla \boldsymbol{u}_\infty^T\right) + \lambda (\operatorname{div} \boldsymbol{u}_\infty) I\right] \nu &= \boldsymbol{0} \quad \text{on } \Gamma_{g_\infty}. \end{aligned}$$

By Theorem 3.4 there exist $c > 0$ and $\beta \in \left(\frac{1}{2}, 1\right)$ such that for all $0 < r < 1$ we have

$$\begin{aligned} \int_{B(\boldsymbol{0},r)\cap A_\infty} |\nabla \boldsymbol{u}_\infty|^2\, dz &\le c r^{2\beta} \int_{A_\infty} \left(|\boldsymbol{u}_\infty|^2 + |\nabla \boldsymbol{u}_\infty|^2\right) dz \\ &\le c r^{2\beta} \int_{A_\infty} |\nabla \boldsymbol{u}_\infty|^2\, dz \le c r^{2\beta}, \end{aligned}$$

where we have used the Poincaré inequality, which holds since $\int_{A_\infty} \boldsymbol{u}_\infty\, dz = 0$, and the fact that $\int_{A_\infty} |\nabla \boldsymbol{u}_\infty|^2\, dz \le \mathcal{L}^2(A_\infty)$. Therefore if τ_0 is such that

$$\tau_0 A_\infty \subseteq (B(\boldsymbol{0}, 1) \cap A_\infty) \setminus \{s\boldsymbol{b}_1 + t\boldsymbol{b}_2 : -1 < s < 1,\ -4L < t < -3L\}$$

we get that for all $0 < \tau \leq \tau_0$,

$$\int_{\tau A_\infty} |\nabla \boldsymbol{u}_\infty|^2 \, dz \leq \int_{B\left(\mathbf{0}, \frac{\tau}{\tau_0}\right) \cap A_\infty} |\nabla \boldsymbol{u}_\infty|^2 \, dz \leq c_2 \tau^{2\beta}.$$

By Step 2 we then have that

$$\lim_{n \to \infty} \int_{\tau A_n} |\nabla \boldsymbol{u}_n|^2 \, dz = \int_{\tau A_\infty} |\nabla \boldsymbol{u}_\infty|^2 \, dz$$

for all $0 < \tau \leq \tau_0$, and so

$$\begin{aligned} \lim_{n \to \infty} \frac{\int_{A(z_0, \tau r_n)} |\nabla \boldsymbol{u}|^2 \, dx}{\int_{A(z_0, r_n)} |\nabla \boldsymbol{u}|^2 \, dx} &= \frac{1}{\mathcal{L}^2(A_\infty)} \lim_{n \to \infty} \int_{\tau A_n} |\nabla \boldsymbol{u}_n|^2 \, dz \\ &\leq \frac{c_2}{\mathcal{L}^2(A_\infty)} \tau^{2\beta}, \end{aligned}$$

which contradicts (5.24), provided we take

$$c_1 \geq 2 \frac{c_2}{\mathcal{L}^2(A_\infty)}.$$

Step 4: Assume next that $l_\infty < \infty$ and $h_0(x_0) > 0$. In this case $e_\infty \equiv e_0$. Define

$$\boldsymbol{v}_\infty(z) := \boldsymbol{u}_\infty(z) - \frac{(e_0 z_1, 0)}{l_\infty}.$$

Then $\boldsymbol{v}_\infty$ is a weak solution of the problem

$$\begin{aligned} \mu \Delta \boldsymbol{v}_\infty + (\lambda + \mu) \nabla (\operatorname{div} \boldsymbol{v}_\infty) = \mathbf{0} \quad &\text{in } A_\infty, \\ \left[\mu \left(\nabla \boldsymbol{v}_\infty + \nabla \boldsymbol{v}_\infty^T\right) + \lambda (\operatorname{div} \boldsymbol{v}_\infty) I\right] \nu = \mathbf{0} \quad &\text{on } \Gamma_{h_{0\infty}} \end{aligned}$$

and thus, as in the previous step, for all $0 < r < 1$ we have

$$\int_{B(\mathbf{0}, r) \cap A_\infty} |\nabla \boldsymbol{v}_\infty|^2 \, dz \leq c r^{2\beta} \int_{A_\infty} \left(|\boldsymbol{v}_\infty|^2 + |\nabla \boldsymbol{v}_\infty|^2\right) dz,$$

from which we obtain that for all $0 < \tau \leq \tau_0$, where τ_0 is the same as in the previous step, there holds

$$\int_{\tau A_\infty} |\nabla \boldsymbol{u}_\infty|^2 \, dz \leq c \tau^{2\beta} \int_{A_\infty} \left(|\nabla \boldsymbol{u}_\infty|^2 + \frac{1}{l_\infty^2}\right) dz \leq c_3 \tau^{2\beta} \left(1 + \frac{1}{l_\infty^2}\right).$$

In turn,

$$\lim_{n\to\infty}\frac{\int_{A(z_0,\tau r_n)}|\nabla \boldsymbol{u}_0|^2\,d\boldsymbol{x}}{\int_{A(z_0,r_n)}\left(|\nabla \boldsymbol{u}_0|^2+1\right)\,d\boldsymbol{x}}=\lim_{n\to\infty}\frac{\int_{A(z_0,\tau r_n)}|\nabla \boldsymbol{u}_0|^2\,d\boldsymbol{x}}{\mathcal{L}^2(A_\infty)\left(l_n^2r_n^2+r_n^2\right)}$$
$$=\lim_{n\to\infty}\frac{\int_{\tau A_n}|\nabla \boldsymbol{u}_n|^2\,d\boldsymbol{z}}{\mathcal{L}^2(A_\infty)\left(1+\frac{1}{l_n^2}\right)}$$
$$\leq\frac{c_3}{\mathcal{L}^2(A_\infty)}\tau^{2\beta},$$

which is again a contradiction provided we take

$$c_1\geq 2\frac{c_3}{\mathcal{L}^2(A_\infty)}.$$

Step 5: Finally we consider the case $l_\infty<\infty$ and $h_0(x_0)=0$. Define

$$\widetilde{e}_\infty(z_2):=\begin{cases} e_0 & \text{if } z_2>0,\\ \gamma & \text{if } z_2<0,\end{cases}$$

for some γ to be determined later, and observe that for every $\varphi\in C_c^1\left(R;\mathbb{R}^2\right)$,

$$\int_{A_\infty}e_\infty(z_2)\frac{\partial\varphi_1}{\partial z_1}\,dz=\int_{A_\infty\cap\{z_2>0\}}e_0\frac{\partial\varphi_1}{\partial z_1}\,dz=\int_{A_\infty\cap\{z_2>0\}}e_0\frac{\partial\varphi_1}{\partial z_1}\,dz$$
$$+\int_{A_\infty\cap\{z_2<0\}}\gamma\frac{\partial\varphi_1}{\partial z_1}\,dz=\int_{A_\infty}\widetilde{e}_\infty(z_2)\frac{\partial\varphi_1}{\partial z_1}\,dz,$$

where we have used the fact that

$$\int_{A_\infty\cap\{z_2<0\}}\frac{\partial\varphi_1}{\partial z_1}\,dz=\int_{A_\infty\cap\{z_2=0\}}\varphi_1\nu_1\,d\mathcal{H}^1(z_1,z_2)=0$$

since $\nu_1=0$. Define

$$\boldsymbol{w}_\infty(\boldsymbol{z}):=e_0\left(z_1,-\min\left\{0,\frac{\lambda z_2}{2\mu+\lambda}\right\}\right),\quad \gamma:=\frac{4\mu(\mu+\lambda)}{(2\mu+\lambda)^2}e_0.$$

A straightforward calculation shows that

$$(2\mu+\lambda)\widetilde{e}_\infty=(2\mu+\lambda)\frac{\partial(\boldsymbol{w}_\infty)_1}{\partial z_1}+\lambda\frac{\partial(\boldsymbol{w}_\infty)_2}{\partial z_2}$$
$$\lambda e_\infty=\lambda\frac{\partial(\boldsymbol{w}_\infty)_1}{\partial z_1}+(2\mu+\lambda)\frac{\partial(\boldsymbol{w}_\infty)_2}{\partial z_2}.$$

Hence for every $\varphi \in C_c^1\left(R;\mathbb{R}^2\right)$,

$$\begin{aligned}
\int_{A_\infty} \boldsymbol{E}(\varphi)\cdot \mathbb{C}\left[\boldsymbol{E}_\infty\right] dz &= (2\mu+\lambda)\int_{A_\infty} e_\infty(z_2)\frac{\partial\varphi_1}{\partial z_1}\,dz \\
&\quad +\lambda\int_{A_\infty} e_\infty(z_2)\frac{\partial\varphi_2}{\partial z_2}\,dz \\
&= (2\mu+\lambda)\int_{A_\infty} \widetilde{e}_\infty(z_2)\frac{\partial\varphi_1}{\partial z_1}\,dz \\
&\quad +\lambda\int_{A_\infty} e_\infty(z_2)\frac{\partial\varphi_2}{\partial z_2}\,dz \\
&= \int_{A_\infty}\left[(2\mu+\lambda)\frac{\partial(\boldsymbol{w}_\infty)_1}{\partial z_1}+\lambda\frac{\partial(\boldsymbol{w}_\infty)_2}{\partial z_2}\right]\frac{\partial\varphi_1}{\partial z_1}\,dz \\
&\quad +\int_{A_\infty}\left[\lambda\frac{\partial(\boldsymbol{w}_\infty)_1}{\partial z_1}+(2\mu+\lambda)\frac{\partial(\boldsymbol{w}_\infty)_2}{\partial z_2}\right]\frac{\partial\varphi_2}{\partial z_2}\,dz \\
&= \int_{A_\infty} \boldsymbol{E}(\varphi)\cdot \mathbb{C}\left[\boldsymbol{E}(\boldsymbol{w}_\infty)\right] dz.
\end{aligned}$$

We can now proceed exactly as in the previous step with the only difference that we now take

$$\boldsymbol{v}_\infty := \boldsymbol{u}_\infty - \frac{1}{l_\infty}\boldsymbol{w}_\infty.$$

Step 6: By Steps 2-5 the estimate (5.22) holds, and we are now in position to prove (5.21). By (5.22) for all $\tau \in (0,\tau_0)$ we have

$$\begin{aligned}
&\int_{A(z_0,\tau r)}\left(1+|\nabla \boldsymbol{u}_0|^2\right)dx \\
&\qquad \le \tau^2 16 L r^2 + c_1\tau^{2\beta}\int_{A(z_0,r)}\left(1+|\nabla \boldsymbol{u}_0|^2\right)dx \\
&\qquad \le (16L+c_1)\,\tau^{2\beta}\int_{A(z_0,r)}\left(1+|\nabla \boldsymbol{u}_0|^2\right)dx
\end{aligned} \tag{5.29}$$

for all $0 < r < r_\tau$. Hence for a fixed $\alpha \in (1/2,\beta)$,

$$(16L+c_1)\,\tau^{2\beta} = (16L+c_1)\,\tau^{2\beta-2\alpha}\tau^{2\alpha} \le \tau^{2\alpha} \tag{5.30}$$

provided τ_0 is sufficiently small.

Fix $0 < r < r_\tau$ and find $k \in \mathbb{N}$ such that

$$\tau^{k+1} r_\tau \le r \le \tau^k r_\tau.$$

By iterating (5.29) and by (5.30) we have

$$\begin{aligned}\int_{B(z_0,r)\cap\Omega_{h_0}} |\nabla \boldsymbol{u}_0|^2\,dx &\le \int_{A(z_0,r)} \left(1+|\nabla \boldsymbol{u}_0|^2\right)dx\\ &\le \int_{A\left(z_0,\tau^k r_\tau\right)} \left(1+|\nabla \boldsymbol{u}_0|^2\right)dx\\ &\le \tau^{2\alpha k}\int_{A(z_0,r_\tau)} \left(1+|\nabla \boldsymbol{u}_0|^2\right)dx\\ &\le \frac{r^{2\alpha}}{(\tau r_\tau)^{2\alpha}}\int_{A(z_0,r_\tau)} \left(1+|\nabla \boldsymbol{u}_0|^2\right)dx,\end{aligned}$$

where we have used the fact that $B(z_0,r)\cap\Omega_{h_0}\subseteq A(z_0,r)$ since $L\ge 1$. This yields (5.21) with

$$c := \frac{1}{(\tau r_\tau)^{2\alpha}}\int_{A(z_0,r_\tau)} \left(1+|\nabla \boldsymbol{u}_0|^2\right)dx$$

and $r_0 := r_\tau$. □

Next we prove that for a local minimizer $(\boldsymbol{u}_0,\Omega)\in X$ the upper boundary $\overline{\Gamma}_{h_0}$ is of class C^1 away from $\Gamma_{\mathrm{cusps}}\cup\Gamma_{\mathrm{cuts}}$.

Theorem 5.12 (C^1 regularity of Γ). *Assume* (3.1) *and* (5.20). *Let* $(\boldsymbol{u}_0,h_0)\in X_0$ *be a minimizer of the functional* $\mathcal{G}$ *defined in* (4.2). *Then* $\overline{\Gamma}_{h_0}\setminus\left(\Gamma_{\mathrm{cusps}}\cup\Gamma_{\mathrm{cuts}}\right)$ *is of class* C^1.

Proof. Since Γ_{cuts} is made of segments, it is enough to prove the regularity of $\overline{\Gamma}_{h_0}\setminus(\Gamma_{\mathrm{cusps}}\cup\Gamma_{\mathrm{cuts}})$. Assume by contradiction that $\overline{\Gamma}_{h_0}$ has a corner at some point $z_0\in\overline{\Gamma}_{h_0}\setminus\left(\Gamma_{\mathrm{cusps}}\cup\Gamma_{\mathrm{cuts}}\right)$.

By Theorem 5.10 and a standard extension argument we may define $\boldsymbol{u}_0$ in a fixed neighborhood of z_0 in such a way that for all $0<r<r_1$,

$$\int_{B(z_0,r)} |\nabla \boldsymbol{u}_0|^2\,dx \le c(L)\int_{B(z_0,r)\cap\Omega} |\nabla \boldsymbol{u}_0|^2\,dx \tag{5.31}$$

for some $r_1>0$, and where the constant $c(L)$ is independent of r and depends only on the Lipschitz constant L of the function given in Theorem 5.10.

Moreover, by Theorem 5.1 there exists $k_0>0$ such that

$$\mathcal{G}(\boldsymbol{u}_0,h_0)=\min\left\{\mathcal{G}(\boldsymbol{u},h)+k_0\left|d-\int_0^b h(x)\,dx\right| : (\boldsymbol{u},h)\in X\right\}. \tag{5.32}$$

We recall also that by Theorem 5.6, $\overline{\Gamma}_{h_0}$ admits a left and a right tangent line at x_0. Assume by contradiction that the two tangent lines are distinct and form an angle $0 < \vartheta < \pi$. For $r > 0$ (sufficiently small) we denote

$$x'_r := \max\{x \in (0,b) : x \le x_0 \text{ and there is } y \text{ such that } (x,y) \in \Gamma_{h_0} \cap \partial B(z_0, r)\},$$

$$x''_r := \min\{x \in (0,b) : x \ge x_0, \text{ and there is } y \text{ such that } (x,y) \in \Gamma_{h_0} \cap \partial B(z_0, r)\},$$

and we let $(x'_r, h_0(x'_r))$ and $(x''_r, h_0(x''_r))$ be the corresponding points on $\overline{\Gamma}_{h_0} \cap \partial B(z_0, r)$. As $r \to 0^+$ we have that $x'_r \to x_0^-$ and $x''_r \to x_0^+$. Construct h_r as the greatest lower semicontinuous function coinciding with h_0 outside $[x'_r, x''_r]$ and with the affine function

$$x \mapsto h_0(x'_r) + \frac{h_0(x''_r) - h_0(x'_r)}{x''_r - x'_r}(x - x'_r)$$

in (x'_r, x''_r). Then for $r > 0$ sufficiently small $(\boldsymbol{u}_0, h_r)$ is admissible for the penalized minimization problem (5.32). Hence,

$$\mathcal{G}(\boldsymbol{u}_0, h_0) \le \mathcal{G}(\boldsymbol{u}_0, h_r) + k_0 \left| d - \int_0^b h_r(x)\, dx \right| \le \mathcal{G}(\boldsymbol{u}_0, h_r) + cr^2,$$

and, in turn, using the estimates (5.21) and (5.31),

$$\begin{aligned} 2r &= \sqrt{(x''_r - x_0)^2 + (h_0(x''_r) - h_0(x_0))^2} \\ &\quad + \sqrt{(x'_r - x_0)^2 + (h_0(x'_r) - h_0(x_0))^2} \\ &\le \text{length}\, \Gamma_{h_0|_{[x'_r, x''_r]}} \qquad (5.33) \\ &\le \int_{x'_r}^{x''_r} \sqrt{1 + (h'_r)^2}\, dx + c \int_{B(z_0,r)} |\nabla \boldsymbol{u}_0|^2\, d\boldsymbol{x} + cr^2 \\ &\le \int_{x'_r}^{x''_r} \sqrt{1 + \left(\frac{h_0(x''_r) - h_0(x'_r)}{x''_r - x'_r} \right)^2}\, dx + cr^{2\alpha} \end{aligned}$$

for r small enough. For $x > x_0$ we have $h_0(x) = h_0(x_0) + (h_0)'_+(x_0)(x - x_0) + o(x - x_0)$. Since

$$\begin{aligned} r &= \sqrt{(x''_r - x_0)^2 + (h_0(x''_r) - h_0(x_0))^2} \\ &= (x''_r - x_0)\sqrt{1 + \left((h_0)'_+(x_0) + \frac{o(x''_r - x_0)}{x''_r - x_0} \right)^2}, \end{aligned}$$

we get that

$$x''_r = x_0 + \frac{r}{\sqrt{1 + \left((h_0)'_+(x_0) + \frac{o(x''_r - x_0)}{x''_r - x_0} \right)^2}}. \qquad (5.34)$$

Similarly, we obtain

$$x'_r = x_0 - \frac{r}{\sqrt{1 + \left((h_0)'_-(x_0) + \frac{o(x'_r - x_0)}{x_0 - x'_r}\right)^2}}. \tag{5.35}$$

Plugging (5.34) and (5.35) in estimate (5.33), dividing both sides by r and letting r go to zero, we get

$$2 \le 2\sin(\vartheta/2),$$

which is impossible. □

Remark 5.13. All the regularity results in this section continue to hold for local minimizers.

Theorem 5.12 can be significantly improved. Indeed, using another blow-up argument it is possible to show that $\overline{\Gamma}_{h_0} \setminus (\Gamma_{\text{cusps}} \cup \Gamma_{\text{cuts}})$ is of class $C^{1,\alpha}$ for all $0 < \alpha < \frac{1}{2}$. In turn, this implies that $\boldsymbol{u}_0$ is of class $C^{1,\beta}$ for some $\beta > 0$ away from the x-axis and from $\Gamma_{\text{cusps}} \cup \Gamma_{\text{cuts}}$. Using a classical bootstrap argument, one can then obtain C^∞ regularity and then use results of Koch, Morini and the author to prove analyticity of $\overline{\Gamma}_{h_0} \setminus (\Gamma_{\text{cusps}} \cup \Gamma_{\text{cuts}})$ away from the x-axis. We refer to [33] for more details.

6 Related literature and open problems

6.1 Voids and anisotropic surface energies

The regularity of optimal configurations is significantly more challenging when the isotropic surface energy ψ in (1.6) is replaced by an anisotropic surface energy $\psi = \psi(\boldsymbol{\nu})$ depending on the outward normal $\boldsymbol{\nu}$ to the profile Γ_h. In this case the main change is that in Theorem 5.2 balls should be replaced by translations of rescaled Wulff sets.

We recall that, given a function $\psi : \partial B(\mathbf{0}, 1) \to (0, \infty)$, where $B(\mathbf{0}, 1) \subset \mathbb{R}^N$, the (open) *Wulff set* is defined by

$$\mathcal{W} := \left\{\boldsymbol{w} \in \mathbb{R}^N : \psi^\circ(\boldsymbol{w}) < 1\right\}, \tag{6.1}$$

where ψ° is the *polar function* of ψ, *i.e.*,

$$\psi^\circ(\boldsymbol{w}) := \max_{|z|=1} \frac{\boldsymbol{z} \cdot \boldsymbol{w}}{\psi(\boldsymbol{z})}, \qquad \boldsymbol{w} \in \mathbb{R}^N. \tag{6.2}$$

It can be shown (see [31,37,81]) that up to translations, the Wulff set is the unique solution of the minimization problem

$$\min\left\{\int_{\partial E} \psi(\boldsymbol{\nu}_E)\, d\mathcal{H}^{N-1} : E \subset \mathbb{R}^N \text{ has finite perimeter}, \mathcal{L}^N(E) = \text{const}\right\}. \tag{6.3}$$

The tools developed in Sections 4 and 5 were used by the author, in collaboration with Fonseca, Fusco, and Millot in [32] to study a variational model proposed by Siegel, Miksis, and Voorhees in [74] to describe the competition between elastic energy and highly anisotropic surface energy for problems involving a material void in a linearly elastic solid. The energy considered in [32] takes the form

$$\mathcal{F}_1(\boldsymbol{u}, V) := \int_{B \setminus V} W(\boldsymbol{E}(\boldsymbol{u}))\, d\boldsymbol{x} + \int_{\partial V} \psi(\boldsymbol{\nu})\, ds, \tag{6.4}$$

where V represents the void, B is a fixed large ball, and $\boldsymbol{\nu}$ is the outward normal to $B \setminus V$. The area of the void is assigned and the mismatch is described by the boundary condition $\boldsymbol{u} = \boldsymbol{u}_0$ a.e. in $\mathbb{R}^2 \setminus B$, for a given function $\boldsymbol{u}_0$.

Problem 6.1. The analysis carried out in [32] was restricted to star-shaped voids. It would be interesting to remove this restriction.

In the *crystalline case*, that is, when the Wulff set is a polygon, under the assumption that *the internal angles of the Wulff set are greater than* $\frac{\pi}{2}$, we proved in [32] that if $(\boldsymbol{u}, V)$ is a minimizer for the relaxed energy $\mathcal{G}_1$ corresponding to $\mathcal{F}_1$ (in the same spirit of Section 4) then ∂V is the union of finitely many Lipschitz graphs. In the absence of the elastic energy and without the restriction that V is star-shaped, we refer to the work of Ambrosio, Novaga, and Paolini [5] and of Novaga and Paolini [67] as well as to the references contained therein.

In the case in which the anisotropy is weak, that is, when the surface energy density ψ is strictly convex, then the Wulff set is of class C^1 and thus many of the arguments obtained in [33] could be adapted to the anisotropic setting, although the proofs in [32] are significantly more involved.

Problem 6.2. What can be said about the regularity of the void V in the crystalline case if the internal angles of the Wulff set are not assumed to be greater than $\frac{\pi}{2}$?

6.2 Cusps and minimality of the flat configuration

In Section 5 we have seen that the boundary $\overline{\Gamma}_h$ corresponding to the optimal profile h is smooth except for a finite number of cusps points.

In the case in which the substrate is rigid, the energy (1.7) should be replaced by

$$\mathcal{F}_2(\boldsymbol{u}, h) := \int_{\Omega_h^+} W(\boldsymbol{E}(\boldsymbol{u}))\, d\boldsymbol{x} + \int_{\Gamma_h} \psi(y)\, ds, \tag{6.5}$$

where we recall (see (1.2)) that

$$\Omega_h^+ := \{\boldsymbol{x} = (x, y) : 0 < x < b,\, 0 < y < h(x)\}, \tag{6.6}$$

and where the mismatch between the film and the substrate is expressed by the Dirichlet boundary condition

$$\boldsymbol{u}(x, 0) = (e_0 x, 0)$$

for some $e_0 > 0$. As in Section 4, when (4.1) holds one has existence of minimizers of the relaxed energy of $\mathcal{F}_2$, given by

$$\mathcal{G}_2(\boldsymbol{u}, h) := \int_{\Omega_h^+} W(\boldsymbol{E}(\boldsymbol{u}))\, d\boldsymbol{x} + \gamma_{\text{film}} \operatorname{length} \Gamma_h, \tag{6.7}$$

The regularity of minimizers and of local minimizers of (6.7) under periodicity conditions was studied by De Maria and Fusco [26] for W of the form (3.1).

In [42] Fusco and Morini, again in the isotropic case (3.1), proved that if the period b is sufficiently small, to be precise, if

$$0 < b \le \frac{\pi}{4} \frac{2\mu + \lambda}{e_0^2 \mu(\mu + \lambda)} := b_1,$$

then the flat configuration

$$h_{\text{flat}}(x) :\equiv \frac{b}{d}, \quad \boldsymbol{u}_{\text{flat}}(x, y) := \left(e_0 x, -\frac{\lambda e_0}{2\mu + \lambda} y\right) \tag{6.8}$$

is an *isolated local minimizer* of $\mathcal{G}_2$, that is, there exists $\delta > 0$ such that

$$\mathcal{G}_2(\boldsymbol{u}_{\text{flat}}, h_{\text{flat}}) \le \mathcal{G}_2(\boldsymbol{u}, h)$$

for all admissible $(\boldsymbol{u}, h)$ with

$$d_{\mathcal{H}}(\overline{\Gamma}_{h_{\text{flat}}}, \overline{\Gamma}_h) \le \delta, \qquad \int_0^b h(x)\, dx = d,$$

with strict inequality whenever $h \neq h_{\text{flat}}$.

On the other hand, if $b > b_1$, let d_b be the unique solution of the equation

$$K\left(\frac{2\pi d_b}{b^2}\right) = \frac{\pi}{4} \frac{2\mu + \lambda}{e_0^2 \mu(\mu + \lambda)} \frac{1}{b},$$

where $K : [0, \infty) \to [0, 1)$ is the strictly increasing function

$$K(y) := \max_{n \in \mathbb{N}} \frac{1}{n} J(ny), \quad J(y) := \frac{y + (3 - 4\nu)\sinh y \cosh y}{4(1-\nu)^2 + y^2 + (3 - 4\nu)\sinh^2 y},$$

where $\nu := \frac{\lambda}{2(\lambda+\mu)}$ is the Poisson modulus. Then Fusco and Morini [42] proved that the flat configuration (6.8) is an isolated local minimizer of $\mathcal{G}_2$ provided the area d of the film is less than or equal the critical value d_b, while if $d > d_b$, then (6.8) is not a local minimizer of $\mathcal{G}_2$ for any $\delta > 0$.

Thus if the area of the film is less than or equal to the critical value d_b, we are in the FM growth mode, while if is is greater than d_b we are in the SK growth mode (see Section 1).

It is important to observe that this result gives the *first rigorous mathematical proof of the existence of island formation* when the area of the film is sufficiently large.

Analogous results were obtained for global minimizers of $\mathcal{G}_2$. We refer to [42] for the precise statements. In particular, it was shown that if the Lamé moduli satisfy the condition $\lambda + \frac{17}{18}\mu \geq 0$ (note that this inequality is slightly stronger than $(3.2)_2$), there exist $b_2 > \frac{2\mu+\lambda}{e_0^2\mu(\mu+\lambda)}$ and $d_0 > 0$ such that if $\frac{2\mu+\lambda}{e_0^2\mu(\mu+\lambda)} \leq b < b_2$, then there is $d(b)$ with the property that for $d(b) \leq d < d(b) + b$, for every global minimizer $(\boldsymbol{u}, h)$ of $\mathcal{G}_2$ the profile h is of class $C^1([0, b])$. Hence, in this range of values *there are no cusp points*.

The results of Fusco and Morini [42] in the isotropic cases have been extended to anisotropic surface energies by Bonacini [10,11] both in the crystalline case and in the smooth case.

The analysis carried out in [10] and [42] is based on sufficient conditions for local minimality, expressed in terms of a suitable notion of second variation of the total energy and is inspired by previous work on the Mumford–Shah functional (see [15] and the references therein, see also [1,12,13,17,29,56] for more recent applications to other problems).

It has been observed in experiments and numerically that high stress concentrations lead to the formation of cusps and cracks (see, *e.g.*, [43, 55,57,87], and the references therein).

Problem 6.3. An analytical proof of cusp formation is a long standing open problem.

Heuristically one would expect cusps and cracks to occur when the elastic energy dominates over the surface energy. In [45] Goldman and Zwicknagl derived the following scaling law for the energy $\mathcal{F}_2$ in (6.5),

$$c_1 \max\left\{1, d, e_0^{2/3} d^{2/3}\right\} \leq \inf_{(\boldsymbol{u},h)} \mathcal{F}_2(\boldsymbol{u}, h) \leq c_2 \max\left\{1, d, e_0^{2/3} d^{2/3}\right\},$$

where c_1 and c_2 are positive constants. Moreover they considered the scaling regimes separately and derived reduced limiting functionals in the spirit of Γ-convergence in the three scaling regimes when $d \to \infty$:

1. If $e_0 \to \infty$ and $\frac{e_0^2}{d} \to 0$, then the surface energy dominates.
2. If $d \simeq \frac{e_0^2}{d}$, then the elastic energy and the surface energy compete against each other.
3. If $e_0 \to \infty$ and $\frac{e_0^2}{d} \to \infty$, then the elastic energy dominates.

It is in the third regime that one expects cusp formation.

The case of unbounded domains $b = \infty$ has also been considered, see the recent paper of Bella, Goldman, and Zwicknagl [7].

6.3 Dynamics

Models for the time evolution – by surface diffusion, of second-phase particles and material voids in a linearly elastic solid – can become ill-posed when the surface energy is highly anisotropic (see, *e.g.*, [27, 74, 78]). One of the first models of the planar motion of grain boundaries using motion by surface diffusion was proposed by Mullins [63] and is based on the relation $V = K$, where V is the normal velocity and K is the curvature of the interfacial curve.

Mullin's theory was generalized by Gurtin [48] (see also [27]) to the case of anisotropic surfaces energies $\psi(\boldsymbol{\nu}) = \hat{\psi}(\theta)$, where θ is defined to be the angle of the outward surface normal $\boldsymbol{\nu}$ with respect to the x-axis. In this case the resulting evolution equation takes the form $aV = \left[\hat{\psi} + \hat{\psi}''\right] K - F$, where $a = a(\theta)$ is a kinetic coefficient and F is a constant expressing the difference in bulk energies. When $\hat{\psi}(\theta) + \hat{\psi}''(\theta) > 0$, this evolution equation is parabolic and it has been studied by several authors (see [49] and the references therein). On the other hand, at angles θ for which $\hat{\psi}(\theta) + \hat{\psi}''(\theta) < 0$, the evolution equation is backward parabolic thus rendering the initial value problem ill-posed and the interfacial energy unstable (see [49]). To overcome the ill-posedness of the evolution equation, one approach that has been used in the literature is to add higher order terms, effectively penalizing the presence of sharp corners. One possible regularization, proposed by Angenent and Gurtin [6] and later studied by DiCarlo, Gurtin and Podio Guidugli in [27], has the form

$$aV = \left[\hat{\psi} + \hat{\psi}''\right] K - F - \varepsilon \left(K_{ss} + \frac{1}{2} K^3 \right),$$

where $\varepsilon > 0$ is a small parameter and s is the arclength. This leads to a fourth-order parabolic equation for the evolution of the interface. This regularization consists in allowing the surface energy to depend not only on the normal to the interface but also on the curvature, to be precise, $\psi = \hat{\psi}(\theta, K)$. This approach was first considered by Herring [51], who considered interfacial energies of the form

$$\hat{\psi}(\theta, K) = \psi_0(\theta) + \varepsilon K^2. \tag{6.9}$$

In this case, sharp corners correspond to $|K| \to \infty$. The introduction of a regularization raises the fundamental problem of the description of equilibrium crystal shapes. A key issue is to understand whether if in the limit as $\varepsilon \to 0^+$ one recovers the Wulff set. Recently, Spencer [78], using matched asymptotic expansions, has shown that, in the static case and in the absence of elastic strain energy and of cusps in ψ, regularized solutions approach the classic sharp-corner equilibrium shape as $\varepsilon \to 0^+$. Siegel, Miksis and Voorhees [74] have studied a related evolution problem numerically in the presence of elastic strain energy in bulk, namely, a 6th order volume-preserving parabolic equation for the surface, coupled with the elliptic system of elasticity in the bulk. To be precise, in term of non-dimensional quantities, they obtained a kinematic equation of the form

$$aV = \frac{d^2}{ds^2}\left\{\left[\hat{\psi} + \hat{\psi}''\right] K - W(\boldsymbol{E}(\boldsymbol{u})) - \varepsilon\left(K_{ss} + \frac{1}{2}K^3\right)\right\}, \tag{6.10}$$

where $W(\boldsymbol{E}(\boldsymbol{u}))$ is the trace of $W(\boldsymbol{E}(\boldsymbol{u}(\cdot,t)))$ on $\Gamma_{h(\cdot,t)}$, and $\boldsymbol{u}(\cdot,t)$ is the elastic equilibrium in $\Omega^+_{h(\cdot,t)}$. Their numerical results indicate that the elastic energy decreases with the corner regularization, but it is not clear if it approaches zero sufficiently fast to be neglected in the zero-regularization limit.

Since the motion flow (6.10) is the gradient flow for the energy

$$\mathcal{G}_\varepsilon(\boldsymbol{u}, h) := \int_{\Omega_h^+} W(\boldsymbol{E}(\boldsymbol{u}))\, dx + \int_{\Gamma_h} \left(\psi(\boldsymbol{\nu}) + \varepsilon K^2\right) d\mathcal{H}^1, \tag{6.11}$$

in the H^{-1} inner product (see the work of Cahn and Taylor [16]), using De Giorgi's minimizing movements approach (see [4]), the author, in collaboration with Fonseca, Fusco, and Morini in [34] proved short time existence, uniqueness, and regularity of solutions.

The study of the gradient flow of (6.11) in the L^2 inner product, which corresponds to the motion by evaporation-condensation, was carried by Piovano in [70].

Problem 6.4. The results in [34] strongly rely on the fact that the interface Γ_h is the graph of a function. It would be important to consider the case in which Γ is a curve.

Problem 6.5. A major open problem is to study what happens when $\varepsilon \to 0^+$.

6.4 The three-dimensional case

In the three-dimensional case the sets Ω_h and Ω_h^+ in (1.1) and (6.6) should be replaced respectively by

$$\Omega_h := \{\boldsymbol{x} = (x, y, z) : (x, y) \in R,\ z < h(x, y)\},$$
$$\Omega_h^+ := \{\boldsymbol{x} = (x, y, z) : (x, y) \in R,\ 0 < z < h(x, y)\},$$

where R is an open rectangle in $\mathbb{R}^2$, and the interface Γ_h is now a two-dimensional surface rather than a curve. Relaxation and existence results corresponding to the ones in Section 4 were proved by Chambolle and Solci [23] in the N-dimensional case for general bulk energy densities W and for isotropic surfaces energies and by Braides, Chambolle, and Solci [14] for anisotropic surfaces energies.

Problem 6.6. Study the (partial) regularity of equilibrium configurations of the functional (1.7) in dimension $N = 3$.

The stability of the flat configuration discussed (see Subsection 6.2) in dimension $N = 3$ and for anisotropic surface energies was studied by Bonacini in [11].

In [38] Fonseca, Pratelli, and Zwicknagl studied a fully faceted variational model introduced by Spencer and Tersoff (see [77, 83]), which takes into account in particular the miscut angle between the substrate and the film. They proved existence of minimizers of the relaxed energy and showed that for small volumes, the minimizers are half-pyramides.

The evolution problem for $N = 3$ was addressed by Fonseca, Fusco, Morini and the author in [36]. As in Subsection 6.3, to regularize the problem, we added to the surface energy density ψ the term $\varepsilon|\boldsymbol{K}|^p$, where $\boldsymbol{K} = \operatorname{div} \boldsymbol{\nu}$ denotes the sum of the principal curvatures of the surface Γ_h and $p > 2$. The H^{-1} gradient flow of the resulting free energy leads to the sixth order evolution equation

$$\begin{aligned} V = \Delta_\Gamma\Big[&\operatorname{div}_\Gamma(\nabla\psi(\boldsymbol{\nu})) + W(\boldsymbol{E}(\boldsymbol{u})) \\ &- \varepsilon\Big(\Delta_\Gamma(|\boldsymbol{K}|^{p-2}\boldsymbol{K}) - \frac{1}{p}|\boldsymbol{K}|^p\boldsymbol{K} + |\boldsymbol{K}|^{p-2}\boldsymbol{K}|\boldsymbol{B}|^2\Big)\Big], \end{aligned} \tag{6.12}$$

where V denotes the outer normal velocity of $\Gamma_{h(\cdot,t)}$, $|\boldsymbol{B}|^2$ is the sum of the squares the principal curvatures of $\Gamma_{h(\cdot,t)}$, $\boldsymbol{u}(\cdot, t)$ is the elastic equilibrium in $\Omega^+_{h(\cdot,t)}$, and $W(\boldsymbol{E}(\boldsymbol{u}))$ is the trace of $W(\boldsymbol{E}(\boldsymbol{u}(\cdot, t)))$ on $\Gamma_{h(\cdot,t)}$. We proved short time existence and regularity of solutions of the evolution equation (6.12). We also studied the Liapunov stability of an admissible flat configuration and showed that if the second variation of the regularized energy

$$\int_{\Omega_h^+} W(\boldsymbol{E}(\boldsymbol{u}))\, d\boldsymbol{x} + \int_{\Gamma_h} (\psi(\boldsymbol{v}) + \varepsilon|\boldsymbol{K}|^p)\, d\mathcal{H}^2 \tag{6.13}$$

computed at h =const. is positive definite, then for any initial configuration sufficiently close to the flat one, solutions of (6.12) exist for all time and approach the flat configuration as $t \to \infty$. Here $\mathcal{H}^2$ denotes the two-dimensional Hausdorff measure.

Problem 6.7. Consider the case $p = 2$ in (6.13).

Problem 6.8. What happens when $\varepsilon \to 0^+$?

6.5 Dislocations

Defects in crystalline solids are usually classified based on their dimension. Zero dimensional or point defects include impurities, vacancies, and self-intersticial atoms. One dimensional or line defects include edge, screw, and mixed dislocations. Finally, planar defects include stacking faults and grain boundaries (see [52,54]).

Edge dislocations were first discovered/hypothesized in crystal plasticity independently by Orowan [69], Polanyi [71], and Taylor [80] in 1934, while the presence of screw dislocations was hypothesized by Frank [39] in crystal growth phenomena in 1949.

At the continuum level, following the classical work of Volterra [86], an infinite straight dislocation can be modeled using a hollow cylinder of elastic material deformed by making a cut it in the z-direction. The resulting two free surfaces are displaced either vertically (screw Volterra dislocation) or horizontally (edge Volterra dislocation) with respect to each other of a distance $b > 0$. The vector $\boldsymbol{b}$ of length b given corresponding to the displacement is called the *Burger vector*. In the case of a screw dislocation it is given by $\boldsymbol{b} = b\boldsymbol{e}_3$, while it lies in the plane $z = 0$ in the case of an edge dislocation.

In the context of epitaxial growth, nucleation of dislocations is a further mode of strain relief for sufficiently thick films (see, for instance, [28,43, 50,55,82]).

Indeed, when a cusp-like morphology is approached as the result of an increasingly greater stress in surface valleys, the local stress may impart on the film a greater energy than that possessed by an equivalent film with a dislocation. When this happens, a dislocation may nucleate in the surface valley. Subsequently, the dislocation migrates to the film/substrate interface and the film surface relaxes towards a planar-like morphology.

In this framework, Volterra dislocations may be viewed as singularities of the strain field $\boldsymbol{H} = \nabla \boldsymbol{u}$. To be precise, given a finite set of points $\{\boldsymbol{x}_1, \dots, \boldsymbol{x}_N\}$ in Ω_h, and a set of vectors $\{\boldsymbol{b}_1, \dots, \boldsymbol{b}_N\}$, with $\boldsymbol{b}_i \in \mathbb{R}^2$, we say that a tensor field $\boldsymbol{H}$ in $\Omega \setminus \{\boldsymbol{x}_1, \dots, \boldsymbol{x}_N\}$ corresponds to a *system of dislocations* located at $\{\boldsymbol{x}_1, \dots, \boldsymbol{x}_N\}$ with Burgers vectors $\{\boldsymbol{b}_1, \dots, \boldsymbol{b}_N\}$, if

$$\begin{cases} \operatorname{curl} \boldsymbol{H} = \sum_{i=1}^{N} \boldsymbol{b}_i \, \delta_{\boldsymbol{x}_i} \\ \operatorname{div} C[\boldsymbol{E}(\boldsymbol{H})] = 0 \end{cases} \quad \text{in } \Omega_h \tag{6.14}$$

in the sense of distributions, where $\boldsymbol{E}(\boldsymbol{H}) = \frac{1}{2}(\boldsymbol{H} + \boldsymbol{H}^\top)$ is the strain associated to $\boldsymbol{H}$ and $\delta_{\boldsymbol{x}}$ denotes the Dirac delta at $\boldsymbol{x}$.

Solutions of (6.14) are not unique even modulo an infinitesimal rigid motion and, moreover, no variational principle may be associated to (6.14) since the elastic energy of a system of Volterra dislocations is not finite. Thus, some kind of regularization is needed. One possible remedy is to remove a core $B(\boldsymbol{x}_i, \varepsilon)$ of radius ε around each dislocation position and associate with $\boldsymbol{H}$ the (finite) elastic energy

$$\int_{\Omega_h \setminus (\cup_{i=1}^{N} B(\boldsymbol{x}_i, \varepsilon))} W\left(\boldsymbol{E}\left(\boldsymbol{H}\right)\right) dx. \tag{6.15}$$

This leads to minimize the energy (6.15) over all fields $\boldsymbol{H}$ for which $\operatorname{curl} \boldsymbol{H} = \boldsymbol{0}$ in $\Omega_h \setminus (\cup_{i=1}^{N} B(\boldsymbol{x}_i, \varepsilon))$ and $\int_{\partial B(\boldsymbol{x}_i, \varepsilon)} \boldsymbol{H}\boldsymbol{t} \, ds = \boldsymbol{b}_i, i = 1, \dots, N$, with $\boldsymbol{t}$ the unit tangent vector to $\partial B(\boldsymbol{x}_i, \varepsilon)$. This classical approach is very common in the engineering literature (see [52, 54]) and more recently it has attracted the attention of several mathematicians in the calculus of variations. For the study of this and of other variational models, although not in the context of epitaxial growth, we refer, *e.g.*, to [9,20,25,30,44,53,61,62,85] and the references contained therein.

In a recent paper [35], in collaboration with Fonseca, Fusco, and Morini, the author considered a variant of this approach, which consists in regularizing the *dislocation measure* $\boldsymbol{\mu} := \sum_{i=1}^{N} \boldsymbol{b}_i \delta_{\boldsymbol{x}_i}$ through a convolution procedure. To be precise, we introduce a convolution kernel $\varphi_\varepsilon := (1/\varepsilon^2)\varphi(\cdot/\varepsilon)$, where φ is a standard mollifier compactly supported in the unit ball, and we require now the compatibility condition

$$\operatorname{curl} \boldsymbol{H} = \boldsymbol{\mu} * \varphi_\varepsilon. \tag{6.16}$$

Here $\varepsilon > 0$ is a fixed constant that may be interpreted as before as the core radius. This alternative formulation is found for instance in [50].

The total energy associated with a profile h, a dislocation measure $\boldsymbol{\mu}$ and a strain field $\boldsymbol{H}$ satisfying the compatibility conditions (6.16) is given by

$$\mathcal{G}_4(\boldsymbol{H}, h, \boldsymbol{\mu}) := \int_{\Omega_h} W\left(\boldsymbol{E}\left(\boldsymbol{H}\right)\right) d\boldsymbol{x} + \gamma_{\text{film}} \,\text{length}\, \Gamma_h. \tag{6.17}$$

Thus, for any given profile h and any given dislocation measure $\boldsymbol{\mu}$, the compatible strain field minimizing the elastic energy is well defined and satisfies the div-curl system

$$\begin{cases} \text{curl}\, \boldsymbol{H} = \boldsymbol{\mu} * \varphi_\varepsilon \\ \text{div}\, C[\boldsymbol{E}(\boldsymbol{H})] = 0 \end{cases} \quad \text{in } \Omega_h. \tag{6.18}$$

Existence and regularity for equilibrium solutions of (6.17) can be obtained as in Sections 4 and 5, although there are many new technical issues due to the presence of dislocations, which require new ideas. In particular, a major difficulty arises in showing that the volume constraint can be replaced by a volume penalization in the case (5.7) (see Theorem 5.1). In the dislocation-free case this was based on a straightforward truncation argument, which fails in the present setting because dislocations cannot be removed in this way. Indeed they act as a sort of obstacle when touching the profile.

In [35] we provide analytical support to the experimental evidence that, after nucleation, dislocations lie at the bottom. In the last part of the paper we study the nucleation of dislocations and we investigate conditions under which it is energetically favorable to create dislocations. To this purpose, we modify the energy (6.17) by adding a term that accounts for the energy dissipated to create dislocations.

Problem 6.9. It would be interesting to study the dynamics of this problem.

The motion of dislocations for (6.15) has been studied by several authors (see, *e.g.*, [2,3,8,18,19,72] and the references therein).

7 Notation

Constants

- c: arbitrary constants that *can change from line to line* and that can be computed in terms of known quantities.

Sets

- X, Y: sets or spaces; E: usually denote a set; ∂E: boundary of E; E°: the interior of E; $\overline{E}$: the closure of E.

- χ_E: characteristic function of the set E.
- dist: distance; diam: diameter.
- d_{reg}: the regularized distance.
- $d_{\mathcal{H}}$: the Hausdorff distance.

Functions

- $f, g, h, \varphi, \psi, \phi, u, v, w$ usually denote functions; supp f: support of the function f; Lip f: Lipschitz constant of the function f; $f * g$: convolution of the functions f and g, f° the polar function of f.

Functions of one variable

- $\mathbb{N}$: the set of positive integers; $\mathbb{N}_0 := \mathbb{N} \cup \{0\}$; $\mathbb{Z}$: the integers; $\mathbb{Q}$: the rational numbers; $\mathbb{R}$: the real line; $\overline{\mathbb{R}} := [-\infty, \infty]$: the extended real line; if $x \in \mathbb{R}$, then $x^+ := \max\{x, 0\}$, $x^- := \max\{-x, 0\}$, $|x| := x^+ + x^-$; $\lfloor x \rfloor$: the integer part of x.
- $\mathcal{L}^1$: the 1-dimensional Lebesgue measure.
- Given a function u, $u^+(x)$ and $u^+(x)$ are the right and left limits of u at a point x, Var u is the pointwise variation of u, $u'_+(x)$ and $u'_-(x)$ are the right and left derivatives of u at a point x
- length γ: the length of a curve γ.

Functions of several variables

- $\mathbb{R}^N$: the N-dimensional Euclidean space, $N \geq 1$, for $\boldsymbol{x} = (x_1, \dots, x_N)$ in $\mathbb{R}^N$,[6]

$$|\boldsymbol{x}| := \sqrt{(x_1)^2 + \ldots + (x_N)^2}.$$

Given $\boldsymbol{x} = (x_1, \dots, x_N) \in \mathbb{R}^N$, for every $i = 1, \dots, N$ we denote by $\boldsymbol{x}_i$ the $(N-1)$-dimensional vector obtained by removing the ith component from $\boldsymbol{x}$ and with an abuse of notation we write

$$\boldsymbol{x} = (\boldsymbol{x}_i, x_i) \in \mathbb{R}^{N-1} \times \mathbb{R}. \tag{7.1}$$

When $i = N$, we will also use the simpler notation

$$\boldsymbol{x} = \left(\boldsymbol{x}', x_N\right) \in \mathbb{R}^{N-1} \times \mathbb{R} \tag{7.2}$$

[6] When there is no possibility of confusion, we will also use x_1, x_2, etc, to denote different points of $\mathbb{R}^N$. Thus, depending on the context, x_i is either a point of $\mathbb{R}^N$ or the ith coordinate of the point $x \in \mathbb{R}^N$.

and

$$\mathbb{R}^N_+ = \left\{x = \left(x', x_N\right) \in \mathbb{R}^{N-1} \times \mathbb{R} : x_N > 0\right\},$$
$$\mathbb{R}^N_- = \left\{x = \left(x', x_N\right) \in \mathbb{R}^{N-1} \times \mathbb{R} : x_N < 0\right\}.$$

- δ_{ij}: the Kronecker delta; that is, $\delta_{ij} := 1$ if $i = j$ and $\delta_{iN} := 0$ otherwise.
- $\boldsymbol{e}_i$, $i = 1, \ldots, N$,: the unit vectors of the standard (or canonical) orthonormal basis of $\mathbb{R}^N$.
- det: determinant of a matrix or a linear mapping.
- Ω: an open set of $\mathbb{R}^N$ (not necessarily bounded); $\boldsymbol{\nu}$: the outward unit normal to $\partial\Omega$; $B_N(\boldsymbol{x}_0, r)$ (or simply $B(\boldsymbol{x}_0, r)$): open ball in $\mathbb{R}^N$ of center $\boldsymbol{x}_0$ and radius r;

$$Q_N(\boldsymbol{x}_0, r) := \boldsymbol{x}_0 + \left(-\frac{r}{2}, \frac{r}{2}\right)^N$$

(or simply $Q(\boldsymbol{x}_0, r)$); S^{N-1}: unit sphere in $\mathbb{R}^N$.
- $\mathcal{L}^N$: the N-dimensional Lebesgue measure;
- $\mathcal{H}^s$: the s-dimensional Hausdorff measure.
- Given $E \subset \mathbb{R}^N$ and $\boldsymbol{u} : E \to \mathbb{R}^M$, $\frac{\partial \boldsymbol{u}}{\partial x_i}(x_0)$ is the partial derivative of $\boldsymbol{u}$ at x_0 with respect to x_i, $\frac{\partial \boldsymbol{u}}{\partial \boldsymbol{n}}(x_0)$ is the directional derivative of $\boldsymbol{u}$ at x_0 in the direction $\boldsymbol{n} \in \partial B(\boldsymbol{0}, 1)$, $\nabla \boldsymbol{u}(x_0)$ is the gradient of $\boldsymbol{u}$ at x_0, and $J\boldsymbol{u}(x_0)$ is the Jacobian of $\boldsymbol{u}$ at x_0 (for $N = M$).
- $\boldsymbol{E}(\boldsymbol{u})$: the symmetrized gradient of $\boldsymbol{u} : \Omega \to \mathbb{R}^N$.
- Δ: the Laplace operator; Δ_Γ: the Laplace–Beltrami operator; div: the divergence operator.
- For a multi-index $\boldsymbol{\alpha} = (\alpha_1, \ldots, \alpha_N) \in (\mathbb{N}_0)^N$,

$$\frac{\partial^{\boldsymbol{\alpha}}}{\partial x^{\boldsymbol{\alpha}}} := \frac{\partial^{|\boldsymbol{\alpha}|}}{\partial x_1^{\alpha_1} \ldots \partial x_N^{\alpha_N}}, \quad |\boldsymbol{\alpha}| := \alpha_1 + \ldots + \alpha_N.$$

- φ_ε: a standard mollifier; $u_\varepsilon := \varphi_\varepsilon * u$: the mollification of u; Ω_ε: the set $\{x \in \Omega : \operatorname{dist}(x, \partial\Omega) > \varepsilon\}$.

Spaces

- $C^m(\Omega)$: the space of all functions that are continuous together with their partial derivatives up to order $m \in \mathbb{N}_0$,

$$C^\infty(\Omega) := \bigcap_{m=0}^{\infty} C^m(\Omega);$$

$C_c^m(\Omega)$ and $C_c^\infty(\Omega)$: the subspaces of $C^m(\Omega)$ and $C^\infty(\Omega)$, respectively, consisting of all functions with compact support.

- $L^p(\Omega)$: L^p space; $\|\cdot\|_{L^p(\Omega)}$: the norm in $L^p(\Omega)$ or in $L^p(\Omega;\mathbb{R}^M)$.
- $\rightharpoonup$: weak convergence in $L^p(\Omega)$.
- $W^{1,p}(\Omega)$: a Sobolev space; $\|\cdot\|_{W^{1,p}(\Omega)}$: the norm in $W^{1,p}(\Omega)$; for $u \in W^{1,p}(\Omega)$, $\frac{\partial u}{\partial x_i}$: the weak (or distributional) partial derivative of u with respect to x_i; ∇u: the weak (or distributional) gradient of u.
- $H^1(\Omega)$: the Hilbert space $W^{1,2}(\Omega)$; $W^{k,p}(\Omega)$: a higher-order Sobolev space.
- $C^{0,\alpha}(\Omega)$: the space of all bounded Hölder continuous functions with exponent α.
- $BV(\Omega)$: the space of functions of bounded variation. Given $u \in BV(\Omega)$, $D_i u$ is the weak (or distributional) partial derivative of u with respect to x_i, Du is the weak (or distributional) gradient of u, $|Du|$ is the total variation measure of the measure Du.
- $\mathrm{P}(E,\Omega)$: the perimeter of a set E in Ω.
- $W^{s,p}(\mathbb{R}^N)$: a fractional Sobolev space.

References

[1] E. ACERBI, N. FUSCO and M. MORINI, *Minimality via second variation for a nonlocal isoperimetric problem*, Communications in Mathematical Physics **322** (2013), 515–557.

[2] R. ALICANDRO, L. DE LUCA, A. GARRONI and M. PONSIGLIONE, *Dynamics of discrete screw dislocations on glide directions*, preprint, 2014.

[3] R. ALICANDRO, L. DE LUCA, A. GARRONI and M. PONSIGLIONE, *Metastability and dynamics of discrete topological singularities in two dimensions: a Γ-convergence approach*, Archive for Rational Mechanics and Analysis **214** (2014), 269–330.

[4] L. AMBROSIO, N. GIGLI and G. SAVARÉ, "Gradient Flows: in Metric Spaces and in the Space of Probability Measures", Springer Science & Business Media, 2008.

[5] L. AMBROSIO, M. NOVAGA and E. PAOLINI, *Some regularity results for minimal crystals*, ESAIM: Control, Optimisation and Calculus of Variations **8** (2002), 69–103.

[6] S. ANGENENT and M. E. GURTIN *Multiphase thermomechanics with interfacial structure 2. Evolution of an isothermal interface*, Archive for Rational Mechanics and Analysis **108** (1989), 323–391.

[7] P. BELLA, M. GOLDMAN and B. ZWICKNAGL *Study of island formation in epitaxially strained films on unbounded domains*, Archive for Rational Mechanics and Analysis (2014), 1–55.

[8] T. BLASS, I. FONSECA, G. LEONI and M. MORANDOTTI, *Dynamics for systems of screw dislocations*, SIAM Journal on Applied Mathematics **75** (2015), 393–419.

[9] T. BLASS and M. MORANDOTTI, *Renormalized energy and Peach-Köhler forces for screw dislocations with antiplane shear*, 2014, arXiv preprint arXiv:1410.6200.

[10] M. BONACINI, *Epitaxially strained elastic films: the case of anisotropic surface energies*, ESAIM: Control, Optimisation and Calculus of Variations **19** (2013), 167–189.

[11] M. BONACINI, *Stability of equilibrium configurations for elastic films in two and three dimensions*, Advances in Calculus of Variations **8** (2015), 117–153.

[12] M. BONACINI and R. CRISTOFERI, *Local and global minimality results for a nonlocal isoperimetric problem on $\mathbb{R}^N$*, SIAM Journal on Mathematical Analysis **46** (2014), 2310–2349.

[13] M. BONACINI and M. MORINI, *Stable regular critical points of the Mumford–Shah functional are local minimizers* In: "Annales de l'Institut Henri Poincare (C) Non Linear Analysis", Elsevier, 2014.

[14] A. BRAIDES, A. CHAMBOLLE and M. SOLCI, *A relaxation result for energies defined on pairs set function and applications*, ESAIM: Control, Optimisation and Calculus of Variations **13** (2007), 717–734.

[15] F. CAGNETTI, M. G. MORA and M. MORINI *A second order minimality condition for the Mumford-Shah functional* Calculus of Variations and Partial Differential Equations **33** (2008), 37–74.

[16] J. W. CAHN and J. E. TAYLOR, *Overview no. 113 surface motion by surface diffusion*, Acta metallurgica et materialia **42** (1994), 1045–1063.

[17] G. M. CAPRIANI, V. JULIN and G. PISANTE, *A quantitative second order minimality criterion for cavities in elastic bodies*, SIAM Journal on Mathematical Analysis **45** (2013), 1952–1991.

[18] P. CERMELLI and T. ARMANO, *Noncrystallographic motion of a dislocation as a fine mixture of rectilinear paths*, SIAM Journal on Applied Mathematics **64** (2004), 2121–2143.

[19] P. CERMELLI and M. E. GURTIN, *The motion of screw dislocations in crystalline materials undergoing antiplane shear: glide, cross-slip, fine cross-slip*, Archive for rational mechanics and analysis **148** (1999), 3–52.

[20] P. CERMELLI and G. LEONI, *Renormalized energy and forces on dislocations*, SIAM journal on mathematical analysis **37** (2005), 1131–1160.

[21] A. CHAMBOLLE and E. BONNETIER, *Computing the equilibrium configuration of epitaxially strained crystalline films*, SIAM Journal on Applied Mathematics **62** (2002), 1093–1121.

[22] A. CHAMBOLLE and C. J. LARSEN, *C^∞ regularity of the free boundary for a two-dimensional optimal compliance problem*, Calculus of Variations and Partial Differential Equations **18** (2003), 77–94.

[23] A. CHAMBOLLE and M. SOLCI, *Interaction of a bulk and a surface energy with a geometrical constraint*, SIAM Journal on Mathematical Analysis **39** (2007), 77–102.

[24] S. CONTI, D. FARACO and F. MAGGI, *A new approach to counterexamples to L^1 estimates: Korn's inequality, geometric rigidity, and regularity for gradients of separately convex functions*, Archive for rational mechanics and analysis **175** (2005), 287–300.

[25] L. DE LUCA, A. GARRONI and M. PONSIGLIONE, *Γ-convergence analysis of systems of edge dislocations: the self energy regime*, Archive for Rational Mechanics and Analysis **206** (2012), 885–910.

[26] B. DE MARIA and N. FUSCO, *Regularity properties of equilibrium configurations of epitaxially strained elastic films*, In: "Topics in Modern Regularity Theory", Springer, 2012, 169–204.

[27] A. DI CARLO, M. E. GURTIN and P. PODIO-GUIDUGLI, *A regularized equation for anisotropic motion-by-curvature*, SIAM Journal on Applied Mathematics **52** (1992), 1111–1119.

[28] K. ELDER, N. PROVATAS, J. BERRY, P. STEFANOVIC and M. GRANT, *Phase-field crystal modeling and classical density functional theory of freezing*, Physical Review B **75** (2007), 064107.

[29] A. FIGALLI, N. FUSCO, F. MAGGI, V. MILLOT and M. MORINI, *Isoperimetry and stability properties of balls with respect to nonlocal energies*, Communications in Mathematical Physics **336** (2015), 441–507.

[30] M. FOCARDI and A. GARRONI, *A 1d macroscopic phase field model for dislocations and a second order γ-limit*, Multiscale Modeling & Simulation **6** (2007),1098–1124.

[31] I. FONSECA, *The Wulff theorem revisited*, **432** (1991), 125–145.

[32] I. FONSECA, N. FUSCO, G. LEONI and V. MILLOT, *Material voids in elastic solids with anisotropic surface energies*, Journal de mathematiques pures et appliquees **96** (2011), 591–639.

[33] I. FONSECA, N. FUSCO, G. LEONI and M. MORINI, *Equilibrium configurations of epitaxially strained crystalline films: existence and regularity results*, Archive for Rational Mechanics and Analysis **186** (2007), 477–537.

[34] I. FONSECA, N. FUSCO, G. LEONI and M. MORINI *Motion of elastic thin films by anisotropic surface diffusion with curvature regularization*, Archive for Rational Mechanics and Analysis **205** (2012), 425–466.

[35] I. FONSECA, N. FUSCO, G. LEONI and M. MORINI, *A model for dislocations in epitaxially strained elastic films*, preprint, 2015.

[36] I. FONSECA, N. FUSCO, G. LEONI and M. MORINI, *Motion of three-dimensional elastic films by anisotropic surface diffusion with curvature regularization*, Analysis & PDE **8** (2015), 373–423.

[37] I. FONSECA and S. MÜLLER, *A uniqueness proof for the Wulff theorem*, Proceedings of the Royal Society of Edinburgh: Section A Mathematics **119** (1991), 125–136.

[38] I. FONSECA, A. PRATELLI and B. ZWICKNAGL, *Shapes of epitaxially grown quantum dots*, Archive for Rational Mechanics and Analysis **214** (2014), 359–401.

[39] F. FRANK and J. H. VAN DER MERWE, *One-dimensional dislocations. I. Static theory* Proceedings of the Royal Society of London. Series A, Mathematical and Physical Sciences (1949), 205–216.

[40] L. B. FREUND and S. SURESH, "Thin Film Materials: Stress, Defect Formation and Surface Evolution", Cambridge University Press, 2004.

[41] K. O. FRIEDRICHS, *On the boundary-value problems of the theory of elasticity and Korn's inequality*, Annals of Mathematics (1947), 441–471.

[42] N. FUSCO and M. MORINI, *Equilibrium configurations of epitaxially strained elastic films: second order minimality conditions and qualitative properties of solutions*, Archive for Rational Mechanics and Analysis **203** (2012), 247–327.

[43] H. GAO and W. D. NIX, *Surface roughening of heteroepitaxial thin films*, Annual Review of Materials Science **29** (1999), 173–209.

[44] A. GARRONI, G. LEONI and M. PONSIGLIONE, *Gradient theory for plasticity via homogenization of discrete dislocations*, J. Eur. Math. Soc. (JEMS) **12** (2010), 1231–1266.

[45] M. GOLDMAN and B. ZWICKNAGL, *Scaling law and reduced models for epitaxially strained crystalline films*, SIAM Journal on Mathematical Analysis **46** (2014), 1–24.

[46] P. GRISVARD, *Singularités en elasticité*, Archive for rational mechanics and analysis **107** (1989), 157–180.

[47] P. GRISVARD, "Elliptic Problems in Nonsmooth Domains", Vol. 69, SIAM, 2011.

[48] M. E. GURTIN, "Multiphase Thermomechanics with Interfacial Structure 1. Heat Conduction and the Capillary Balance Law", Springer, 1999.

[49] M. E. GURTIN, H. M. SONER and P. E. SOUGANIDIS, *Anisotropic motion of an interface relaxed by the formation of infinitesimal wrinkles*, Journal of differential equations **119** (1995), 54–108.

[50] M. HAATAJA, J. MÜLLER, A. RUTENBERG and M. GRANT, *Dislocations and morphological instabilities: Continuum modeling of misfitting heteroepitaxial films*, Physical Review B **65** (2002), 165414.

[51] C. HERRING, *Some theorems on the free energies of crystal surfaces*, Physical Review **82** (1951), 87.

[52] J. P. HIRTH and J. LOTHE, "Theory of Dislocations", John Wiley\ & Sons, 1982.

[53] T. HUDSON and C. ORTNER, *Existence and stability of a screw dislocation under anti-plane deformation*, Archive for Rational Mechanics and Analysis **213** (2014), 887–929.

[54] D. HULL and D. J. BACON, "Introduction to Dislocations", Butterworth-Heinemann, 2001.

[55] D. JESSON, S. PENNYCOOK, J.-M. BARIBEAU and D. HOUGHTON, *Direct imaging of surface cusp evolution during strained-layer epitaxy and implications for strain relaxation*, Physical review letters **71** (1993), 1744.

[56] V. JULIN and G. PISANTE, *Minimality via second variation for microphase separation of diblock copolymer melts*, Journal für die reine und angewandte Mathematik (Crelles Journal), 2013.

[57] P. KOHLERT, K. KASSNER and C. MISBAH, *Large-amplitude behavior of the Grinfeld instability: a variational approach*, The European Physical Journal B-Condensed Matter and Complex Systems **35** (2003), 493–504.

[58] R. KUKTA and L. FREUND, *Minimum energy configuration of epitaxial material clusters on a lattice-mismatched substrate*, Journal of the Mechanics and Physics of Solids **45** (1997), 1835–1860.

[59] G. LEONI, "A First Course in Sobolev Spaces", Vol. 105, American Mathematical Society Providence, RI, 2009.

[60] G. LIEBERMAN, *Regularized distance and its applications*, Pacific journal of Mathematics **117** (1985),329–352.

[61] M. G. MORA, M. PELETIER and L. SCARDIA, *Convergence of interaction-driven evolutions of dislocations with Wasserstein dissipation and slip-plane confinement*, 2014, preprint arXiv:1409.4236.

[62] S. Müller, L. Scardia and C. I. Zeppieri, *Geometric rigidity for incompatible fields and an application to strain-gradient plasticity*, Indiana University Mathematics Journal, 2014.

[63] W. W. Mullins, *Theory of thermal grooving*, Journal of Applied Physics **28** (1957), 333–339.

[64] S. Nicaise, *About the Lamé system in a polygonal or a polyhedral domain and a coupled problem between the Lamé system and the plate equation. I: Regularity of the solutions*, Annali della Scuola Normale Superiore di Pisa-Classe di Scienze **19** (1992), 327–361.

[65] S. Nicaise and A.-M. Sändig, *General interface problems: I*, Mathematical Methods in the Applied Sciences **17** (1994), 395–429.

[66] J. A. Nitsche, *On Korn's second inequality*, RAIRO-Analyse numérique **15** (1981), 237–248.

[67] M. Novaga and E. Paolini, *Regularity results for boundaries in $\mathbb{R}^2$ with prescribed anisotropic curvature*, Annali di Matematica Pura ed Applicata **184** (2005), 239–261.

[68] D. Ornstein, *A non-inequality for differential operators in the L^1 norm*, Archive for Rational Mechanics and Analysis **11** (1962), 40–49.

[69] E. Orowan, *Zur kristallplastizität. iii*, Zeitschrift für Physik **89** (1934), 634–659.

[70] P. Piovano, *Evolution of elastic thin films with curvature regularization via minimizing movements*, Calculus of Variations and Partial Differential Equations **49** (2014), 337–367.

[71] M. Polanyi, *Über eine art gitterstörung, die einen kristall plastisch machen könnte*, Zeitschrift für Physik **89** (1934), 660–664.

[72] C. Reina and S. Conti, *Kinematic description of crystal plasticity in the finite kinematic framework: a micromechanical understanding of* $\mathbf{F} = \mathbf{F}^e\mathbf{F}^p$, Journal of the Mechanics and Physics of Solids **67** (2014), 40–61.

[73] A. Rössle, *Corner singularities and regularity of weak solutions for the two-dimensional Lamé equations on domains with angular corners*, Journal of elasticity **60** (2000), 57–75.

[74] M. Siegel, M. Miksis and P. Voorhees, *Evolution of material voids for highly anisotropic surface energy*, Journal of the Mechanics and Physics of Solids **52** (2004), 1319–1353.

[75] B. Spencer, *Asymptotic derivation of the glued-wetting-layer model and contact-angle condition for Stranski-Krastanow islands*, Physical Review B **59** (1999), 2011.

[76] B. Spencer and J. Tersoff, *Equilibrium shapes and properties of epitaxially strained islands*, Physical Review Letters **79** (1997), 4858.

[77] B. SPENCER and J. TERSOFF, *Asymmetry and shape transitions of epitaxially strained islands on vicinal surfaces*, Applied Physics Letters **96** (2010), 073114.

[78] B. J. SPENCER, *Asymptotic solutions for the equilibrium crystal shape with small corner energy regularization*, Physical Review E **69** (2004), 011603.

[79] E. M. STEIN, "Singular Integrals and Differentiability Properties of Functions", Vol. 2. Princeton University Press, 1970.

[80] G. I. TAYLOR, *The mechanism of plastic deformation of crystals. Part I. Theoretical*. Proceedings of the Royal Society of London, Series A, Containing Papers of a Mathematical and Physical Character, 1934, 362–387.

[81] J. E. TAYLOR, *Crystalline variational problems*, Bulletin of the American Mathematical Society **84** (1978), 568–588.

[82] J. TERSOFF and F. LEGOUES, *Competing relaxation mechanisms in strained layers*, Physical review letters **72** (1994), 3570.

[83] J. TERSOFF and R. TROMP, *Shape transition in growth of strained islands: spontaneous formation of quantum wires*, Physical review letters **70** (1993), 2782.

[84] T. W. TING, *Generalized Korn's inequalities*, Tensor **25** (1972), 295–302.

[85] P. VAN MEURS, A. MUNTEAN and M. PELETIER, *Upscaling of dislocation walls in finite domains*, European Journal of Applied Mathematics **25** (2014), 749–781.

[86] V. VOLTERRA, *Sur l'équilibre des corps élastiques multiplement connexes*, In: "Annales scientifiques de l'Ecole Normale Superieure", Vol. 24, Société mathématique de France, 1907, 401–517.

[87] W. YANG and D. SROLOVITZ, *Cracklike surface instabilities in stressed solids*, Physical review letters **71** (1993), 1593.

Local and global minimality results for an isoperimetric problem with long-range interactions

Massimiliano Morini

1 Introduction

In this paper we review some recent results concerning the following *non-local isoperimetric problem*:

$$\min\left\{ P_\Omega(\{u=1\}) + \gamma \int_\Omega\int_\Omega G_\Omega(z_1,z_2)u(z_1)u(z_2)\,dz_1dz_2 : \right. \\ \left. u \in BV(\Omega;\{-1,1\}),\ \fint_\Omega u\,dz = m \right\}, \tag{1.1}$$

where $m \in (-1,1)$ is given and prescribes the volume of the two phases $\{u=1\}$ and $\{u=-1\}$. Here P_Ω stands for the perimeter relative to Ω (in the sense of Caccioppoli-De Giorgi) and $BV(\Omega;\{-1,1\})$ denotes the space of functions of bounded variation taking values in $\{-1,1\}$. We refer to [4] and [33] for a rather complete account on the properties of functions of bounded variations and sets of finite perimeter.

The function G_Ω appearing in the double integral term is the Laplacian's Green function associated with Ω under Neumann boundary conditions. Precisely, denoting by δ_{z_1}, $z_1 \in \Omega$, the Dirac measure at z_1, $G_\Omega(z_1,\cdot)$ is the unique solution to

$$\begin{cases} -\Delta_{z_2} G_\Omega(z_1,\cdot) = \delta_{z_1} - \frac{1}{|\Omega|} & \text{in } \Omega, \\ \nabla_{z_2} G_\Omega(z_1,\cdot)\cdot n_\Omega = 0 & \text{on } \partial\Omega, \\ \int_\Omega G_\Omega(z_1,z_2)\,dz_2 = 0\,. \end{cases} \tag{1.2}$$

Here and in the following, given a (sufficiently smooth) open subset $\Omega \subset \mathbb{R}^N$, we denote by n_Ω the outward unit normal vector to $\partial\Omega$. We will also consider periodic boundary conditions; in this case $\Omega = \mathbb{T}^N$, where $\mathbb{T}^N$ denotes the N-dimensional flat torus, and $G_{\mathbb{T}^N}$ is defined as in (1.2), but with periodic boundary conditions in place of the homogeneous Neumann ones.

Problem (1.1) arises as the *sharp interface limit* of the following ε-diffuse energies, which were proposed by Ohta and Kawasaki [42] to model the behavior of a class of two-phase materials called *diblock coplymers*:

$$OK_\varepsilon(u) := \varepsilon \int_\Omega |\nabla u|^2 \, dz + \frac{1}{\varepsilon} \int_\Omega (u^2 - 1)^2 \, dz \\ + \gamma_0 \int_\Omega \int_\Omega G_\Omega(z_1, z_2) u(z_1) u(z_2) dz_1 \, dz_2 \tag{1.3}$$

where u is an $H^1(\Omega)$ phase parameter describing the density distribution of the components ($u = -1$ stands for one phase, $u = +1$ for the other), G_Ω is as before, and the constants $\gamma_0 > 0$ and $\varepsilon > 0$ are characteristic of the material (the latter usually very small). According to the Ohta-Kawasaki approach observable configurations are described by the global (or local) minimizers of the above energy under a volume constraint $⨍_\Omega u = m$. In order to describe the main features of the problem, we recall that diblock copolymers are macromolecules composed of two distinct polymer chains: a monomer specie A, corresponding roughly to the region where the order parameter $u \approx 1$, and a second monomer specie B, corresponding to the region where $u \approx -1$. The two subchains are chemically bonded and thermodynamically incompatible. The incompatibility between the different subchains drives the system

Figure 1.1. A diblock copolymer.

towards phase separation, whereas the chemical bonds between the different species makes it impossible to achieve a complete macroscopic phase separation, which can only occur at a mesoscale. We refer to [9] for a rigorous derivation of the Ohta-Kawasky density functional theory from first principles.

As the result of the two competing trends, the observed configurations may display complex patterns and depend very much on the volume fraction of one phase with respect to the other. For instance, when the volume fraction is small the minority phase arranges in many small almost round connected components (droplets) periodically distributed. On the contrary, when the volume fraction is approximately one half, then a *lamellar pattern* is experimentally observed (see Figure 1.2). For intermediate volume fractions more complicated patterns, such as triply periodic structures, may appear. See the phase diagram below (Figure 1.3).

Figure 1.2. An example of lamellar pattern observed when the volume fraction is approximately one half.

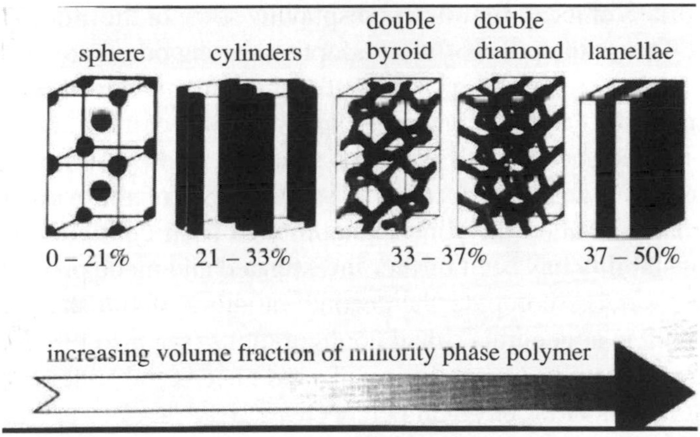

Figure 1.3. The typical patterns that are observed according to value of the volume fraction. The picture is taken from Edwin Thomas' talk at MSRI 1999.

From the mathematical point of view, the main feature is a competition between the short-range interfacial energy and the nonlocal Green's function, which acts as a long-range repulsive interaction of Coulombic type. Indeed, the perimeter term favors the formation of large regions of pure phases with minimal interface area, whereas the Green's term prefers configurations in which several connected components nucleate and try to separate from each other as much as possible, due to the repulsive nature of the interaction. In general, highly oscillating patterns, with the two phases finely intertwined, have smaller nonlocal energy. The observed configuration is the nontrivial result of this competition.

As observed in the literature and displayed in Figure 1.3, the domain structures in phase-separated diblock copolymers closely resemble *periodic* surfaces with constant mean curvature, see for instance [54].

A related challenging mathematical problem is to analytically validate the expectation that global minimizers of (1.1) are (almost) periodic. This is still a largely open issue, which seems to escape the current mathematical grasp of the nonlocal isoperimetric problem. Up to now, periodicity has been established only in one dimension, see [37,45]. In higher dimensions, the best result in this direction is due to Alberti, Choksi and Otto, who were able to show that the energy of global minimizers is uniformly distributed in space, see [2]. See also [51] for an extension of this result to the global minimizers of (1.3). We refer also to [7,13,21,22,39,52,55] for other results on the structure of global minimizers. However, all these works deal with some asymptotic regimes, namely the small γ and the small mass regimes (see Subsections 5.2 and 5.3).

A mathematically more affordable and yet highly nontrivial task is to exhibit a class of local minimizers displaying some of the aforementioned patterns. This is the point of view adopted, among others, by Ren & Wei and by Choksi & Sternberg. The first authors in a series of papers (see for instance [44,46–49]) construct explicit critical configurations for the sharp interface energy with lamellar, spherical and cylindrical patterns. Moreover, they are able to identify some ranges of the parameters involved that guarantee the (linear) stability of such configurations. The notion of stability has been further investigated and made precise in [11], where the authors compute the second variation of the sharp interface energy in (1.1) at general critical configurations (see also [38,40], where related linear stability/instability results have been established for the first time, but from a more physical perspective).

The aforementioned works give rise to some natural questions:

(i) Are the explicit stable critical configurations constructed by Ren & Wei true local minimizer of the energy?
(ii) Is it possible to approximate such configurations by local minimizers of the ε-diffuse energies (1.3)?
(iii) To what extent are strictly stable (in the sense of positive definite second variation) critical configurations of the sharp interface energy local minimizers?
(iv) Can we exhibit a class of explicit global minimizers of (1.1)?

Such issues have been successfully addressed in a series of recent papers and it is the main purpose of this review to present the results of such investigations.

The plan of the paper is the following: In Section 2 we give the precise formulation of the problem, compute the first variation of the energy and provide the second variation formula computed by Choksi & Sternberg. In Sections 3 and 4 we establish the main local minimality criterion; namely, we show that every critical configuration with positive definite second variation is an isolated local minimizer of the sharp interface energy, with respect to the L^1-metric. This results was established in [1] in the periodic setting and also in the Neumann setting under some restrictions. The methods of [1] have been subsequently adapted in [27] to settle the general Neumann case. Section 5 is devoted to some applications of the theory developed in the previous sections to establish several local and global minimality results. Among them, we highlight the connection between strictly stable critical configurations of the sharp interface energy and local minimizers of (1.3) (see Subsection 5.1) and the results contained in Subsection 5.4, where we display a cascade of global minimizers of lamellar type with more and more interfaces in sufficiently thin rectangles. In such a setting, we can fully appreciate the impact of the nonlocal term on the structure of global minimizers. The results of this subsection were established in [36].

The focus of this presentation will be on the main ideas and techniques behind the proofs of the results. Therefore, while trying to keep it as self-contained as possible, we will from time to time skip some arguments of more technical nature or impose additional simplifying assumptions, when this is instrumental to make the overall strategy and the "essence" of a proof more transparent. The interested reader may anyhow find the missing details in the aforementioned original references.

ACKNOWLEDGEMENTS. The author warmly thanks Aldo Pratelli and Nicola Fusco for organizing the ERC School on Free Discontinuity Problems, July 7th-11th, 2014 and the De Giorgi Center of Pisa for the hospitality.

2 The first and the second variation of the nonlocal functional

We start by briefly recalling the notation and the main definitions concerning sets of finite perimeter relative to an open set $\Omega \subset \mathbb{R}^N$, referring to [1,33] for a complete treatment of the theory. Given a measurable set $E \subset \Omega$, we set

$$P_\Omega(E) := \sup\left\{\int_E \operatorname{div}\varphi\, dz \,:\, \varphi \in C^1_c(\Omega;\mathbb{R}^N),\ \|\varphi\|_\infty \le 1\right\}.$$

If $P_\Omega(E) < +\infty$, then we say that E has finite perimeter in Ω. It turns out that $P_\Omega(E) < +\infty$ if and only if the distributional derivative of the characteristic function χ_E is a Radon measure, denoted by μ_E, with finite total variation $|\mu_E|(\Omega) = P_\Omega(E)$. The celebrated De Giorgi's Structure

Theorem states that if E is of finite perimeter in Ω, then there exists a suitable subset of the topological boundary $\partial E \cap \Omega$ relative to Ω, called the *reduced boundary* and denoted by $\partial^* E$, which is $(N-1)$-*countably rectifiable* and such that the following generalized Divergence Theorem holds:

$$\int_E \operatorname{div}\varphi \, dz = \int_{\partial^* E} \varphi \cdot n_E \, d\mathcal{H}^{N-1} \qquad \text{for all } \varphi \in C^1_c(\Omega; \mathbb{R}^N). \tag{2.1}$$

Here n_E denotes the generalized outer normal to E, which is well defined on $\partial^* E$. Formula (2.1) implies in particular that $\mu_E = -n_E \mathcal{H}^{N-1} \llcorner \partial^* E$. In turn, if $E = F \cap \Omega$ for some set $F \subset \mathbb{R}^N$ of class C^1, then $\partial^* E = \partial E \cap \Omega$ and $P_\Omega(E) = \mathcal{H}^{N-1}(\partial E \cap \Omega)$. We will make frequent use of the following important compactness property of sets of finite perimeters.

Theorem 2.1 (Compactness of sets of finite perimeters). *Let $\Omega \subset \mathbb{R}^N$ be a bounded open Lipschitz set, and let $(E_n)_n$ be a sequence of subsets of Ω such that*

$$\sup_n P_\Omega(E_n) < +\infty .$$

Then, there exists $E \subset \Omega$ with finite perimeter in Ω and a subsequence $(E_{n_k})_k$ such that

$$\chi_{E_{n_k}} \to \chi_E \qquad \text{in } L^1(\Omega)$$

as $k \to \infty$.

With a slight abuse of notation, in the following we will write $E_n \to E$ in $L^1(\Omega)$ instead of $\chi_{E_n} \to \chi_E$.

We now rewrite the functional in (1.1) in a more convenient way. First of all, we adopt a more geometric point of view and we regard it as a functional defined on sets, rather than on functions. Indeed, any set $E \subset \Omega$ may be thought as the region where the associated function

$$u_E := \chi_E - \chi_{\Omega \setminus E}$$

takes the value 1. Note that E is of finite perimeter in Ω if and only if $u_E \in BV(\Omega; \{-1, 1\})$. We may now define the potential v_E associated with E as

$$v_E(z) := \int_\Omega G_\Omega(z, z_2) u_E(z_2) \, dz_2 ,$$

where G_Ω is the Green functions introduced in (1.2). By the properties of the Green function, v_E is the unique solution to

$$\begin{cases} -\Delta v_E = u_E - m & \text{in } \Omega, \\ \nabla v_E \cdot n_\Omega = 0 & \text{on } \partial\Omega, \\ \displaystyle\int_\Omega v_E \, dz = 0 , \end{cases} \tag{2.2}$$

where $m := \fint_\Omega u_E\,dz = 2|E|/|\Omega| - 1$. We can now rewrite the double integral term in (1.1) (for $u = u_E$) in the following way:

$$\int_\Omega\int_\Omega G_\Omega(z_1, z_2)u_E(z_1)u_E(z_2)\,dz_1dz_2 = \int_\Omega v_E\,u_E\,dz$$
$$= \int_\Omega v_E(u_E - m)\,dz = \int_\Omega |\nabla v_E|^2\,dz\,,$$

where the last equality follows immediately from the weak formulation of (2.2). Motivated by the above observations, for every subset $E \subset \Omega$ we set

$$J(E) := P_\Omega(E) + \gamma \int_\Omega |\nabla v_E|^2\,dz\,,$$

where γ is a fixed positive constant. It is now clear that (1.1) is equivalent to

$$\min\{J(E) :\ E \subset \Omega,\ |E| = d\}\,, \tag{2.3}$$

where $d := (m + 1)|\Omega|/2$.

Remark 2.2. It is not difficult to check that

$$\int_\Omega |\nabla v_E|^2 dz = \sup\left\{\left(\int_\Omega \varphi(u_E - m)\,dz\right)^2 : \varphi \in H^1(\Omega),\ \|\nabla\varphi\|_{L^2(\Omega)} = 1\right\}.$$

Thus, the nonlocal term in the functional J may be regarded as the square of H^{-1}-norm of $u_E - m$.

The following remark will be used several times throughout the paper.

Remark 2.3. Notice that if Ω is of class C^2, then by standard elliptic regularity $v_E \in W^{2,p}(\Omega)$ for all $p \in [1, +\infty)$. More precisely, given $p > 1$, there exists a constant $C = C(p, \Omega)$ such that

$$\|v_E\|_{W^{2,p}(\Omega)} \leq C\|u_E - m\|_{L^p(\Omega)} \leq C(2|\Omega|)^{1/p} \qquad \text{for all } E \subset \Omega.$$

The above estimate implies in particular that the nonlocal term is continuous with respect to the L^1-convergence of sets. This observation, the fact that the perimeter functional P_Ω is lower semicontinuous with respect to the L^1-convergence of sets, and Theorem 2.1 yield the existence of a global minimizer to (2.3), through a standard application of the Direct Method of the Calculus of Variations.

We now briefly describe the periodic setting. To this end, we recall that the (unit) flat torus $\mathbb{T}^N$ is the quotient of $\mathbb{R}^N$ under the equivalence relation $x \sim y \iff x - y \in \mathbb{Z}^N$. Thus, the functional space $W^{k,p}(\mathbb{T}^N)$, $k \in \mathbb{N}$, $p \geq 1$, can be identified with the subspace of $W^{k,p}_{\rm loc}(\mathbb{R}^N)$ of functions that are one-periodic with respect to all coordinate directions. Similarly, $C^{k,\alpha}(\mathbb{T}^N)$, $\alpha \in (0,1)$ may be identified with the space of one-periodic functions in $C^{k,\alpha}(\mathbb{R}^N)$. Finally, a set $E \subset \mathbb{T}^N$ will said of class C^k (or smooth) if its one-periodic extension to $\mathbb{R}^N$ is of class C^k (or smooth).

In the periodic setting, the perimeter of a set $E \subset \mathbb{T}^N$ is defined as

$$P_{\mathbb{T}^N}(E) := \sup\left\{\int_E \operatorname{div}\varphi \, dz : \varphi \in C^1(\mathbb{T}^N; \mathbb{R}^N), \|\varphi\|_\infty \le 1\right\}.$$

The generalized Divergence Theorem for sets of finite perimeter in $\mathbb{T}^N$ now reads

$$\int_E \operatorname{div}\varphi \, dz = \int_{\partial^* E} \varphi \cdot n_E \, d\mathcal{H}^{N-1} \qquad \text{for all } \varphi \in C^1\left(\mathbb{T}^N; \mathbb{R}^N\right).$$

All the structure properties seen before and the compactness property stated in Theorem 2.1 continue to hold.

Given $E \subset \Omega$, we consider the associated potential

$$v_E(z) := \int_{\mathbb{T}^N} G_{\mathbb{T}^N}(z, z_2) u_E(z_2) \, dz_2 , \tag{2.4}$$

where $u_E := \chi_E - \chi_{\mathbb{T}^N \setminus E}$ and $G_{\mathbb{T}^N}$ is the Laplacian's Green function in the torus; that is, for $z_1 \in \mathbb{T}^N$, $G_{\mathbb{T}^N}(z_1, \cdot)$ is the unique solution to

$$\begin{cases} -\Delta_{z_2} G_{\mathbb{T}^N}(z_1, \cdot) = \delta_{z_1} - 1 & \text{in } \mathbb{T}^N, \\ \int_{\mathbb{T}^N} G_{\mathbb{T}^N}(z_1, z_2) \, dz_2 = 0 . \end{cases}$$

The function v_E satisfies (2.2) in $\mathbb{T}^N$, with periodic boundary conditions in place of the homogeneous Neumann ones. Clearly, the elliptic estimates stated in Remark 2.3 are still true in the periodic case.

Finally, for every set $E \subset \mathbb{T}^N$ we denote

$$J(E) := P_{\mathbb{T}^N}(E) + \gamma \int_{\mathbb{T}^N} |\nabla v_E|^2 \, dz$$

and consider the corresponding *periodic nonlocal isoperimetric problem*

$$\min\left\{J(E) : E \subset \mathbb{T}^N, \ |E| = d\right\} ,$$

where $d \in (0,1)$ is the given volume constraint.

We are ready to define and compute the FIRST VARIATION of the nonlocal energy functional J with respect to volume-preserving variations. The procedure is very classical in the context of minimal surfaces. We need the following definition.

Definition 2.4. (i) (NEUMANN CASE). Let $\Omega \subset \mathbb{R}^N$ be a bounded smooth set and let $E \subset \Omega$ be *relatively smooth*; that is, $E = \hat{E} \cap \Omega$ for some smooth (C^∞) set $\hat{E} \subset \mathbb{R}^N$. We say that a vector-field $X \in C^\infty(\overline{\Omega}; \mathbb{R}^N)$ is admissible for E if $X_{|\partial\Omega}$ is tangential to $\partial\Omega$, that is,

$$X \cdot n_\Omega = 0 \quad \text{on } \partial\Omega,$$

and the associated flow $\Phi : (-1, 1) \times \Omega \to \mathbb{R}^N$, defined by

$$\begin{cases} \dfrac{\partial \Phi}{\partial t} = X(\Phi) \\ \Phi(0, z) = z \end{cases} \tag{2.5}$$

satisfies

$$|E_t| = |E| \qquad \text{for all } t \in (0, 1), \tag{2.6}$$

where we set $E_t := \Phi(t, E)$.

(ii) (PERIODIC CASE). Let $E \subset \mathbb{T}^N$ be a smooth set. We say that a vector-field $X \in C^\infty(\mathbb{T}^N; \mathbb{R}^N)$ is admissible for E if the associated flow defined by (2.5) satisfies (2.6).

Given an admissible field X and the associated Φ, the *first variation of J at E with respect to the flow Φ* is nothing but

$$\frac{d}{dt} J(E_t)\Big|_{t=0}.$$

Before computing the first variation, we need a bit of notation: Given a vector X, its tangential part on some smooth $(N-1)$-manifold $\mathcal{M}$ is defined as $X_\tau := X - (X \cdot n)n$, with n being a unit normal vector to $\mathcal{M}$. In particular, we will denote by D_τ the tangential gradient operator given by $D_\tau \varphi := (D\varphi)_\tau$. Finally $\operatorname{div}_\tau X$ will stand for the *tangential divergence* of X on $\mathcal{M}$ defined as $\operatorname{div}_\tau X := \operatorname{div} X - \partial_n X \cdot n$.

Theorem 2.5 (First variation of J). (i) (NEUMANN CASE). *Let Ω, E, and X be as in Definition* 2.4-(i). *Then, setting $\mathcal{F} := \overline{\partial E \cap \Omega} \cap \partial\Omega$,*

$$\begin{aligned} \frac{d}{dt} J(E_t)\Big|_{t=0} &= \int_{\partial E \cap \Omega} (H_{\partial E} + 4\gamma v_E) X \cdot n_E \, d\mathcal{H}^{N-1} \\ &\quad + \int_{\mathcal{F}} X \cdot \eta \, d\mathcal{H}^{N-2}, \end{aligned} \tag{2.7}$$

where $H_{\partial E}$ denotes the sum of the principal curvatures of ∂E and coincides with the tangential divergence $\operatorname{div}_\tau n_E$ *of n_E, and η stands for the* outward co-normal *of $\partial E \cap \Omega$ on $\mathcal{F}$.*

(ii) (PERIODIC CASE). *Let E and X be as in Definition* 2.4-(ii). *Then,*

$$\frac{d}{dt}J(E_t)\Big|_{t=0} = \int_{\partial E}(H_{\partial E} + 4\gamma v_E)X \cdot n_E \, d\mathcal{H}^{N-1}.$$

Proof. We only deal with the Neumann case, the other one being analogous. We start by computing the derivative of the perimeter. Note that by the AREA FORMULA (see for instance [4]) we have

$$P_\Omega(E_t) = \int_{\partial E \cap \Omega} J^{N-1}\Phi(t,z)\, d\mathcal{H}^{N-1}(z),$$

where $J^{N-1}\Phi(t,\cdot)$ denotes the $(N-1)$-Jacobian of $\Phi(t,\cdot)$ and it is given by

$$J^{N-1}\Phi(t,\cdot) := \sqrt{\det\big(D_\tau^T\Phi(t,\cdot)D_\tau\Phi(t,\cdot)\big)}.$$

Recall now that for any $(N-1)\times(N-1)$ matrix A we have the expansion

$$\det(I_{N-1} + tA) = 1 + t\,\mathrm{trace}(A) + o(t), \tag{2.8}$$

with I_{N-1} denoting the $(N-1)\times(N-1)$ identity matrix. Moreover, from (2.5) we can expand Φ and $D\Phi$ in t to get $D\Phi = I_N + tDX + o(t)$; in turn, $D_\tau\Phi = I_{N-1} + tD_\tau X + o(t)$ and thus

$$D_\tau^T\Phi\, D_\tau\Phi = I_{N-1} + t\,(D_\tau^T X + D_\tau X) + o(t).$$

Using (2.8), we deduce

$$\det\big(D_\tau^T\Phi\, D_\tau\Phi\big) = 1 + 2t\,\mathrm{div}_\tau X + o(t)$$

and finally

$$J^{N-1}\Phi = \sqrt{1 + 2t\,\mathrm{div}_\tau X + o(t)} = 1 + t\,\mathrm{div}_\tau X + o(t).$$

Thus,

$$\begin{aligned}\frac{d}{dt}P_\Omega(E_t)\Big|_{t=0} &= \int_{\partial E\cap\Omega}\frac{d}{dt}J^{N-1}\Phi(t,z)\Big|_{t=0}\, d\mathcal{H}^{N-1}(z)\\ &= \int_{\partial E\cap\Omega}\mathrm{div}_\tau X\, d\mathcal{H}^{N-1} \\ &= \int_{\partial E\cap\Omega} H_{\partial E}\, X\cdot n_E\, d\mathcal{H}^{N-1} + \int_{\mathcal{F}} X\cdot\eta\, d\mathcal{H}^{N-2},\end{aligned} \tag{2.9}$$

where the last equality follows from the Divergence Theorem on Manifolds (see for instance [43, Equation (7.6)]).

In order to compute the t-derivative of the nonlocal term, we set for notational convenience

$$v(t,\cdot) := v_{E_t}(\cdot) = \int_\Omega G_\Omega(\cdot,z)u_{E_t}(z)\,dz = \int_{E_t} G_\Omega(\cdot,z)\,dz - \int_{E_t^c} G_\Omega(\cdot,z)\,dz\,.$$

Then,

$$\begin{aligned}\frac{d}{dt}\left(\int_\Omega |\nabla v_{E_t}|^2\,dz\right)\Big|_{t=0} &= \frac{d}{dt}\left(\int_\Omega |\nabla_z v(t,z)|^2\,dz\right)\Big|_{t=0}\\ &= 2\int_\Omega v_E(z)\frac{\partial}{\partial t}(\nabla_z v(t,z))\Big|_{t=0}\,dz\\ &= 2\int_\Omega (u_E - m)\frac{\partial}{\partial t}v\Big|_{t=0}\,dz\,,\end{aligned}\tag{2.10}$$

where the last equality follows by integration by parts, recalling that $\Delta v_E = u_E - m$ subject to homogeneous Neumann conditions. Write now

$$\frac{\partial}{\partial t}v\Big|_{t=0} = \frac{\partial}{\partial t}\left(\int_{E_t} G_\Omega(\cdot,z)\,dz\right)\Big|_{t=0} - \frac{\partial}{\partial t}\left(\int_{E_t^c} G_\Omega(\cdot,z)\,dz\right)\Big|_{t=0}\,.\tag{2.11}$$

By a change of variable

$$\frac{\partial}{\partial t}\left(\int_{E_t} G_\Omega(\cdot,z)\,dz\right)\Big|_{t=0} = \frac{\partial}{\partial t}\left(\int_E G_\Omega(\cdot,\Phi(t,z))J\Phi(t,z)\,dz\right)\Big|_{t=0}\,,$$

where, using (2.8) with N in place of $N-1$,

$$J\Phi = \det D\Phi = \det(I_N + tDX + o(t)) = 1 + t\,\mathrm{div}X + o(t)\,.$$

Thus,

$$\begin{aligned}\frac{\partial}{\partial t}\left(\int_{E_t} G_\Omega(\cdot,z)dz\right)\Big|_{t=0} &= \frac{\partial}{\partial t}\left(\int_E G_\Omega(\cdot,\Phi(t,z))J\Phi(t,z)\,dz\right)\Big|_{t=0}\\ &= \int_E \big(\nabla_z G_\Omega(\cdot,z)\cdot X(z) + G_\Omega(\cdot,z)\mathrm{div}X(z)\big)dz\\ &= \int_E \mathrm{div}_z\big(G_\Omega(\cdot,z)X(z)\big)\,dz\\ &= \int_{\partial E\cap\Omega} G_\Omega(\cdot,z)X(z)\cdot n_E(z)\,d\mathcal{H}^{N-1}(z)\,,\end{aligned}\tag{2.12}$$

where the last equality follows from an application of the Divergence Theorem together with the fact that $X \cdot n_\Omega = 0$ on $\partial\Omega$. By similar computations we also get

$$-\frac{\partial}{\partial t}\left(\int_{E_t^c} G_\Omega(\cdot, z)\,dz\right)\Big|_{t=0} = \int_{\partial E\cap\Omega} G_\Omega(\cdot, z)X(z)\cdot n_E(z)\,d\mathcal{H}^{N-1}(z)\,. \quad (2.13)$$

Combining (2.10)–(2.13), we obtain

$$\begin{aligned}
&\frac{d}{dt}\left(\int_\Omega |\nabla v_{E_t}|^2\,dz\right)\Big|_{t=0}\\
&= 4\int_\Omega (u_E(z_1)-m)\left(\int_{\partial E\cap\Omega} G_\Omega(z_1, z_2)X(z_2)\cdot n_E(z_2)\,d\mathcal{H}^{N-1}(z_2)\right)dz_1\\
&= 4\int_{\partial E\cap\Omega}\underbrace{\left(\int_\Omega G_\Omega(z_1, z_2)(u_E(z_1)-m)dz_1\right)}_{=v_E(z_2)} X(z_2)\cdot n_E(z_2)\,d\mathcal{H}^{N-1}(z_2)\\
&= 4\int_{\partial E\cap\Omega} v_E\,X\cdot n_E\,d\mathcal{H}^{N-1}\,. \qquad (2.14)
\end{aligned}$$

Combining (2.9) and (2.14), we finally obtain (2.7). □

It is clear that every smooth *volume-constrained (local) minimizer* set E will satisfy the CRITICALITY CONDITION

$$\begin{aligned}
&\int_{\partial E\cap\Omega} (H_{\partial E} + 4\gamma v_E)X\cdot n_E\,d\mathcal{H}^{N-1}\\
&\qquad + \int_{\mathcal{F}} X\cdot\eta\,d\mathcal{H}^{N-2} = 0 \text{ for all admissible } X.
\end{aligned} \qquad (2.15)$$

Note now that if X is admissible, then $|E_t| = |E|$ for all $t \in [0, 1]$ and thus, arguing as in the proof of Theorem 2.5,

$$0 = \frac{d}{dt}|E_t|\Big|_{t=0} = \int_E \frac{d}{dt}J\Phi\Big|_{t=0} = \int_E \operatorname{div}X\,dx = \int_{\partial E\cap\Omega} X\cdot n_E\,d\mathcal{H}^{N-1}\,,$$

that is, the normal component $X\cdot n_E$ has zero average on $\partial E\cap\Omega$. In fact, it is possible to prove (see [1] and [27] for the details) that the set

$$\Big\{(X\cdot n_E)_{|\partial E\cap\Omega} : X \text{ admissible in the sense of Definition 2.4}$$
$$\text{and supp } X \Subset \partial E\cap\Omega\Big\}$$

is dense in

$$\left\{\varphi \in L^2(\partial E \cap \Omega) : \int_{\partial E\cap\Omega} \varphi \, d\mathcal{H}^{N-1} = 0\right\}.$$

Thus, (2.15) implies

$$\int_{\partial E\cap\Omega} (H_{\partial E} + 4\gamma v_E)\varphi \, d\mathcal{H}^{N-1} = 0$$
$$\text{for all } \varphi \in L^2(\partial E \cap \Omega) \text{ s.t. } \int_{\partial E\cap\Omega} \varphi \, d\mathcal{H}^{N-1} = 0,$$

which is equivalent to saying that there exists a constant $\lambda \in \mathbb{R}$ such that

$$H_{\partial E} + 4\gamma v_E = \lambda \qquad \text{on } \partial E \cap \Omega. \tag{2.16}$$

In turn, if $\mathcal{F} \neq \emptyset$, by (2.15) we have

$$\int_{\mathcal{F}} X \cdot \eta \, d\mathcal{H}^{N-2} = 0 \qquad \text{for all admissible } X,$$

from which, recalling that $X \cdot n_\Omega = 0$ on $\partial\Omega$, one can deduce that $\eta = n_\Omega$ on $\mathcal{T}$; that is,

$$\partial E \cap \Omega \text{ and } \partial\Omega \text{ meet orthogonally.} \tag{2.17}$$

Summarizing, we have discovered that (2.15) is equivalent to (2.16) if $\mathcal{F} = \emptyset$, and to (2.16)-(2.17) otherwise.

Remark 2.6. Note that λ may be interpreted as a *Lagrange multiplier* associated with the volume constraint. When $\gamma = 0$ we recover the *classical constant mean curvature condition* for minimal surfaces.

The above discussion motivates the following definition.

Definition 2.7 (smooth critical sets). (i) (NEUMANN CASE). Let $\Omega \subset \mathbb{R}^N$ be a bounded smooth set. We say that $E \subset \Omega$ is a *smooth critical set* if E is relatively smooth (see Definition 2.4), (2.16) holds and either $\partial E \cap \Omega \Subset \Omega$ or $\partial E \cap \Omega$ meets $\partial\Omega$ orthogonally.
(ii) (PERIODIC CASE). We say that $E \subset \mathbb{T}^N$ is a smooth critical set if ∂E is smooth and (2.16) holds.

We now consider the second variation of J. Given a relatively smooth subset $E \subset \Omega$ ($E \subset \mathbb{T}^N$ in the periodic case) and an admissible vector-field X (in the sense of Definition 2.4), the *second variation of J at E with respect to the associated flow* Φ is nothing but

$$\frac{d^2}{dt^2} J(E_t)\Big|_{t=0}.$$

The derivation of the explicit second variation formula is in the same spirit of Theorem 2.5, although the computations are more involved. Here we only state the final result of such a derivation, referring the interested reader to the literature for the details. Before giving the precise statement, we fix a bit of notation: Given a (relatively) smooth subset $E \subset \Omega$, the second fundamental form $B_{\partial E}$, defined on $\partial E \cap \Omega$, is given by $D_\tau n_E$. The square $|B_{\partial E}|^2$ of its Euclidean norm coincides with the sum of the squares of the principal curvatures of $\partial E \cap \Omega$.

Theorem 2.8. (i) (NEUMANN CASE). *Let Ω, E, X, and $\mathcal{F}$ be as in Theorem* 2.5-(i). *Then,*

$$
\begin{aligned}
&\frac{d^2}{dt^2} J(E_t)\Big|_{t=0} \\
&= \int_{\partial E \cap \Omega} \Big(|D_\tau (X \cdot n_E)|^2 - |B_{\partial E}|^2 (X \cdot n_E)^2 \Big)\, d\mathcal{H}^{N-1} \\
&\quad - \int_{\mathcal{F}} B_{\partial E}(n_E, n_E)(X \cdot n_E)^2\, d\mathcal{H}^{N-2} \\
&\quad + 8\gamma \int_{\partial E \cap \Omega} \int_{\partial E \cap \Omega} G_\Omega(z_1, z_2)(X \cdot n_E)(z_1)(X \cdot n_E)(z_2)\, d\mathcal{H}^{N-1} \\
&\quad \times (z_1)\, d\mathcal{H}^{N-1}(z_2) \\
&\quad + 4\gamma \int_{\partial E \cap \Omega} \nabla v_E \cdot n_E (X \cdot n_E)^2\, d\mathcal{H}^{N-1} - R\,,
\end{aligned}
\tag{2.18}
$$

where the remainder R is defined as

$$
\begin{aligned}
R := &- \int_{\partial E \cap \Omega} (4\gamma v_E + H_{\partial E}) \mathrm{div}_\tau \big(X_\tau (X \cdot n_E) \big)\, d\mathcal{H}^{N-1} \\
&+ \int_{\partial E \cap \Omega} (4\gamma v_E + H_{\partial E})(\mathrm{div} X)(X \cdot n_E)\, d\mathcal{H}^{N-1}\,.
\end{aligned}
\tag{2.19}
$$

(ii) (PERIODIC CASE). *Let E and X be as in Theorem* 2.5-(ii). *Then*

$$
\begin{aligned}
&\frac{d^2}{dt^2} J(E_t)\Big|_{t=0} \\
&= \int_{\partial E} \Big(|D_\tau (X \cdot n_E)|^2 - |B_{\partial E}|^2 (X \cdot n_E)^2 \Big)\, d\mathcal{H}^{N-1} \\
&\quad + 8\gamma \int_{\partial E} \int_{\partial E \cap \Omega} G_{\mathbb{T}^N}(z_1, z_2)(X \cdot n_E)(z_1)(X \cdot n_E)(z_2)\, d\mathcal{H}^{N-1} \\
&\quad \times (z_1)\, d\mathcal{H}^{N-1}(z_2) \\
&\quad + 4\gamma \int_{\partial E} \nabla v_E \cdot n_E (X \cdot n_E)^2\, d\mathcal{H}^{N-1} + R\,,
\end{aligned}
\tag{2.20}
$$

where the remainder R is defined as in (2.19), *with ∂E in place of $\partial E \cap \Omega$.*

Remark 2.9. (i) The above second variation formulas were derived in [11] when E is a critical set. Note that in this case the remainder term R vanishes (see below). The expression of R was found in [1], by revisiting the Choksi-Sternberg derivation in the case of non critical sets. Both periodic and Neumann boundary conditions are considered in [1], but the latter only under the additional assumption that $\partial E \cap \Omega$ does not meet $\partial\Omega$. The general Neumann case has been treated in [27].

(ii) Note that the remainder R vanishes at critical sets. Indeed, using (2.16) (and (2.17) in the Neumann case), we have

$$
\begin{aligned}
R = -\lambda \int_{\partial E \cap \Omega} & \operatorname{div}_\tau \big(X_\tau (X \cdot n_E)\big)\, d\mathcal{H}^{N-1} \\
+ \lambda \int_{\partial E \cap \Omega} & (\operatorname{div} X)(X \cdot n_E)\, d\mathcal{H}^{N-1} = 0\,,
\end{aligned}
$$

thanks to the Divergence Theorem on Manifolds and the fact that

$$
0 = \frac{d^2}{dt^2} |E_t| \Big|_{t=0} = \int_{\partial E \cap \Omega} (\operatorname{div} X)(X \cdot n_E)\, d\mathcal{H}^{N-1}\,. \tag{2.21}
$$

For a proof of the last equality see for instance [11, Equation (2.30)].

In view of the previous remark, the second variation formula at a smooth critical set E is a quadratic form on the normal component $X \cdot n_E$ only. For this reason, we denote

$$
\widetilde{H}(\partial E \cap \Omega) := \left\{ \varphi \in H^1(\partial E \cap \Omega) : \int_{\partial E \cap \Omega} \varphi\, \mathcal{H}^{N-1} = 0 \right\}, \tag{2.22}
$$

and consider the quadratic form $\partial^2 J(E) : \widetilde{H}(\partial E \cap \Omega) \to \mathbb{R}$ defined as

$$
\begin{aligned}
\partial^2 J(E)[\varphi] = & \int_{\partial E \cap \Omega} \Big(|D_\tau \varphi|^2 - |B_{\partial E}|^2 \varphi^2 \Big)\, d\mathcal{H}^{N-1} \\
& - \int_{\mathcal{F}} B_{\partial E}(n_E, n_E) \varphi^2\, d\mathcal{H}^{N-2} \\
& + 8\gamma \int_{\partial E \cap \Omega} \int_{\partial E \cap \Omega} G_\Omega(z_1, z_2) \varphi(z_1) \varphi(z_2)\, d\mathcal{H}^{N-1}(z_1)\, d\mathcal{H}^{N-1}(z_2) \\
& + 4\gamma \int_{\partial E \cap \Omega} \nabla v_E \cdot n_E\, \varphi^2\, d\mathcal{H}^{N-1}\,.
\end{aligned} \tag{2.23}
$$

In the periodic case, we define $\widetilde{H}(\partial E)$ as in (2.22) with $\partial E \cap \Omega$ replaced by ∂E, and consider the quadratic form

$$\partial^2 J(E)[\varphi] = \int_{\partial E} \left(|D_\tau \varphi|^2 - |B_{\partial E}|^2 \varphi^2 \right) d\mathcal{H}^{N-1} \\ + 8\gamma \int_{\partial E} \int_{\partial E} G_{\mathbb{T}^N}(z_1, z_2)\varphi(z_1)\varphi(z_2)\, d\mathcal{H}^{N-1}(z_1)\, d\mathcal{H}^{N-1}(z_2) \quad (2.24) \\ + 4\gamma \int_{\partial E} \nabla v_E \cdot n_E\, \varphi^2\, d\mathcal{H}^{N-1}$$

again defined for all $\varphi \in \widetilde{H}(\partial E)$.

Remark 2.10. (i) Note that when $\gamma = 0$ the above quadratic forms reduce to the standard second variation of the Area Functional.
(ii) Setting for $\varphi \in \widetilde{H}(\partial E \cap \Omega)$

$$v_\varphi(\cdot) := \int_{\partial E \cap \Omega} G_\Omega(\cdot, z)\varphi(z)\, d\mathcal{H}^{N-1}(z)$$

and $\mu_\varphi := \varphi\, \mathcal{H}^{N-1} \llcorner (\partial E \cap \Omega)$, by the properties of the Green's function (see for instance [31, Chapter 18]) we have that v_φ satisfies

$$\begin{cases} -\Delta v_\varphi = \mu_\varphi & \text{in } \Omega, \\ \partial_n v_\varphi = 0 & \text{on } \partial\Omega, \end{cases}$$

or, equivalently,

$$\int_\Omega \nabla v_\varphi \cdot \nabla \psi\, dz = \int_{\partial E \cap \Omega} \varphi\, \psi\, d\mathcal{H}^{N-1} \qquad \text{for all } \psi \in H^1(\Omega). \quad (2.25)$$

Thus,

$$\int_{\partial E \cap \Omega} \int_{\partial E \cap \Omega} G_\Omega(z_1, z_2)\varphi(z_1)\varphi(z_2)\, d\mathcal{H}^{N-1}(z_1)\, d\mathcal{H}^{N-1}(z_2) \\ = \int_{\partial E \cap \Omega} \varphi\, v_\varphi\, d\mathcal{H}^{N-1} = \int_\Omega |\nabla v_\varphi|^2\, dz\,,$$

where the last equality follows from (2.25). Therefore, the double integral term in (2.23) is always nonnegative. The same holds for the double integral term in (2.24).

We conclude this section by noticing that if E is a (relatively smooth) volume constrained local minimizer of J, then for any admissible vectorfield X (and the associated flow Φ) we clearly have

$$\frac{d^2}{dt^2} J(E_t)\Big|_{t=0} = \partial^2 J(E)[X \cdot n_E] \geq 0\,.$$

By a density argument (see [1] or [27] for details), we infer that

$$\partial^2 J(E)[\varphi] \geq 0 \qquad \text{for all } \varphi \in \widetilde{H}(\partial E \cap \Omega). \tag{2.26}$$

Thus, the non-negative definiteness condition (2.26) is necessary for the (local) minimality of E. Clearly, the same holds in the periodic setting.

3 Strict stability and $W^{2,p}$-local minimality

In the previous section we have established the stability condition (2.26) as a necessary condition for local minimality. In this section we start investigating to what extent the *strict stability*, namely the positive definiteness of the quadratic form in (2.26) at a critical set, is a *sufficient condition* for minimality. In order to give the precise definition of stability in the periodic setting, we have to take into account that the functional is translation invariant. Therefore, if one considers the (admissible) flow $\Phi(t, z) = z + t\,\eta, \eta \in \mathbb{R}^N$, then $E_t = E + t\eta$ and thus $J(E_t) = J(E)$. It follows that

$$0 = \frac{d^2}{dt^2} J(E_t)\Big|_{t=0} = \partial^2 J(E)[\eta \cdot n_E] \qquad \text{for all } \eta \in \mathbb{R}^N.$$

We conclude that the quadratic form $\partial^2 J(E)$ always vanishes on the finite dimensional subspace $T(\partial E) \subset \widetilde{H}(\partial E)$ (see (2.22)) defined by

$$T(\partial E) := \left\{ \eta \cdot n_E \,:\, \eta \in \mathbb{R}^N \right\}.$$

Thus, strict stability can only hold in the orthogonal space

$$T^{\perp}(\partial E) := \left\{ \varphi \in \widetilde{H}(\partial E) \,:\, \int_{\partial E} \varphi\, n_E \, d\mathcal{H}^{N-1} = 0 \right\}. \tag{3.1}$$

The above observation motivates the second part of the following definition.

Definition 3.1 (Strictly stable smooth critical sets).
(i) (NEUMANN CASE.) Let $\Omega \subset \mathbb{R}^N$ be a bounded smooth set and let $E \subset \Omega$ be a smooth critical set in the sense of Definitiion 2.7. We say that E is *strictly stable* if

$$\partial^2 J(E)[\varphi] > 0 \qquad \text{for all } \varphi \in \widetilde{H}(\partial E \cap \Omega) \setminus \{0\}.$$

(ii) (PERIODIC CASE.) Let $E \subset \mathbb{T}^N$ be a smooth critical set in the sense of Definitiion 2.7. We say that E is strictly stable if

$$\partial^2 J(E)[\varphi] > 0 \qquad \text{for all } \varphi \in T^{\perp}(\partial E) \setminus \{0\},$$

where $T^{\perp}$ is the subspace defined in (3.1).

In order to state the stability results in the periodic setting, we need again to take into account the translation invariance of the energy and consider a suitable L^1-distance on the quotient of $L^1(\mathbb{T}^N)$ with respect to the relation

$$f \sim g \iff f(\cdot) = g(\cdot + \eta) \quad \text{for some } \eta \in \mathbb{T}^N. \tag{3.2}$$

Restricted to sets such a distance, denoted by α, reduces to

$$\alpha(E, F) := \min_{\eta \in \mathbb{T}^N} |E \Delta (\eta + F)|. \tag{3.3}$$

Theorem 3.2 (Strict stability implies $W^{2,p}$-minimality).
Fix $p > \max\{N-1, 2\}$.
(i) (NEUMANN CASE.) *Let* $\Omega \subset \mathbb{R}^N$ *be a bounded smooth set and let* $E \subseteq \Omega$ *be a* smooth strictly stable critical set *in the sense of Definition* 3.1-(i). *Then, there exists* $\delta_1 > 0$, $C_1 > 0$ *such that:*

$$J(F) \geq J(E) + C_1 |E \Delta F|^2$$

whenever $|F| = |E|$ *and* $F = \Phi(E)$ *for some diffeomorphism* $\Phi : \Omega \to \Omega$ *of class* $W^{2,p}$, *with* $\|\Phi - Id\|_{W^{2,p}} \leq \delta_1$. *Here and in the following* Id *denotes the identity mapping.*
(ii) (PERIODIC CASE.) *Let* $E \subset \mathbb{T}^N$ *be a smooth strictly stable critical set in the sense of Definition* 3.1-(ii). *Then, there exist* $\delta_1 > 0$ *and* $C_1 > 0$ *with the following properties:*

$$J(F) \geq J(E) + C_1 \alpha(E, F)^2 \text{whenever } \partial F = \{z + \psi(z) n_E(z) : z \in \partial E\}, \\ \text{with } \|\psi\|_{W^{2,p}(\partial E)} \leq \delta_1, \text{ and } |F| = |E|.$$

We will prove only part (i) of the statement, under the additional assumption

$$\partial E \cap \Omega \Subset \Omega. \tag{3.4}$$

By exchanging E with E^c if necessary, we may in fact reduce (3.4) to

$$\partial E \Subset \Omega. \tag{3.5}$$

The proof of (ii) is similar but technically more involved since one has to deal with the translation invariance of the nonlocal energy; for the details see [1]. Part (i) of the theorem under (3.5) and part (ii) were established in [1], while the general Neumann case has been treated in [27].

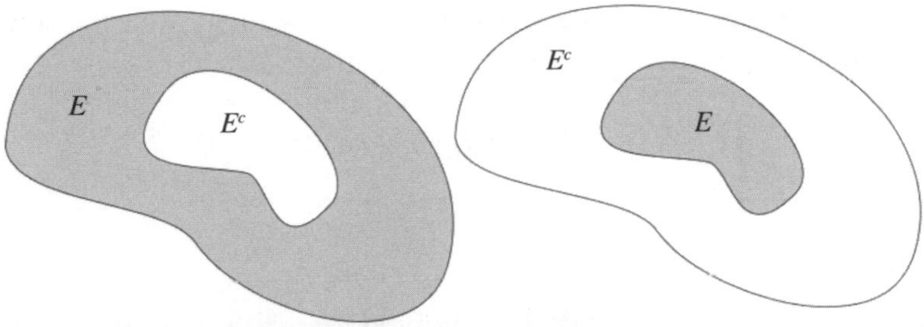

Figure 3.1. Two cases where the additional assumption (3.4) holds.

Remark 3.3. Observe that under (3.5) an application of the Implicit Function Theorem shows that the conclusion of part (i) is equivalent to:

$$J(F) \geq J(E) + C_1|E\Delta F|^2 \text{ whenever } \partial F = \{x + \psi(x)n_E(x) : x \in \partial E\},$$
$$\text{with } \|\psi\|_{W^{2,p}(\partial E)} \leq \delta_1, \text{ and } |F| = |E| \quad (3.6)$$

(for a possibly smaller δ_1).

Our next aim is to provide a way of connecting two sets that are sufficiently $W^{2,p}$-close through a suitable volume preserving flow. Given $E \subset \Omega$ smooth and satisfying (3.5), we denote by $\mathcal{N}_\varepsilon$ the tubular neighborhood of ∂E of thickness 2ε. In the following proposition we fix $\varepsilon_0 > 0$ so small that the signed distance d_E from E and the projection π_E on ∂E are smooth in $\mathcal{N}_{\varepsilon_0}$.

Proposition 3.4. *Let $E \subset \Omega$ be a smooth set satisfying (3.5), fix ε_0 as above and let $p > N-1$. There exist $\varepsilon_1 \in (0, \varepsilon_0)$ and a constant $C > 0$ with the following property: If $\psi \in C^\infty(\partial E)$ and $\|\psi\|_{W^{2,p}(\partial E)} \leq \varepsilon_1$ then there exists a field $X \in C^\infty$ with $\operatorname{div} X = 0$ in $\mathcal{N}_{\varepsilon_0}$ such that its flow*

$$\Phi(0, z) = z\,, \qquad \frac{\partial \Phi}{\partial t} = X(\Phi) \tag{3.7}$$

satisfies $\Phi(1, z) = z + \psi(z)n_E(z)$ for $z \in \partial E$. Moreover, for every $t \in [0,1]$

$$\|\Phi(t,\cdot) - \mathrm{Id}\|_{W^{2,p}(\partial E)} \leq C\|\psi\|_{W^{2,p}(\partial E)}\,. \tag{3.8}$$

If in addition E_1 has the same volume as E, then for every $t \in [0,1]$ we have $|E_t| = |E|$ and

$$\int_{\partial E_t} X \cdot n_{E_t}\, d\mathcal{H} = 0\,.$$

Proof. We start by constructing a C^∞ vector field $\tilde{X} : \mathcal{N}_{\varepsilon_0} \to \mathbb{R}^N$ such that

$$\operatorname{div}\tilde{X} = 0 \quad \text{in } \mathcal{N}_{\varepsilon_0}, \qquad \tilde{X} = n_E \quad \text{on } \partial E. \tag{3.9}$$

To this aim, for every $z \in \partial E$ consider $f_z : (-\varepsilon_0, \varepsilon_0) \to \mathbb{R}$ the solution to the following Cauchy problem

$$\begin{cases} (f_z)'(t) + f_z(t)\Delta d_E(z + t n_E(z)) = 0 \\ f_z(0) = 1\,, \end{cases}$$

and set

$$\zeta(z + t n_E(z)) := f_z(t) = \exp\Big(-\int_0^t \Delta d_E(z + s n_E(z))\, ds\Big)\,.$$

The vector field $\tilde{X}$ can be taken of the form $\tilde{X} := \zeta \nabla d_E$. For every smooth ψ we have $\|\psi\|_{L^\infty(\partial E)} \leq C_E \|\psi\|_{W^{2,p}(\partial E)}$: take $\varepsilon_1 < \varepsilon_0 / C_E$ so that $z + \psi(z) n_E(z) \in \mathcal{N}_{\varepsilon_0}$ for all $z \in \partial E$, and let X be any C^∞ admissible vector field (in the sense of Definition 2.4) such that

$$X(z) := \left(\int_0^{\psi(\pi_E(z))} \frac{ds}{\zeta(\pi_E(z) + s n_E(\pi_E(z)))}\right) \zeta(z) \nabla d_E(z)\,, \quad \text{for every } z \in \mathcal{N}_{\varepsilon_0}. \tag{3.10}$$

Let Φ be the associated flow. Notice that the integral appearing in (3.10) represents the time needed to go from $\pi_E(z)$ to $\pi_E(z) + \psi(\pi_E(z)) n_E(\pi_E(z)))$ along the trajectory of the vector field $\tilde{X}$. Therefore the above definition ensures that the time needed to go from $\pi_E(z)$ to $\pi_E(z) + \psi(\pi_E(z)) n_E(\pi_E(z)))$ along the trajectory of the modified vector field X is always one. This is equivalent to saying that for all $z \in \partial E$ we have $\Phi(1, z) = z + \psi(z) n_E(z)$. Moreover, since $\mathrm{div}(\zeta \nabla_E d) = 0$ in $\mathcal{N}_{\varepsilon_0}$ and the function

$$z \mapsto \int_0^{\psi(\pi_E(z))} \frac{ds}{\zeta(\pi_E(z) + s n_E(\pi_E(z)))}$$

is constant along the trajectories of $\zeta \nabla d_E$, we deduce that also the modified field X is divergence free in the same tubular neighborhood.

Finally, note that from (3.10) it follows

$$\|X\|_{W^{2,p}(\mathcal{N}_{\varepsilon_0})} \leq C \|\psi\|_{W^{2,p}(\partial E)} \tag{3.11}$$

for a constant $C > 0$ depending on E. To establish (3.8), observe that the closeness of Φ to Id in L^∞ follows from (3.7) and (3.11). By differentiating (3.7) and solving the resulting equation, and since $p > N - 1$, one easily gets

$$\|\nabla_z \Phi - I\|_{L^\infty(\mathcal{N}_{\varepsilon_0})} \leq C(\varepsilon_0) \|\nabla X\|_{L^\infty(\mathcal{N}_{\varepsilon_0})} \leq C(\varepsilon_0) \|\psi\|_{W^{2,p}(\partial E)} \leq C(\varepsilon_0) \varepsilon_1\,.$$

In particular, if ε_1 is small enough the $(N-1)$-dimensional Jacobian of $\Phi(t,\cdot)$ on ∂E is uniformly close to 1. Using this information and by differentiating again (3.7), we have

$$\|\nabla_z^2\Phi(t,\cdot)\|_{L^p(\partial E)} \le C(\varepsilon_0)\|\nabla^2 X\|_{L^p(\mathcal{N}_{\varepsilon_0})},$$

whence (3.8) follows.

Assume now that $|E_1| = |E|$ and recall that by [11, Equation (2.30)]

$$\frac{d^2}{dt^2}|E_t| = \int_{\partial E_t} (\operatorname{div} X)(X\cdot n_{E_t}) = 0 \qquad \text{for all } t\in[0,1].$$

Hence the function $t\mapsto |E_t|$ is affine in $[0,1]$ and since $|E_0| = |E| = |E_1|$ it is constant. Therefore

$$0 = \frac{d}{dt}|E_t| = \int_{E_t} \operatorname{div} X\, dz = \int_{\partial E_t} X\cdot n_{E_t}\, d\mathcal{H} \qquad \text{for all } t\in[0,1].$$

This concludes the proof of the proposition. □

Sketch of the Proof of Theorem 3.2-(i). As already mentioned we assume (3.5) and we recall that under this assumption the conclusion of the theorem is equivalent to (3.6). Since all estimates will depend only on $\|\psi\|_{W^{2,p}(\partial E)}$, by an approximation argument we may assume that ψ is of class C^∞. We split the proof into three steps.

Step 1. We claim that

$$m_0 := \inf\Big\{\partial^2 J(E)[\varphi] : \varphi\in \widetilde{H}(\partial E),\, \|\varphi\|_{H^1(\partial E)} = 1\Big\} > 0. \qquad (3.12)$$

Indeed, let φ_h be a minimizing sequence for the infimum in (3.12) and assume that $\varphi_h \rightharpoonup \varphi_0 \in \widetilde{H}(\partial E)$ weakly in $H^1(\partial E)$. If $\varphi_0\neq 0$, by (2.23) it follows that

$$m_0 = \lim_h \partial^2 J(E)[\varphi_h] \ge \partial^2 J(E)[\varphi_0] > 0,$$

where the last inequality follows from the strict stability assumption. If $\varphi_0 = 0$, then

$$m_0 = \lim_h \partial^2 J(E)[\varphi_h] = \lim_h \int_{\partial E} |D_\tau\varphi_h|^2\, d\mathcal{H} = 1.$$

Step 2. We claim that there exists $\bar\delta > 0$ such that if $\partial F = \{z + \psi(z) n_E(z) : z\in\partial E\}$ with $|F| = |E|$ and $\|\psi\|_{W^{2,p}(\partial E)} \le \bar\delta$, then

$$\inf\Big\{\partial^2 J(F)[\varphi] : \varphi\in\widetilde{H}(\partial F),\, \|\varphi\|_{H^1(\partial F)} = 1\Big\} \ge \frac{m_0}{2}, \qquad (3.13)$$

where m_0 is defined in (3.12). We argue by contradiction assuming that there exist a sequence F_h, with $\partial F_h = \{z + \psi_h(z) n_E(z) \,:\, z \in \partial E\}$, $|F_h| = |E|$ and $\|\psi_h\|_{W^{2,p}(\partial E)} \to 0$, and a sequence $\varphi_h \in \widetilde{H}(\partial F_h)$, with $\|\varphi_h\|_{H^1(\partial F_h)} = 1$ and such that

$$\partial^2 J(F_h)[\varphi_h] < \frac{m_0}{2} \,. \tag{3.14}$$

The idea now is to regard φ_h as a function on ∂E. To this end, for $z \in \partial E$ set

$$\tilde{\varphi}_h(z) := \varphi_h\big(z + \psi_h(z) n_E(z)\big) - \fint_{\partial E} \varphi_h(z + \psi_h(z) n_E(z))\, d\mathcal{H}^{N-1}(z) \,.$$

Note that

$$\tilde{\varphi}_h \in \widetilde{H}(\partial E) \qquad \text{and} \qquad \|\tilde{\varphi}_h\|_{H^1(\partial E)} \to 1 \,, \tag{3.15}$$

due to the fact that $\psi_h \to 0$ in $W^{2,p}(\partial E)$. Moreover, the $W^{2,p}$ convergence of F_h to E and Remark 2.3 imply

$$\begin{aligned} &B_{\partial F_h}\big(\cdot + \psi_h(\cdot) n_E(\cdot)\big) \to B_{\partial E} \text{ in } L^p(\partial E), \\ &v_{F_h} \to v_E \text{ in } C^{1,\beta}\big(\overline{\Omega}\big) \text{ for all } \beta < 1. \end{aligned} \tag{3.16}$$

We now observe that

$$\begin{aligned} &\int_{\partial F_h}\int_{\partial F_h} G_\Omega(z_1, z_2)\varphi_h(z_1)\varphi_h(z_2)\, d\mathcal{H}^{N-1}(z_1) d\mathcal{H}^{N_1}(z_2) \\ &\quad - \int_{\partial E}\int_{\partial E} G_\Omega(z_1, z_2)\tilde{\varphi}_h(z_1)\tilde{\varphi}_h(z_2)\, d\mathcal{H}^{N-1}(z_1) d\mathcal{H}^{N_1}(z_2) \to 0 \end{aligned} \tag{3.17}$$

as $h \to \infty$. Indeed, thanks to Remark 2.10-(ii) this is equivalent to

$$\int_\Omega \big(|\nabla z_h|^2 - |\nabla \tilde{z}_h|^2\big)\, dz \to 0 \,, \tag{3.18}$$

where

$$-\Delta z_h = \mu_h := \varphi_h \mathcal{H} \lfloor \partial F_h \,, \qquad -\Delta \tilde{z}_h = \tilde{\mu}_h := \tilde{\varphi}_h \mathcal{H} \lfloor \partial E \,,$$

under homogeneous Neumann conditions. In fact, it is not difficult to show that

$$\mu_h - \tilde{\mu}_h \to 0 \quad \text{in } H^{-1}(\Omega), \tag{3.19}$$

we refer to [1, Proof of Theorem 3.9] for the details. In turn, the convergence (3.19) clearly implies (3.18). Finally, we observe that since $p > \max\{2, N-1\}$, the Sobolev Embedding theorem and the $W^{2,p}$-convergence of ∂F_h to ∂E imply

$$\int_{\partial F_h} |B_{\partial F_h}|^2 \varphi_h^2 \, d\mathcal{H} - \int_{\partial E} |B_{\partial E}|^2 \tilde{\varphi}_h^2 \, d\mathcal{H} \to 0 \,. \tag{3.20}$$

Collecting (3.16), (3.17), and (3.20) we may conclude that all terms in the expression (2.23) of $\partial^2 J(F_h)[\varphi_h]$ are asympotically close to the corresponding terms of $\partial^2 J(E)[\tilde{\varphi}_h]$; that is,

$$\partial^2 J(F_h)[\varphi_h] - \partial^2 J(E)[\tilde{\varphi}_h] \to 0 \,.$$

The above convergence, together with (3.14), contradicts the conclusion of Step 1. This establishes (3.13).

Step 3. Let us fix F such that $|F| = |E|$ and

$$\partial F = \{z + \psi(z) n_E(z) : z \in \partial E\} \,,$$

with ψ smooth and $\|\psi\|_{W^{2,p}(\partial E)} \leq \delta_1$, where $\delta_1 \in (0, \bar{\delta} \wedge \varepsilon_1)$, where $\bar{\delta}$ is the constant provided by Step 2 and ε_1 the one provided by Proposition 3.4. Let X and Φ be the field and the associated flow constructed in Proposition 3.4. Since $|E_1| = |F| = |E|$, we have $|E_t| = |E|$ for all $t \in (0, 1)$, and thus

$$0 = \frac{d}{dt}|E_t| = \int_{\partial E_t} X \cdot n_{E_t} \, d\mathcal{H}^{N-1} \,.$$

In particular $X \cdot n_{E_t} \in \widetilde{H}(\partial E_t)$. By taking δ_1 smaller and recalling (3.8), we may assume that

$$\partial E_t = \{z + \psi_t(z) n_E(z) : z \in \partial E\} \,,$$

with $\|\psi_t\|_{W^{2,p}(\partial E)} \in (0, \bar{\delta} \wedge \varepsilon_1)$ for all $t \in (0, 1)$. Thus, by Step 2 we have

$$\partial^2 J(E_t)[X \cdot n_{E_t}] \geq \frac{m_0}{2} \|X \cdot n_{E_t}\|_{H^1(\partial E_t)} \qquad \text{for all } t \in (0, 1). \tag{3.21}$$

Recalling that by criticality $\frac{d}{dt}J(E_t)\Big|_{t=0} = 0$, and using (2.18), (2.19), (2.23), and the fact that by construction $\operatorname{div} X = 0$ in a tubular neighborhood of ∂E (where all the ∂E_t, $t \in (0,1)$, lie), we have

$$J(F) - J(E) = J(E_1) - J(E) = \int_0^1 (1-t)\frac{d^2}{dt^2}J(E_t)\,dt$$

$$= \int_0^1 (1-t)\Big(\partial^2 J(E_t)[X\cdot n_{E_t}]$$

$$- \int_{\partial E_t} (4\gamma v_{E_t} + H_{\partial E_t})\operatorname{div}_{\tau_t}\big(X_{\tau_t}(X\cdot n_{E_t})\big)\,d\mathcal{H}^{N-1}\Big)\,dt \quad (3.22)$$

$$= \int_0^1 (1-t)\Big(\partial^2 J(E_t)[X\cdot n_{E_t}]$$

$$- \int_{\partial E_t} (4\gamma v_{E_t} + H_{\partial E_t} - \lambda)\operatorname{div}_{\tau_t}\big(X_{\tau_t}(X\cdot n_{E_t})\big)\,d\mathcal{H}^{N-1}\Big)\,dt\,,$$

where $\operatorname{div}_{\tau_t}$ stands for the tangential divergence on ∂E_t, $X_{\tau_t} := X - (X\cdot n_{E_t})n_{E_t}$, and λ is the constant value of $4\gamma v_E + H_{\partial E}$ on ∂E (see (2.16)). Note that in the last equality we used the fact that $\int_{\partial E_t}\operatorname{div}_{\tau_t}\big(X_{\tau_t}(X\cdot n_{E_t})\big)\,d\mathcal{H}^{N-1} = 0$ by the Divergence Theorem on Manifolds. We will make use of the following two inequalities:

$$\|\operatorname{div}_{\tau_t}\big(X_{\tau_t}(X\cdot n_{E_t})\|_{L^{\frac{p}{p-1}}(\partial E_t)} \le C'\|X\cdot n_{E_t}\|^2_{H^1(\partial E_t)} \qquad \text{for all } t\in(0,1),$$

$$\|X\cdot n_{E_t}\|_{H^1(\partial E_t)} \ge \tfrac{1}{2}\|\psi\|_{H^1(\partial E)} \qquad (3.23)$$

for a suitable constant $C' > 0$ and taking δ_1 smaller if needed. Such inequalities can be deduced from the explicit construction of the field X given in Proposition 3.4; we refer to [1, Proof of Theorem 3.9] for the details. Now choose ε so small that

$$\varepsilon C' \le \frac{1}{4}m_0 \qquad (3.24)$$

and observe that by taking δ_1 smaller if needed, we may guarantee

$$\|4\gamma v_{E_t} + H_{\partial E_t} - \lambda\|_{L^p(\partial E_t)} \le \varepsilon \qquad \text{for all } t\in(0,1). \qquad (3.25)$$

We can continue the estimate (3.22) and deduce:

$$
\begin{aligned}
&J(F) - J(E)\\
&\overset{(3.21)}{\geq} \int_0^1 (1-t)\Big(\frac{m_0}{2}\|X\cdot n_{E_t}\|^2_{H^1(\partial E_t)} - \|4\gamma v_{E_t} + H_{\partial E_t}\\
&\qquad\qquad - \lambda\|_{L^p(\partial E_t)}\|\mathrm{div}_{\tau_t}(X_{\tau_t}(X\cdot n_{E_t})\|_{L^{\frac{p}{p-1}}(\partial E_t)}\Big)\,dt\\
&\overset{(3.23)\text{ and }(3.25)}{\geq} \int_0^1 (1-t)\Big(\frac{m_0}{2}\|X\cdot n_{E_t}\|^2_{H^1(\partial E_t)} - \varepsilon\, C'\|X\cdot n_{E_t}\|^2_{H^1(\partial E_t)}\Big)\,dt\\
&\overset{(3.24)}{\geq} \frac{m_0}{4}\int_0^1 (1-t)\|X\cdot n_{E_t}\|^2_{H^1(\partial E_t)}\,dt\\
&\overset{(3.23)-(ii)}{\geq} \frac{m_0}{8}\int_0^1 (1-t)\|\psi\|^2_{H^1(\partial E)}\,dt = \frac{m_0}{16}\|\psi\|^2_{H^1(\partial E)}\\
&\geq \frac{m_0}{16}C''|E\Delta F|^2,
\end{aligned}
$$

where we used the obvious estimate

$$
C''|E\Delta F|^2 \leq \|\psi\|^2_{L^2(\partial E)}
$$

for some constant $C'' > 0$ depending only on E and δ_1. This concludes the proof of the theorem. □

4 Strict stability and L^1-local minimality

In this section we show that smooth strictly stable critical sets are local minimizers of the nonlocal energy in a much stronger sense; namely, with respect to the L^1 topology. More precisely, we will establish the following.

Theorem 4.1. (i) (NEUMANN CASE) *Let $\Omega \subset \mathbb{R}^N$ be a bounded smooth set and let $E \subseteq \Omega$ be a* smooth strictly stable critical set *in the sense of Definition 3.1-(i). Then, there exists $\delta_0 > 0$, $C_0 > 0$ such that:*

$$
J(F) \geq J(E) + C_0|E\Delta F|^2 \qquad \textit{whenever } |E\Delta F| \leq \delta_0 \textit{ and } |F| = |E|.
$$

(ii) (PERIODIC CASE) *Let $E \subseteq \mathbb{T}^N$ be a smooth strictly stable critical set in the sense of Definition* 3.1-(ii). *Then, there exists $\delta_0, C_0 > 0$ such that:*

$$
J(F) \geq J(E) + C_0\alpha(E,F)^2 \qquad \textit{whenever } \alpha(E,F) \leq \delta_0 \textit{ and } |F| = |E|.
$$

Regarding the smoothness assumption on E, see Remark 4.10 below.

Remark 4.2. (i) The strict stability inequalities established in Theorem 4.1 may be regarded as sort of local *quantitative isoperimetric inequalities* for the nonlocal energy J. More precisely, by considering the case $\gamma = 0$ (to which the theorem applies), we obtain a quantitative inequality, which holds in an L^1-neighborhood of E and has essentially the same form of the global quantitative Isoperimetric Inequality established in [18] (see also [12,17]) for a ball B (of course in the latter case, in order to have a global inequality one has to suitably truncate α, taking for instance $\tilde{\alpha} := \alpha \wedge 2|B|$).
(ii) As already mentioned, Theorem 4.1 applies also to the standard Isoperimetric Problem corresponding to the case $\gamma = 0$. Weaker minimality results were previously established in [57] (see also [23]). More recently, Morgan & Ros in [35] were able to prove an L^1-local minimality result for *strictly stable constant mean curvature surfaces*, but only in dimensions $N \leq 7$ and without the quantitative estimates provided by Theorem 4.1.

The main result will be achieved by showing that any weak $W^{2,p}$-local minimizer for the Nonlocal Isoperimetric Problem, in the sense of Theorem 3.2, is in fact an L^1-local minimizer. The idea of first using a second variation argument to establish a minimality result with respect to regular perturbations, and then of deducing the desired local minimality result via penalization and regularity arguments (see the proof of Theorem 4.3 below) was already used in [57] (for the L^∞-local minimality of stable minimal surfaces) and in [19] (for a free-discontinuity functional related to the epitaxial growth of elastic films). In the context of the standard Isoperimetric Problem such a strategy was first used in [12].

This is made precise by following theorem, in which the second variation plays no role.

Theorem 4.3. (i) (NEUMANN CASE) *Let $\Omega \subset \mathbb{R}^N$ be a bounded smooth open set and fix $p > N - 1$. Let $E \subseteq \Omega$ be such that ∂E is smooth and compactly contained in Ω. Assume also that E is a weak $W^{2,p}$-local minimizer for the Nonlocal Isoperimetric Problem;* i.e., *there exist $\delta_1 > 0$ and $C_1 > 0$ with the following properties:*

$$J(F) \geq J(E) + C_1 |E \Delta F|^2 \text{ whenever } \partial F = \{x + \psi(x) n_E(x) : x \in \partial E\}, \\ \text{with } \|\psi\|_{W^{2,p}(\partial E)} \leq \delta_1, \text{ and } |F| = |E|. \quad (4.1)$$

Then, there exists δ_0 such that:

$$J(F) \geq J(E) + \frac{C_1}{2} |E \Delta F|^2 \qquad \text{whenever } |E \Delta F| \leq \delta_0 \text{ and } |F| = |E|.$$

(ii) (PERIODIC CASE) *Let $p > N - 1$ and let $E \subseteq \mathbb{T}^N$ be such that ∂E is smooth. Assume also that E is weak $W^{2,p}$-local minimizer for the Nonlocal Isoperimetric Problem;* i.e., *there exist $\delta_1 > 0$ and $C_1 > 0$ with the following properties:*

$$J(F) \geq J(E) + C_1 \alpha(E, F)^2 \text{ whenever } \partial F = \{x + \psi(x) n_E(x) : x \in \partial E\}, \\ \text{with } \|\psi\|_{W^{2,p}(\partial E)} \leq \delta_1, \text{ and } |F| = |E|.$$

Then, there exists δ_0 such that:

$$J(F) \geq J(E) + \frac{C_1}{2} \alpha(E, F)^2 \quad \text{whenever } \alpha(E, F) \leq \delta_0 \text{ and } |F| = |E|.$$

We recall that the the quantity $\alpha(E, F)$ appearing in the above statement stands for the distance up to translations defined in (3.3). Throughout the section we will focus on the Neumann case only. The periodic case can be treated in a completely analogous fashion, see [1].

Remark 4.4. As already observed in the previous section, the simplifying assumption $\partial E \Subset \Omega$ can be removed at the expense of some technical effort but without affecting the overall strategy and the main ideas of [1]. This has been accomplished in [27], to which we refer the interested reader.

We postpone the proof of Theorem 4.3 until after introducing several auxiliary results. In all the following statements we implicitly assume that the hypotheses of Theorem 4.3-(i) are satisfied.

Lemma 4.5 (Lipschitz continuity of the nonlocal energy). *There exists a constant $C = C(\Omega) > 0$ such that*

$$\left| \int_\Omega |\nabla v_F|^2 \, dz - \int_\Omega |\nabla v_E|^2 \, dz \right| \leq C |E \Delta F|$$

for all $E, F \subset \Omega$, where v_E and v_F denote as usual the potentials associated with E and F, respectively (under Neumann conditions).

Proof. Write

$$\begin{aligned} \int_\Omega |\nabla v_F|^2 \, dz - \int_\Omega |\nabla v_E|^2 \, dz &= \int_\Omega |\nabla v_F - \nabla v_E|^2 \, dz \\ &\quad + 2 \int_\Omega \nabla v_E \cdot (\nabla v_F - \nabla v_E) \, dz \\ &= \int_\Omega |\nabla v_F - \nabla v_E|^2 \, dz \\ &\quad + 2 \int_\Omega v_E \cdot (u_F - u_E + m_E - m_F) \, dz \, , \end{aligned} \tag{4.2}$$

where in the last equality we used integration by parts and the fact tha $v_F - v_E$ solves

$$\begin{cases} -\Delta(v_F - v_E) = u_F - u_E + m_E - m_F & \text{in } \Omega, \\ \partial_n(v_F - v_E) = 0 & \text{on } \partial\Omega, \end{cases}$$

where, we recall, $u_E = \chi_E - \chi_{\Omega\setminus E}$ and $m_E = \fint_\Omega u_E$ (and analogously for u_F, m_F). By the classical Calderon-Zygmund L^p-estimates for elliptic equations (see for instance [20]), we have that for all $q > 1$

$$\begin{aligned} \|v_F - v_E\|^q_{W^{2,q}} &\leq C(q,\Omega)\|u_F - u_E + m_E - m_F\|^q_{L^q} \\ &\leq C'(q,\Omega)|E\Delta F|\,. \end{aligned} \tag{4.3}$$

By the same argument , for any set $F \subset \Omega$ we have

$$\|v_F\|^q_{W^{2,q}} \leq C(q,\Omega)\|u_F - m_F\|^q_{L^q} \leq C'(q,\Omega)\,.$$

In particular, by the Sobolev Embedding, the L^∞-norm of v_E (and of v_F) is bounded by a universal constant independent of E. Using this piece of information and (4.3) (with $q = 2$) to estimate the righ-hand side of (4.2), we deduce the thesis of the lemma. □

In the next lemma we show that the volume constraint can be replaced by a suitable volume penalization.

Lemma 4.6 (volume penalization). *Let $\delta_0 > 0$, $\eta_0 \in (0, |\Omega|/2)$, $M > 0$, and let V be a set-functional satisfying the Lipschitz continuity condition:*

$$|V(F_1) - V(F_2)| \leq C|F_1\Delta F_2| \qquad \textit{for all } F_1, F_2 \subset \Omega \tag{4.4}$$

for some constant $C > 0$. Let $E \subset \Omega$ be a solution to the volume-constrained *minimization problem*

$$\min\{P_\Omega(F) + V(F) : F \subset \Omega,\ |F| = \eta,\ |F\Delta E| \leq \delta\}\,, \tag{4.5}$$

with $\delta \in [\delta_0, +\infty]$ and $\eta \in [\eta_0, |\Omega| - \eta_0]$, and assume further that $P_\Omega(E) \leq M$. Then, there exists $\lambda_0 = \lambda_0(C, \Omega, \delta_0, \eta_0, M) > 0$ such that E is also an unconstrained *minimizer of*

$$\min\left\{P_\Omega(F) + V(F) + \lambda\big||F| - |E|\big| : F \subset \Omega,\ |F\Delta E| \leq \frac{\delta}{2}\right\}, \tag{4.6}$$

for all $\lambda \geq \lambda_0$.

Proof. We adapt an argument taken from [15, Section 2] (see also [1, Proposition 2.7]) and we indicate below only the relevant changes. In order to prove the lemma it is enough to show that there exists $\lambda_0 = \lambda_0(C, \Omega, \delta_0, \eta_0, M) > 0$ such that if E solves (4.5) (with η, δ, and V satisfying the assumption of the statement), then any set $\hat{F}$ that solves

$$\min\left\{P_\Omega(F) + V(F) + \lambda\big||F| - |E|\big| : F \subset \Omega,\ |F\Delta E| \leq \frac{\delta}{2}\right\},$$

with $\lambda \geq \lambda_0$, satisfies $|\hat{F}| = |E|$. To this aim, we argue by contradiction assuming that there exist $\lambda_n \to \infty$, $\delta_n \in [\delta_0, +\infty]$, functionals V_n satisfying the Lipschitz continuity condition (4.4), and sets $E_n \subset \Omega$ satisfying

$$E_n \in \operatorname{argmin}\{P_\Omega(F)+V_n(F) : F \subset \Omega,\ |F|=|E_n|,\ |F\Delta E| \leq \delta_n\}, \tag{4.7}$$

$|E_n| \in [\eta_0, |\Omega| - \eta_0]$, $P_\Omega(E_n) \leq M$, and such that if

$$F_n \in \operatorname{argmin}\left\{P_\Omega(F)+V_n(F)+\lambda_n\big||F|-|E_n|\big| : |F\Delta E_n| \leq \frac{\delta_n}{2}\right\}, \tag{4.8}$$

then $|F_n| \neq |E_n|$. Without loss of generality we may assume that $|F_n| < |E_n|$ (the other case being similar). Let us now set for notational convenience

$$J_n(F) := P_\Omega(F) + V_n(F) + \lambda_n\big||F| - |E_n|\big|.$$

Using the minimality of F_n, the Lipschitz continuity of V_n and the fact that $P_\Omega(E_n) \leq M$, we have

$$\begin{aligned} &P_\Omega(F_n) + \lambda_n\big||F_n| - |E_n|\big| = J_n(F_n) - V_n(F_n) \\ &\leq J_n(E_n) - V_n(F_n) = P_\Omega(E_n) + V_n(E_n) - V_n(F_n) \leq M + C|\Omega|. \end{aligned}$$

Thus, by Theorem 2.1 we may extract a (not relabeled) subsequence such that and

$$F_n \to F_\infty \qquad \text{in } L^1,$$

with $|F_\infty| \in [\eta_0, |\Omega| - \eta_0]$. Moreover,

$$\big||F_n| - |E_n|\big| \to 0. \tag{4.9}$$

Arguing as in Step 1 of [15], for any given small $\varepsilon > 0$ we can find $r > 0$ and a point $x_0 \in \Omega$ such that

$$|F_n \cap B_{r/2}(x_0)| < \varepsilon r^N, \quad |F_n \cap B_r(x_0)| > \frac{\omega_N r^N}{2^{N+2}}$$

for all n sufficiently large. In the following, to simplify the notation we assume that $x_0 = 0$ and we write B_r instead of $B_r(0)$. For a sequence $0 < \sigma_n < 1/2^N$ to be chosen, we introduce the following bilipschitz maps:

$$\Phi_n(x) := \begin{cases} (1-\sigma_n(2^N-1))x & \text{if } |x| \le \frac{r}{2}, \\ x + \sigma_n\Big(1 - \frac{r^N}{|x|^N}\Big)x & \frac{r}{2} \le |x| < r, \\ x & |x| \ge r. \end{cases}$$

Setting $\widetilde{F}_n := \Phi_n(F_n)$, we have as in Step 3 of [15]

$$P_{B_r}(F_n) - P_{B_r}(\widetilde{F}_n) \ge -2^N N P_{B_r}(F_n)\sigma_n \,. \tag{4.10}$$

Moreover, as in Step 4 of [15] we have

$$|\widetilde{F}_n| - |F_n| \ge \sigma_n r^N \Big[c\frac{\omega_N}{2^{N+2}} - \varepsilon(c + (2^N-1)N) \Big]$$

for a suitable constant c depending only on the dimension N. If we fix ε so that the negative term in the square bracket does not exceed half the positive one, then we have

$$|\widetilde{F}_n| - |F_n| \ge \sigma_n r^N C_1 \,, \tag{4.11}$$

with $C_1 > 0$ depending on N. In particular from this inequality it is clear that we can choose σ_n so that $|\widetilde{F}_n| = |E_n|$; this implies $\sigma_n \to 0$, thanks to (4.9).

Now, arguing as in [1, Equations (2.12) and (2.13)], we obtain

$$|\widetilde{F}_n \Delta F_n| \le C_3 \sigma_n P_{B_r}(F_n) \,. \tag{4.12}$$

In particular, since $\sigma_n \to 0$ and $\delta_n \ge \delta_0$, we have that $|\widetilde{F}_n \Delta E_n| \le \delta_n$ for n sufficiently large. Finally, recall that by the Lipschitz continuity assumption on V_n

$$|V_n(\widetilde{F}_n) - V_n(F_n)| \le C|\widetilde{F}_n \Delta F_n| \,. \tag{4.13}$$

Combining (4.10), (4.11), (4.12), and (4.13), we conclude that for n sufficiently large

$$\begin{aligned} J_n(\widetilde{F}_n) &\le J_n(F_n) + \sigma_n\big[(2^N N + CC_3)P_{B_r}(F_n) - \lambda_n r^N C_1\big] \\ &< J_n(F_n) \le J_n(E_n) = P_\Omega(E_n) + V_n(E_n) \,, \end{aligned}$$

a contradiction to (4.7), since $|\widetilde{F}_n| = |E_n|$ and $|\widetilde{F}_n \Delta E_n| \le \delta_n$. □

Remark 4.7. Let $E \subset \Omega$ be a volume-constrained L^1-local minimizer of J; *i.e.*, there exists $\delta > 0$ such that

$$J(F) \geq J(E) \quad \text{for all } F \subseteq \Omega, \text{with } |F| = |E| \text{ and } |E \Delta F| \leq \delta.$$

By applying Lemma 4.6 with $V(F) := \gamma \int_\Omega |\nabla v_F|^2 \, dz$ (and taking into account Lemma 4.5), we may infer the existence of $\lambda > 0$ such that

$$J(F) + \lambda \big| |F| - |E| \big| \geq J(E) \qquad \text{if } |E \Delta F| \leq \frac{\delta}{2}. \tag{4.14}$$

Thus, if $r > 0$ is such that $|B_r| \leq \delta/2$ and F is a set such that $F \Delta E \subset B_r(x_0)$ for some $x_0 \in \Omega$, then by (4.14) we have

$$\begin{aligned} P_\Omega(E) &\leq P_\Omega(F) + \gamma \left(\int_\Omega |\nabla v_F|^2 \, dz - \int_\Omega |\nabla v_E|^2 \, dz \right) + \lambda \big| |F| - |E| \big| \\ &\leq P_\Omega(F) + (\gamma C + \lambda) |E \Delta F| , \end{aligned}$$

where C is the constant provided by Lemma 4.5.

The previous remark is telling us that any volume-constrained L^1-local minimizer of the nonlocal functional J is in fact a STRONG QUASI-MINIMIZER OF THE PERIMETER, according to the following definition.

Definition 4.8. We say that a set $E \subseteq \Omega$ is a *strong quasi-minimizer* of P_Ω with constants (ω, r) if

$$P_\Omega(E) \leq P_\Omega(F) + \omega |E \Delta F| \quad \text{provided } F \subseteq \Omega \text{ and } E \Delta F \subset B_r(x_0) \text{ for some } x_0 \in \Omega.$$

The above notion of quasi-minimality is useful since it allows one to apply the well-established regularity theory developed by De Giorgi and his School. We recall the main results.

Theorem 4.9 (see [53]). *Let E be a strong quasi-minimizer of P_Ω in the sense of Definition* 4.8. *Then $\partial E = \partial^* E \cup \Sigma$, where the reduced boundary $\partial^* E$ is a $(N-1)$-manifold of class $C^{1,\alpha}$ for all $\alpha \in (0, 1)$ and the* singular set Σ *has Hausdorff dimension less than or equal to $N - 8$.*

Remark 4.10. In particular, in low dimensions ($N \leq 7$) the singular set Σ is empty and thus ∂E is fully $C^{1,\alpha}$-regular. Thanks to Remark 4.7, the full regularity result holds for all volume constrained local minimizers of the nonlocal functional in dimensions $N \leq 7$. In turn, higher regularity can be gained from the Euler-Lagrange condition

$$H_{\partial E} + 4\gamma v_E = \text{const} \qquad \text{on } \partial E. \tag{4.15}$$

In fact, it can be shown that the initial $C^{1,\alpha}$-regularity provided by Theorem 4.9 implies the C^∞-regularity of $\partial^* E$, thanks to (4.15). We refer the reader to [27] for the details. This shows that the C^∞-regularity assumption on E in Theorem 4.1 is not very restrictive and, in fact, necessary in low dimensions.

We finally recall the following important consequence of the regularity theory for quasi-minimizers of the perimeter. Roughly speaking, it says that any sequence of UNIFORM (*i.e.*, with the same constants) strong quasi-minimizers of P_Ω L^1-converging to a smooth set E must eventually inherit the regularity of the limiting set and the convergence holds in fact in a much stronger sense. This is made precise by the following statement, which is well-know to the experts (see for instance [57]). A clear proof can be found in [12].

Theorem 4.11 (Improved convergence). *Let $(E_n)_n$ be a sequence of strong quasi minimizers of P_Ω, with constants (ω, r) independent of n, such that $E_n \to E$ in L^1, where E is a set of class C^1 with $\partial E \Subset \Omega$. Then, for n sufficiently large ∂E_n is of class $C^{1,\alpha}$ for all $\alpha \in (0,1)$ and, in fact,*

$$\partial E_n \to \partial E \qquad \text{in } C^{1,\alpha} \text{ for all } \alpha \in (0,1).$$

The latter convergence can be rephrased in the following way: for n large enough

$$\partial E_n = \{x + \psi_n(x) n_E(x) : x \in \partial E\},$$

where $\psi_n \to 0$ in $C^{1,\alpha}(\partial E)$ for all $\alpha \in (0,1)$, as $n \to \infty$.

Remark 4.12. Once again, the assumption $\partial E \Subset \Omega$ in the previous statement can be removed (see for instance [27]).

A further intermediate result needed to prove Theorem 4.3 is the following L^∞-local minimality property.

Proposition 4.13. *Under the assumptions of Theorem 4.3 the set E is a local minimizer with respect to the Hausdorff distance. Precisely, there exists a tubular neighborhood $\mathcal{N}$ of ∂E such that*

$$J(F) \geq J(E) \qquad \text{if } |F| = |E| \text{ and } \partial F \subset \mathcal{N}.$$

We postpone the proof of Proposition 4.13 until the end of the section. We are now in a position to prove Theorem 4.3.

Proof of Theorem 4.3-(i). We argue by contradiction assuming the existence of a sequence $E_n \to E$ in L^1 such that $|E_n| = |E|$ and

$$J(E_n) < J(E) + \frac{C_1}{2}|E_n \Delta E|^2 \qquad \text{for all } n \in \mathbb{N}. \tag{4.16}$$

Let us set $\varepsilon_n := |E_n \Delta E|$. Clearly $\varepsilon_n \to 0$ as $n \to \infty$. We now replace (E_n) by a new sequence (F_n), where

$$F_n \in \operatorname{argmin}\left\{ J(F)+\Lambda_1\sqrt{(|F\Delta E|-\varepsilon_n)^2+\varepsilon_n} : F\subset\Omega,\ |F|=|E|\right\}, \quad (4.17)$$

with $\Lambda_1 > 0$ to be chosen later. Note that

$$\begin{aligned} J(F_n) &\le J(F_n) + \Lambda_1\left(\sqrt{(|F_n\Delta E| - \varepsilon_n)^2 + \varepsilon_n} - \sqrt{\varepsilon_n}\right) \\ &\le J(E_n) < J(E) + \frac{C_1}{2}\varepsilon_n^2 , \end{aligned} \quad (4.18)$$

where the second inequality follows from the minimality of F_n, while the last one from (4.16). In particular,

$$\sup_n P_\Omega(F_n) \le \sup_n J(F_n) < +\infty$$

and thus, by the Compactness Theorem 2.1, we may extract a (not relabeled) subsequence and find a set $F_\infty \subset \Omega$ of finite perimeter such that

$$F_n \to F_\infty \qquad \text{in } L^1$$

as $n \to \infty$. We now split the rest of the proof into several steps.

Step 1. (F_∞ solves the limiting problem) Take any subset F, with $|F| = |E|$. Then, by minimality we have

$$J(F_n)+\Lambda_1\sqrt{(|F_n\Delta E| - \varepsilon_n)^2 + \varepsilon_n} \le J(F)+\Lambda_1\sqrt{(|F\Delta E| - \varepsilon_n)^2 + \varepsilon_n} .$$

Using the semicontinuity of J with respect to the L^1- convergence and recalling that $\varepsilon_n \to 0$, we can pass to the limit in both sides of the above inequality to deduce that

$$J(F_\infty) + \Lambda_1|F_\infty\Delta E| \le J(F) + \Lambda_1|F\Delta E| .$$

This shows that

$$F_\infty \in \operatorname{argmin}\{J(F) + \Lambda_1|F\Delta E| :\ F \subset \Omega,\ |F| = |E|\} . \quad (4.19)$$

Step 2. ($F_\infty = E$) We now claim that for Λ_1 sufficiently large the set E itself is the UNIQUE minimizer of (4.19) and thus $F_\infty = E$. To this aim, fix a smooth vector-field X such that $\operatorname{supp} X \Subset \Omega$, which agrees with n_E on ∂E and satisfies $\|X\|_\infty = 1$. Then, for any set $F \subset \Omega$ we have

$$\begin{aligned} P_\Omega(F) - P_\Omega(E) &\ge \int_{\partial^* F} X \cdot n_F \, d\mathcal{H}^{N-1} - \int_{\partial E} X \cdot n_E \, d\mathcal{H}^{N-1} \\ &= \int_F \operatorname{div} X \, dz - \int_E \operatorname{div} X \, dz \ge -\|\operatorname{div} X\|_\infty |E\Delta F| . \end{aligned}$$

In turn, by Lemma 4.5, we deduce

$$\begin{aligned} J(F) - J(E) &= P_\Omega(F) - P_\Omega(E) + \gamma \left(\int_\Omega |\nabla v_F|^2 \, dz - \int_\Omega |\nabla v_E|^2 \, dz \right) \\ &\geq -\big(\|\mathrm{div}\, X\|_\infty + \gamma\, C\big)|E \Delta F| \,. \end{aligned}$$

If

$$\Lambda_1 > \|\mathrm{div}\, X\|_\infty + \gamma\, C \,, \tag{4.20}$$

then

$$J(E) < J(F) + \Lambda_1 |E \Delta F| \qquad \text{whenever } |E \Delta F| \neq 0.$$

Therefore, with this choice of Λ_1, E is the unique minimizer of the problem in (4.19). Thus, $F_\infty = E$ and

$$F_n \to E \qquad \text{in } L^1. \tag{4.21}$$

Step 3. (Improved $C^{1,\alpha}$-convergence) Set

$$V_n(F) := \gamma \int_\Omega |\nabla v_F|^2 \, dz + \Lambda_1 \sqrt{(|F \Delta E| - \varepsilon_n)^2 + \varepsilon_n} \,.$$

By Lemma 4.5 and the fact that the maps $t \mapsto \Lambda_1 \sqrt{(t - \varepsilon_n)^2 + \varepsilon_n}$ are equi-Lipschitz continuous, one can easily check that

$$|V_n(F_1) - V_n(F_2)| \leq C' |F_1 \Delta F_2| \qquad \text{for all } F_1, F_2 \subset \Omega, \tag{4.22}$$

with $C' > 0$ independent of n. Thus, by Lemma 4.6 (with $V = V_n$), we may find $\Lambda_2 > 0$ independent of n such that

$$F_n \in \operatorname{argmin} \left\{ J(F) + \Lambda_1 \sqrt{(|F \Delta E| - \varepsilon_n)^2 + \varepsilon_n} + \Lambda_2 \big| |F| - |F_n| \big| \,:\, F \subset \Omega \right\}.$$

In turn, taking into account again (4.22) and arguing as in Remark 4.7, we deduce that

$$P_\Omega(F_n) \leq P_\Omega(F) + (C' + \Lambda_2) |F \Delta F_n| \qquad \text{for all } F \subset \Omega.$$

In other words, the F_n's are uniform strong quasi-minimizers of P_Ω. Recalling (4.21) and since ∂E is smooth, we may apply Theorem 4.11 to deduce that for n large enough

$$\begin{gathered} \partial F_n = \{x + \psi_n(x) n_E(x) : x \in \partial E\}, \\ \text{with } \psi_n \in C^{1,\alpha}(\partial E) \text{ for all } \alpha \in (0,1) \end{gathered} \tag{4.23}$$

and

$$\psi_n \to 0 \qquad \text{in } C^{1,\alpha}(\partial E) \text{ for all } \alpha \in (0,1). \tag{4.24}$$

Step 4. We claim that

$$\frac{\varepsilon_n}{|F_n \Delta E|} \to 1 \qquad \text{as } n \to \infty. \tag{4.25}$$

To this aim, observe that (4.24) implies in particular that $\partial F_n \to \partial E$ in the Hausdorff metric. Therefore, for n large enough

$$J(F_n) \geq J(E), \tag{4.26}$$

thanks to Proposition 4.13. For all such n we have

$$\begin{aligned} J(F_n) &+ \Lambda_1 \sqrt{(|F_n \Delta E| - \varepsilon_n)^2 + \varepsilon_n} \\ &\overset{(\text{minimality of } F_n)}{\leq} J(E_n) + \Lambda_1 \sqrt{\varepsilon_n} \\ &\overset{(4.16)}{<} J(E) + \frac{C_1}{2}\varepsilon_n^2 + \Lambda_1 \sqrt{\varepsilon_n} \\ &\overset{(4.26)}{\leq} J(F_n) + \frac{C_1}{2}\varepsilon_n^2 + \Lambda_1 \sqrt{\varepsilon_n}\,. \end{aligned}$$

Thus,

$$\Lambda_1 \left(\sqrt{(|F_n \Delta E| - \varepsilon_n)^2 + \varepsilon_n} - \sqrt{\varepsilon_n} \right) \leq \frac{C_1}{2}\varepsilon_n^2\,.$$

It is now an elementary exercise to check that the above inequality remains valid as $\varepsilon_n \to 0$ if and only if (4.25) holds true.

Step 5. (First variations) Because of the $C^{1,\alpha}$-regularity of ∂F_n, recalling (4.17) and arguing as in the previous section, we may write the *stationarity condition* for F_n in the following form:

$$H_{\partial F_n} + 4\gamma v_{F_n} - \lambda_n = g_n \qquad \text{on } \partial F_n, \tag{4.27}$$

where the left-hand side comes from the first variation of J with respect to volume preserving flows (λ_n are the constant Lagrange multipliers associated with the volume constraint), while the functions g_n on the left-hand side represent the extra contribution coming from the added volume penalization. More precisely, since $\big||F_n \Delta E| - \varepsilon_n\big| << \varepsilon_n$ by (4.25) and using the fact that the map $t \mapsto \sqrt{(t-\varepsilon_n)^2 + \varepsilon_n}$ has vanishing derivative at ε_n, one can show that

$$\|g_n\|_{L^\infty(\partial F_n)} \to 0 \qquad \text{as } n \to \infty, \tag{4.28}$$

we refer the reader to [1, page 546] for the details. Note that (4.27) and elliptic regularity imply in particular that ∂F_n is of class $W^{2,p}$ for all $p > 1$, see Lemma 4.14 below.

Step 6. (Improved $W^{2,p}$-convergence) We start by showing that

$$\lambda_n \to \lambda\,, \tag{4.29}$$

where λ is the constant such that

$$H_{\partial E} + 4\gamma v_E = \lambda \qquad \text{on } \partial E, \tag{4.30}$$

see (2.16). To this aim, recall that

$$v_{F_n} \to v_E \qquad \text{in } C^1(\overline{\Omega}), \tag{4.31}$$

thanks to the elliptic estimates recalled in the proof of Lemma 4.5 (together with the Sobolev Embedding Theorem). Fix now a coordinate frame and a cylinder $C := B' \times (-L, L)$, where $B' \subset \mathbb{R}^{N-1}$ is a ball centered at the origin, and find functions $f_n, f \in W^{2,p}(B'; (-L, L))$ for all $p > 1$ such that $E \cap C = \{(x', x_n) \in B' \times (-L, L) : x_n < f(x')\}$,

$$F_n \cap C = \{(x', x_n) \in B' \times (-L, L) : x_n < f_n(x')\} \tag{4.32}$$

for all $n \in \mathbb{N}$ sufficiently large. Recall that by Step 3 we also have

$$f_n \to f \qquad \text{in } C^{1,\alpha}(\overline{B'}) \text{ for all } \alpha \in (0,1). \tag{4.33}$$

Then, by (4.27) we have

$$\begin{aligned}
&\lambda_n \mathcal{H}^{N-1}(B') - 4\gamma \int_{B'} v_{F_n}(x', f_n(x'))\,dx' + \int_{B'} g_n(x', f_n(x'))\,dx' \\
&= -\int_{B'} \operatorname{div}\left(\frac{\nabla_{x'} f_n}{\sqrt{1+|\nabla_{x'} f_n|^2}}\right) dx' \\
&= -\int_{\partial B'} \frac{\nabla_{x'} f_n}{\sqrt{1+|\nabla_{x'} f_n|^2}} \cdot \frac{x'}{|x'|}\, d\mathcal{H}^{N-2}\,.
\end{aligned}$$

Since by (4.33) and (4.30)

$$\begin{aligned}
-\int_{\partial B'} &\frac{\nabla_{x'} f_n}{\sqrt{1+|\nabla_{x'} f_n|^2}} \cdot \frac{x'}{|x'|}\, d\mathcal{H}^{N-2} \\
&\to -\int_{\partial B'} \frac{\nabla_{x'} f}{\sqrt{1+|\nabla_{x'} f|^2}} \cdot \frac{x'}{|x'|}\, d\mathcal{H}^{N-2} \\
&= -\int_{B'} \operatorname{div}\left(\frac{\nabla_{x'} f}{\sqrt{1+|\nabla_{x'} f|^2}}\right) dx' \\
&= \lambda \mathcal{H}^{N-1}(B') - 4\gamma \int_{B'} v_E(x', f(x'))\,dx'\,,
\end{aligned}$$

we deduce that

$$\lambda_n \mathcal{H}^{N-1}(B') - 4\gamma \int_{B'} v_{F_n}(x', f_n(x'))\, dx' + \int_{B'} g_n(x', f_n(x'))\, dx' \longrightarrow \lambda \mathcal{H}^{N-1}(B') - 4\gamma \int_{B'} v_E(x', f(x'))\, dx' .$$

In turn, taking into account (4.28) and (4.31), we may conclude that (4.29) holds.

Combining now (4.24), (4.27), (4.28), (4.29), and (4.30), we arrive at

$$H_{\partial F_n}(\cdot + \psi_n(\cdot) n_E(\cdot)) \to H_{\partial E}(\cdot) \qquad \text{in } L^\infty(\partial E). \tag{4.34}$$

In turn, Lemma 4.14 below yields

$$\psi_n \to 0 \qquad \text{in } W^{2,p}(\partial E) \text{ for all } p > 1. \tag{4.35}$$

Step 7. (Conclusion) By the assumption (4.1) and by (4.35), we have

$$J(F_n) \geq J(E) + C_1 |E \Delta F_n|^2 \tag{4.36}$$

for n large enough. On the other hand, for n large enough we also have

$$J(F_n) \overset{(4.18)}{<} J(E) + \frac{C_1}{2}\varepsilon_n^2 \overset{(4.25)}{\leq} J(E) + \frac{3C_1}{4}|E\Delta F_n|^2 ,$$

which clearly contradicts (4.36). This concludes the proof of the theorem. □

The following lemma is well-known to the experts. We provide a sketch of the proof for the reader's convenience.

Lemma 4.14. *Let E be a set of class C^2 and let F_n be a sequence of sets of class $C^{1,\alpha}$ for some $\alpha \in (0, 1)$ such that*

$$\partial F_n = \{x + \psi_n(x) n_E(x) : x \in \partial E\} ,$$

where $\psi_n \to 0$ in $C^{1,\alpha}(\partial E)$. Assume also that $H_{\partial F_n} \in L^p(\partial F_n)$ and that

$$H_{\partial F_n}\big(\cdot + \psi_n(\cdot) n_E(\cdot)\big) \to H_{\partial E}(\cdot) \qquad \text{in } L^p(\partial E). \tag{4.37}$$

Then $\psi_n \to 0$ in $W^{2,p}(\partial E)$.

Proof. By localization we may reduce to the case when both ∂E and all ∂F_n are graphs of functions f, f_n in a cylinder of the $C = B' \times (-L, L)$, where $B' \subset \mathbb{R}^{N-1}$ is a ball centered at the origin. By assumption $f \in C^2(\overline{B}')$ and

$$f_n \to f \in C^{1,\alpha}(\overline{B}'),\ \operatorname{div}\left(\frac{\nabla_{x'} f_n}{\sqrt{1+|\nabla_{x'} f_n|^2}}\right) \to \operatorname{div}\left(\frac{\nabla_{x'} f}{\sqrt{1+|\nabla_{x'} f|^2}}\right) \text{ in } L^p(B').$$

Standard elliptic regularity (see for instance [4, Proposition 7.56]) gives that f_n is bounded in $W^{2,2}$ in a smaller ball B'', thus we may carry out the differentiation and using the C^1 convergence of f_n to f we are led to

$$A_{ij}(x')\nabla^2_{ij} f_n \to A_{ij}(x')\nabla^2_{ij} f \quad \text{in } L^p(B'') ,$$

where

$$A = I - \frac{\nabla_{x'} f \otimes \nabla_{x'} f}{1 + |\nabla_{x'} f|^2} .$$

Again the standard L^p-elliptic estimates imply the strong (local) convergence in L^p of $\nabla^2_{x'} f_n$. □

We conclude the section by proving Proposition 4.13.

Proof of Proposition 4.13. We only sketch the main steps of the proof. The details can be filled arguing as in the proof of Theorem 4.3.

Recalling that d_E denotes the signed distance to E, we define for all $\delta \in \mathbb{R}$

$$\mathcal{I}_\delta(E) = \{x : d_E(x) < \delta\} .$$

We argue by contradiction assuming that there exist two sequences $\delta_n \to 0$ and $E_n \subset \Omega$ such that $|E_n| = |E|, \mathcal{I}_{-\delta_n}(E) \subset E_n \subset \mathcal{I}_{\delta_n}(E)$, and

$$J(E_n) < J(E) \tag{4.38}$$

for all n. For later use, we fix a vector-field $X \in C^\infty_c(\Omega; \mathbb{R}^N)$ such that

$$\|X\|_\infty = 1 \quad \text{and} \quad X = \nabla d_E \text{ in a tubular neighborhood of } \partial E. \tag{4.39}$$

As in the proof of Theorem 4.3 we replace the sequence $(E_n)_n$ with a new sequence $(F_n)_n$, where each F_n is the solution of a suitable obstacle problem; precisely,

$$F_n \in \operatorname{argmin}\{J(F) + \Lambda\big||F| - |E|\big| : \mathcal{I}_{-\delta_n}(E) \subset F \subset \mathcal{I}_{\delta_n}(E)\}, \tag{4.40}$$

where

$$\Lambda > \|\operatorname{div} X\|_\infty + \gamma C , \tag{4.41}$$

and C is the Lipschitz constant provided by Lemma 4.5. Clearly the inclusions in the above minimization problem must be understood in the a.e. sense.

Step 1. We claim that for n sufficiently large

$$|F_n| = |E|\,. \tag{4.42}$$

Indeed, assume by contradiction that $|F_n| < |E|$. We consider the competitor

$$\widetilde{F}_n := F_n \cup \mathcal{I}_{\tau_n}(E)$$

for some $\tau_n \in (-\delta_n, \delta_n)$ chosen in such a way to ensure $|\widetilde{F}_n| = |E|$. Since $\partial^* \widetilde{F}_h$ is the disjoint union of a part contained in $\partial^* F_h \setminus \partial I_{\tau_n}(E)$, another one contained in $\partial I_{\tau_n}(E) \setminus \partial^* F_n$, and $\{x \in \partial^* F_n \cap \partial I_{\tau_n}(E) : n_{F_n}(x) = n_{\mathcal{I}_{\tau_h}(E)}(x)\}$, and using also the fact that $n_{\mathcal{I}_{\tau_n}(E)} = X$ for n large enough (see (4.39)), for all such n we have

$$P_\Omega(\widetilde{F}_n) - P_\Omega(F_n) \le \int_{\partial^* \widetilde{F}_n} X \cdot n_{\widetilde{F}_n}\, d\mathcal{H}^{N-1} \quad \int_{\partial^* F_n} X \cdot n_{F_n}\, d\mathcal{H}^{N-1}\,.$$

Hence, also by Lemma 4.5,

$$\begin{aligned}
&J(\widetilde{F}_n) + \Lambda\big||\widetilde{F}_n| - |E|\big| - J(F_n) - \Lambda\big||F_n| - |E|\big| \\
&\quad = P_\Omega(\widetilde{F}_n) - P_\Omega(F_n) \\
&\qquad + \gamma \int_\Omega \big(|\nabla v_{\widetilde{F}_n}|^2 - |\nabla v_{F_n}|^2\big)\, dz - \Lambda\big(|\widetilde{F}_n| - |F_n|\big) \\
&\quad \le \int_{\partial^* \widetilde{F}_n} X \cdot n_{\widetilde{F}_n}\, d\mathcal{H}^{N-1} \\
&\qquad - \int_{\partial^* F_n} X \cdot n_{F_n}\, d\mathcal{H}^{N-1} + (\gamma C - \Lambda)\big(|\widetilde{F}_n| - |F_n|\big) \\
&\quad \le \int_{\widetilde{F}_n \Delta F_n} |\operatorname{div} X|\, dz + (\gamma C - \Lambda)\big(|\widetilde{F}_n| - |F_n|\big) \\
&\quad \le (\|\operatorname{div} X\|_\infty + \gamma C - \Lambda)\big(|\widetilde{F}_n| - |F_n|\big) < 0\,,
\end{aligned} \tag{4.43}$$

where the last inequality follows from (4.41) and contradicts the minimality of F_n. If $|F_n| > |E|$, we argue similarly.

Step 2. For any set $F \subset \Omega$ define its "truncation" as

$$\mathcal{K}_n(F) := (F \cup \mathcal{I}_{-\delta_n}(E)) \cap \mathcal{I}_{\delta_n}(E)\,.$$

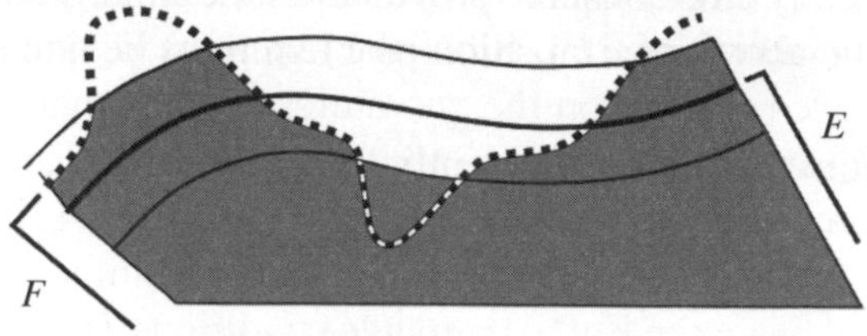

Figure 4.1. The dotted line denotes the boundary of F, while the shaded region correspond to the truncated set $\mathcal{K}_n(F)$.

We claim that F_n solves a suitable penalized problem (without obstacle); preciselsy,

$$F_n \in \operatorname{argmin}\{J(F)+\Lambda\big||F|-|E|\big|+2\Lambda|F\Delta\mathcal{K}_n(F)| : F \subset \Omega\}. \quad (4.44)$$

Indeed, let $\widetilde{J}$ denote the functional in (4.44). Using Lemma 4.5 and arguing as in (4.43), we obtain by the minimality of F_n

$$\begin{aligned}\widetilde{J}(F) - \widetilde{J}(F_n) &= J(\mathcal{K}_n(F)) + \Lambda\big||\mathcal{K}_n(F)| - |E|\big| \\ &\quad - \big[J(F_n) + \Lambda\big||F_n| - |E|\big|\big] \\ &\quad + \big[P_\Omega(F) - P_\Omega(\mathcal{K}_n(F))\big] + \gamma\int_\Omega |\nabla v_F|^2\,dz - \gamma\int_\Omega |\nabla v_{\mathcal{K}_n(F)}|^2\,dz \\ &\quad + \Lambda\big(\big||F| - |E|\big| - \big||\mathcal{K}_n(F)| - |E|\big|\big) + 2\Lambda|F\Delta\mathcal{K}_n(F)| \\ &\geq -\|\operatorname{div} X\|_\infty |F\Delta\mathcal{K}_n(F)| - \gamma C|F\Delta\mathcal{K}_n(F)| + \Lambda|F\Delta\mathcal{K}_n(F)| > 0,\end{aligned}$$

where in the last inequality we used (4.41).
Step 3. We can now argue as in Step 3 of the proof of Theorem 4.3 to deduce from (4.44) that

$$P_\Omega(F_n) \leq P_\Omega(F) + 4\Lambda|F\Delta F_n| \qquad \text{for all } F \subset \Omega \qquad (4.45)$$

and in turn, by Theorem 4.11, that

$$\partial F_n = \{x + \psi_n(x) n_E(x) : x \in \partial E\},$$

with $\psi_n \to 0$ in $C^{1,\alpha}(\partial E)$ for all $\alpha \in (0, 1)$.
Step 4. Here we show that $\psi_n \to 0$ in $W^{2,p}(\partial E)$ for all $p > 1$. To this aim, we first observe that since F_n is of class C^1 and satisfies (4.45), a standard variation argument (see Step 2 of the proof of Proposition 7.41 in [4]) yields

$$\|H_{\partial F_n}\|_{L^\infty(\partial F_n)} \leq 4\Lambda, \qquad (4.46)$$

for all n sufficiently large, where $H_{\partial F_h}$ denotes the (distributional) sum of the principal curvatures of ∂F_n. Since the functions ψ_n are equibounded in $C^{1,\alpha}$, the above estimate on the curvatures implies that for all $p > 1$ the functions ψ_n are equibounded in $W^{2,p}(\partial E)$, by elliptic regularity (see the proof of Lemma 4.14). Observe now that, due to (4.42), each F_n is a solutions of the obstacle problem (4.40) under the volume constraint. We can then easily deduce the Euler-Lagrange conditions $H_{\partial F_n} = f_n$ on ∂F_n, where, setting

$$\mathcal{N}_{\delta_n}(\partial E) := \{-\delta_n < d_E < \delta_n\},$$

$$f_h := \begin{cases} \lambda_n - 4\gamma v_{F_n} & \text{in } \partial F_n \cap \mathcal{N}_{\delta_n}(\partial E), \\ \lambda - 4\gamma v_E + r_n & \text{otherwise}, \end{cases} \tag{4.47}$$

λ_n and λ are the Lagrange multipliers corresponding to F_n and E, respectively, and

$$r_n = H_{\partial F_n} - H_{\partial E}(\cdot - \psi_n(\cdot)n_E(\cdot)) + 4\gamma\big[v_E - v_E\big(\cdot - \psi_n(\cdot)n_E(\cdot)\big)\big]$$

is a remainder clearly satisfying $\|r_n\|_{L^\infty(\partial F_n)} \to 0$.

We claim that (4.34) holds. If $\mathcal{H}^{N-1}(\partial F_n \cap \mathcal{N}_{\delta_n}(\partial E)) \to 0$, the claim follows immediately from (4.47), (4.31), and the fact the the λ_n's are bounded (thanks to (4.46)). Otherwise, we can apply a straightforward variant of the argument used in Step 6 of the proof of Theorem 4.3, to deduce first that $\lambda_n \to \lambda$ and then that (4.34) holds. In turn, we can conclude that $\psi_n \to 0$ in $W^{2,p}(\partial E)$ for all $p > 1$.

Thus, by (4.1) we have that $J(F_n) \geq J(E)$ for n sufficiently large, a contradiction to (4.38). □

Remark 4.15. The idea of solving the obstacle problem (4.40) to improve the convergence was already used in [57] in the context of the minimal surfaces.

5 Applications

In this section we apply the theory developed so far to establish several global and local minimality results. We start by investigating, in Subsection 5.1, the connections between the sharp-interface and diffuse Ohta-Kawasaki energies. In Subsection 5.2 we establish global and local minimality results in the small γ and small mass regimes for some simple configurations. Then, in Subsection 5.4, we consider lamellar configurations with more than one stripe and study their local and global minimality in two-dimensional thin rectangles.

5.1 Local minimizers of the diffuse Ohta-Kawasaki energy

In this subsection we explore the connection between local minimizers of the sharp interface Ohta-Kawasaki and ε-diffuse one defined in (1.3). Roughly speaking we will show that any strictly stable critical set for the sharp interface energy J can be approximated by local minimizers of the energy OK_ε for ε sufficiently small. This will be a consequence of Theorem 4.1 and of the Γ-convergence of OK_ε to J with respect to the L^1-distance. In the following, we briefly recall the definition of Γ-convergence and some of the properties we will need.

Definition 5.1 (Γ-convergence). Let X be a metric space endowed with the distance d. Let $F_n, F : X \to \mathbb{R} \cup \{+\infty\}$ be functionals defined in X. We say that the sequence $(F_n)_n$ Γ-converges to F with respect to the metric d, and we write

$$F_n \xrightarrow{\Gamma(d)} F\,,$$

if the following conditions are satisfied:

(i) (Γ-lim inf *inequality*) for all $x_n \xrightarrow{d} x$ in X, we have

$$F(x) \leq \liminf_n F_n(x_n)\,;$$

(ii) (Γ-lim sup *inequality*) for all $x \in X$ there exists a *recovery sequence* $(x_n)_n \subset X$ such that $x_n \xrightarrow{d} x$ and

$$F(x) \geq \limsup_n F_n(x_n)\,.$$

Given a one-parameter family of functionals $F_\varepsilon : X \to \mathbb{R} \cup \{+\infty\}$, $\varepsilon \in (0, \varepsilon_0)$, we say that $F_\varepsilon \xrightarrow{\Gamma(d)} F$ as $\varepsilon \to 0^+$, if $F_{\varepsilon_n} \xrightarrow{\Gamma(d)} F$ for every sequence $\varepsilon_n \to 0^+$.

Remark 5.2. Due to property (i), the inequality in (ii) is in fact equivalent to

$$F(x) = \lim_n F_n(x_n)\,.$$

The main property of Γ-convergence is stated in the following proposition, whose easy proof is omitted (see [6] or [14]).

Proposition 5.3. *Let (X, d) be a metric space. Assume that $F_n \xrightarrow{\Gamma(d)} F$ and that for all $n \in \mathbb{N}$ there exists $x_n \in X$ such that*

$$F_n(x_n) = \min_X F_n\,.$$

Assume also that there exist $x \in X$ and a subsequence such that $x_{n_k} \xrightarrow{d} x$. Then

$$F(x) = \min_X F\,.$$

We conclude this digression by recalling a *bridge principle* between *isolated* local minimizers of the Γ-limit functional and local minimizers of the approximating functionals, which was first observed in [30]. We provide the easy proof for the reader's convenience. To this end, given a sequence of functionals $F_n : X \to \mathbb{R} \cup \{+\infty\}$, we say that they are *equi-coercive* if for every $M > 0$ there exists a set $K \subset X$, compact with respect to the metric d, such that

$$\{x \in X : F_n(x) \leq M\} \subseteq K \qquad \text{for all } n \in \mathbb{N}.$$

The definition obviously extends to one-parameter families of functionals.

Theorem 5.4. *Let (X, d) be a metric space and let $F_n : X \to \mathbb{R} \cup \{+\infty\}$ be a sequence of lower semicontinuous and equi-coercive functionals such that $F_n \xrightarrow{\Gamma(d)} F$ for some functional F. Assume also that $\bar{x} \in X$ is an* isolated *local minimizer (with respect to the metric d) of F. Then, there exists a sequence $x_n \xrightarrow{d} \bar{x}$ such that x_n is a local minimizer of F_n for all n sufficiently large.*

Proof. By assumption there exists $\eta > 0$ such that $\bar{x}$ is the *unique* global minimizer of F in the closed ball $\overline{B_\eta(\bar{x})} := \{x \in X : d(x, \bar{x}) \leq \eta\}$; that is,

$$\{\bar{x}\} = \operatorname*{argmin}_{\overline{B_\eta(\bar{x})}} F\,. \tag{5.1}$$

Now, for every $n \in \mathbb{N}$ choose

$$x_n \in \operatorname*{argmin}_{\overline{B_\eta(\bar{x})}} F_n\,.$$

Note that the existence of a minimizer of F_n in $\overline{B_\eta(\bar{x})}$ follows from an application of the Direct Method of the Calculus of Variations, since by assumptions F_n is coercive and lower semicontinuous. Recall now that from the very definition of Γ-convergence we may find a recovery sequence $\bar{x}_n \xrightarrow{d} \bar{x}$ such that $F_n(\bar{x}_n) \to F(\bar{x})$. Thus, for n sufficiently large, say $n \geq n_0$, we have $\bar{x}_n \in B_\eta(\bar{x})$ and $F_n(\bar{x}_n) \leq F(\bar{x}) + 1$. In turn, by minimality and the equi-coercivity assumption, there exists a compact set K such that

$$x_n \in \{x \in X : F_n(x) \leq F(\bar{x}) + 1\} \subseteq K \qquad \text{for all } n \geq n_0.$$

We may therefore find $\bar{y} \in \overline{B_\eta(\bar{x})}$ and a subsequence (x_{n_k}) such that

$$x_{n_k} \xrightarrow{d} \bar{y}$$

By the minimality of x_n and the Γ-liminf inequality, we have

$$F(\bar{y}) \leq \liminf_k F_{n_k}(x_{n_k}) \leq \lim_k F_{n_k}(\bar{x}_{n_k}) = F(\bar{x}).$$

Recalling (5.1), we deduce $\bar{y} = \bar{x}$. Since the limit does not depend on the subsequence, we infer that the whole sequence converges; that is, $x_n \xrightarrow{d} \bar{x}$. In particular, for n sufficiently large, x_n belongs to the open ball $B_\eta(\bar{x})$ and thus is a local minimizer of F_n. □

Remark 5.5. (i) The above theorem obviously extends to one-parameter families of equi-coercive Γ-converging functionals.
(ii) The proof of the theorem shows in particular that for every $\eta' \in (0, \eta)$ we have

$$x_n \in \operatorname*{argmin}_{B_{\eta'}(x_n)} F_n ,$$

provided that n is large enough.

We now state the precise Γ-convergence result of the ε-diffuse energies OK_ε to J, which is a straightforward consequence of the Modica result [34]. We consider the Neumann case. Let $\Omega \subset \mathbb{R}^N$ be a bounded and Lipschitz open set and fix $d \in (0, |\Omega|)$. For $\varepsilon > 0$ define $OK_\varepsilon : L^1(\Omega) \to \mathbb{R} \cup \{+\infty\}$ as

$$\begin{aligned} OK_\varepsilon(u) := \varepsilon \int_\Omega |\nabla u|^2 \, dz + \frac{1}{\varepsilon} \int_\Omega (u^2 - 1)^2 \, dz \\ + \gamma_0 \int_\Omega \int_\Omega G_\Omega(z_1, z_2) u(z_1) u(z_2) \, dz_1 \, dz_2 \end{aligned}$$

if $u \in H^1(\Omega)$ and

$$\int_\Omega u = 2d - |\Omega| \tag{5.2}$$

and

$$OK_\varepsilon(u) := +\infty \qquad \text{otherwise in } L^1(\Omega).$$

Theorem 5.6. *The functionals OK_ε are equi-coercive and Γ-converge with respect to the L^1-distance to the function $\hat{J} : L^1(\Omega) \to \mathbb{R} \cup \{+\infty\}$*

defined as

$$\hat{J}(u) := \begin{cases} P_\Omega(E) \\ +\gamma \int_\Omega \int_\Omega G_\Omega(z_1, z_2) u_E(z_1) u_E(z_2)\, dz_1\, dz_2 & \text{if } u = u_E, \\ & \text{for some } E \\ & \text{with } |E| = d, \\ +\infty & \text{otherwise} \\ & \text{in } L^1(\Omega), \end{cases}$$

where $u_E := \chi_E - \chi_{\Omega \setminus E}$ *and* $\gamma := 3\gamma_0/16$.

The result follows from the Γ-convergence result of [34], since the nonlocal perturbation (*i.e.*, the double integral term) is continuous with respect to the L^1-convergence.

We are now in a position to state the main result of this section, which shows that any strictly stable critical set can be approximated by local minimizers of the ε-diffuse energy.

Theorem 5.7. *Let* $\Omega \subset \mathbb{R}^N$ *be a bounded smooth set and let* $E \subseteq \Omega$ *be a* smooth strictly stable critical set *in the sense of Definition* 3.1-(i). *Then there exists* $u_\varepsilon \to u_E$ *in* $L^1(\Omega)$ *such that* u_ε *is an* L^1*-local minimizer for* OK_ε *(under the volume constraint* (5.2)*) for all* ε *sufficiently small.*

Proof. Since Theorem 4.1 implies that u_E is an isolated L^1-local minimizer for the functional $\hat{J}$, the result follows from Theorem 5.6 and Theorem 5.4. □

Remark 5.8 (The periodic case). The analog of Theorem 5.6 holds in the periodic case, with $L^1(\Omega)$ and the L^1-distance replaced by the space X obtained as the quotient of $L^1(\Omega)$ with respect to the relation (3.2), and the distance

$$\alpha(u, v) = \min_{\tau \in \mathbb{T}^N} \|u(\cdot + \tau) - v\|_{L^1(\mathbb{T}^N)},$$

respectively. In this way any strictly stable critical set (in the sense of Definition 3.1-(ii)) is an isolated α-local minimizer. Therefore, by Theorem 5.4 the function u_E is the α-limit of α-local minimizers of the ε-diffuse energies (see [10] for similar observations in the context of the standard Periodic Isoperimetric Problem).

We conclude this subsection by observing that Theorem 5.7 (and its periodic analog) apply to a wealth of examples of strictly stable critical configurations for the sharp interface functional scattered in the literature. Among these, we mention the many droplet and spherical patterns proved to be strictly stable in [48] and [49], for some range of the parameters involved (see also [46,47]). Theorem 5.7 implies that for small values of ε there are local minimizers of the diffuse energies which are close (in an L^1-sense) to such configurations.

5.2 Minimality results in the small γ regime

In this subsection, we stress the dependence of the nonlocal functional upon the parameter γ by setting

$$J_\gamma(E) := P_{\mathbb{T}^N}(E) + \gamma \int_{\mathbb{T}^N} |\nabla v_E|^2 \, dz \,,$$

for all subsets $E \subset \mathbb{T}^N$ of finite perimeter. We recall that v_E is the potential associated with the set E according to (2.2). As usual, for any given $d \in (0, 1)$, we are interested in the periodic nonlocal isoperimetric problem

$$\min\{J_\gamma(E) : E \subset \mathbb{T}^N, \, |E| = d\} \,. \tag{5.3}$$

The goal is to show that for γ sufficiently small, $N = 2, 3$, and d properly chosen, the single lamella configuration $L := \mathbb{T}^{N-1} \times [0, d]$ is the unique (up to translations and relabeling of the coordinates) global minimizer of (5.3). The results described below are contained in [1]. In order

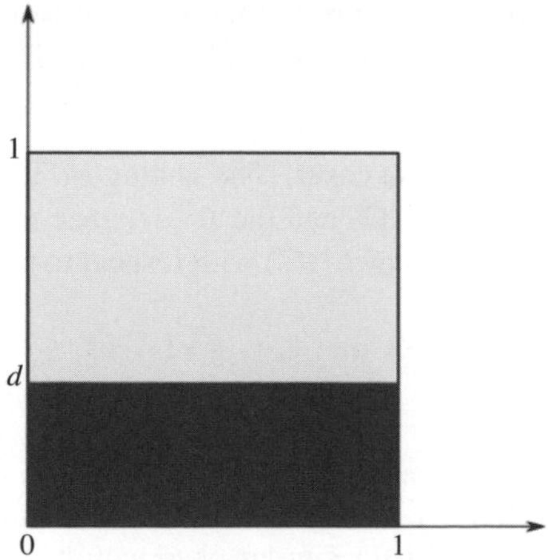

Figure 5.1. The single lamella with volume d.

to achieve our goal, we establish the following general criterion, linking the non local periodic isoperimetric problem to the (local) periodic isoperimetric problem (corresponding to $\gamma = 0$).

Theorem 5.9. *Assume that L is the unique, up to translations and relabeling of coordinates, global minimizer of the (local) periodic isoperimetric problem. Then the same set is also the unique global minimizer of* (5.3), *provided γ is sufficiently small.*

Proof. We argue by contradiction assuming that there exist a sequence $\gamma_n \searrow 0$ and a sequence of global minimizers E_n of

$$\min\{J_{\gamma_n}(E) : E \subset \mathbb{T}^N, |E| = d\}$$

such that E_n is NOT a single lamella for all n. Note that

$$\sup_n J_{\gamma_n}(E_n) \leq \sup_n J_{\gamma_n}(L) \leq J_{\gamma_1}(L) < +\infty .$$

Thus, by the Compactness Theorem 2.1, up to a (not relabeled) subsequence, we have $E_n \to E$ in L^1. Moreover, by the (easily proved) Γ-convergence of J_{γ_n} to the perimeter, we have that E is a global minimizer of the periodic isoperimetric problem. Thus, by assumption and up to translations and relabeling of coordinates, we may assume

$$E_n \to L \qquad \text{in } L^1(\mathbb{T}^N). \tag{5.4}$$

Note now that in $\mathbb{T}^N$ the boundary ∂L is made up of the two flat interfaces $\mathbb{T}^{N-1} \times \{0\}$ and $\mathbb{T}^{N-1} \times \{d\}$. Thus, the second variation $\partial^2 J_{\gamma_n}(L)$ of J_{γ_n} at L may be written as

$$\begin{aligned} \partial^2 J_{\gamma_n}(L)[\varphi] = & \int_{\partial L} |D_\tau \varphi|^2 \, d\mathcal{H}^{N-1} \\ & + 8\gamma_n \int_{\partial L}\int_{\partial L} G(z_1, z_2)\varphi(z_1)\varphi(z_2) d\mathcal{H}^{N-1}(z_1)\, d\mathcal{H}^{N-1}(z_2) \\ & + 4\gamma_n \int_{\partial L} \partial_n v_L \varphi^2 \, d\mathcal{H}^{N-1} \\ \geq & \int_{\partial L} |D_\tau \varphi|^2 \, d\mathcal{H} - 4\gamma_n \|\nabla v_L\|_{L^\infty} \int_{\partial L} \varphi^2 \, d\mathcal{H}^{N-1}, \end{aligned} \tag{5.5}$$

for all $\varphi \in H^1(\partial L)$. Since $\varphi \in T^\perp(\partial L)$ if and only if φ is a function in $H^1(\partial L)$ with zero average on each connected component of ∂L, using Poincaré inequality we deduce from (5.5) that

$$\partial^2 J_{\gamma_n}(L)[\varphi] \geq \frac{1}{2} \int_{\partial L} |D_\tau \varphi|^2 \, d\mathcal{H}^{N-1},$$

provided that n is large enough, say $n \geq n_0$. By Theorem 4.1, there exists $\delta > 0$ such that

$$J_{\gamma_{n_0}}(F) > J_{\gamma_{n_0}}(L), \qquad \text{for all } |F| = |L|, \text{ with } 0 < \alpha(F, L) < \delta . \tag{5.6}$$

We claim that the same holds true for all $n > n_0$. Indeed, assume towards a contradiction that for some $n > n_0$ and for some set F, with $|F| = d$ and $0 < \alpha(F, L) < \delta$, we have

$$J_{\gamma_n}(F) \leq J_{\gamma_n}(L) . \tag{5.7}$$

Since L is the unique global minimizer of the perimeter and thus $P_{\mathbb{T}^N}(F) > P_{\mathbb{T}^N}(L)$, it follows from (5.7) that

$$\int_{\mathbb{T}^N} |\nabla v_L|^2 \, dz - \int_{\mathbb{T}^N} |\nabla v_F|^2 \, dz > 0 .$$

But then, again by (5.7), we also have

$$\begin{aligned} P_{\mathbb{T}^N}(F) - P_{\mathbb{T}^N}(L) &\leq \gamma_n \left(\int_{\mathbb{T}} |\nabla v_L|^2 \, dz - \int_{\mathbb{T}} |\nabla v_F|^2 \, dz \right) \\ &< \gamma_{n_0} \left(\int_{\mathbb{T}} |\nabla v_L|^2 \, dz - \int_{\mathbb{T}} |\nabla v_F|^2 \, dz \right), \end{aligned}$$

which contradicts (5.6). Thus, we have proved that for all $n \geq n_0$

$$J_{\gamma_n}(F) > J_{\gamma_n}(L), \quad \text{for all } F \subset \mathbb{T}^N, \text{with } |F| = d \text{ and } 0 < \alpha(F, L) < \delta.$$

Recalling (5.4), for n large enough we also have $J_{\gamma_n}(E_n) > J_{\gamma_n}(L)$, contradicting the minimality of E_n. □

The application of the previous criterion combined with some classical results on the periodic isoperimetric problem yields the following corollaries.

Corollary 5.10 (Minimality of the single lamella in $\mathbb{T}^2$). *Let $N = 2$ and assume $d \in (0, 1 - 2/\pi)$. Then, for $\gamma > 0$ sufficiently small the single lamella L is the unique global minimizer of*

$$\min \left\{ P_{\mathbb{T}^2}(E) + \gamma \int_{\mathbb{T}^2} |\nabla v_E|^2 \, dx \, : \, E \subset \mathbb{T}^2, \, |E| = d \right\} .$$

Uniqueness is, as usual, up to translations and relabeling of coordinates.

Proof. As proven in [25] (see also [10]), for $d \in (0, 1 - 2/\pi)$ the single lamella L is the unique solution of the (local) periodic isoperimetric problem. Thus, the conclusion follows by applying Theorem 5.9. □

Remark 5.11. The result stated in Corollary 5.10 was obtained first in [52], with a different purely two-dimensional proof. See also [55] for the analogue on the two-sphere.

Corollary 5.12 (Minimality of the single lamella in $\mathbb{T}^3$). *Let $N = 3$. There exists $\eta > 0$ such that if $d \in (1/2 - \eta, 1/2 + \eta)$, then the single lamella L is the unique (in the usual sense) solution to*

$$\min\left\{ P_{\mathbb{T}^3}(E) + \gamma \int_{\mathbb{T}^3} |\nabla v_E|^2 \, dx \, : \, E \subset \mathbb{T}^3, \ |E| = d \right\},$$

provided that γ is small enough.

Proof. If $d = 1/2$ it was proven in [24] (see also [10, Theorem 4.3]) that the single lamella is the unique solution to the (local) periodic isoperimetric problem. A perturbation argument shows that the same is true for all d ranging in a small interval centered at $1/2$, see [1, Theorem 5.3] for the details. The conclusion follows again by applying Theorem 5.9. □

We conclude this subsection by showing a connection between sets whose boundary is a strictly stable (periodic) constant mean curvature hypersurface and isolated local minimizers of the non local functional J_γ, for small γ's. Precisely, we have:

Theorem 5.13. *Let $E \subset \mathbb{T}^N$ be a smooth strictly stable set for the perimeter functional; that is, there exists $\lambda \in \mathbb{R}$ such that*

$$H_{\partial E} = \lambda \qquad \text{on } \partial E$$

and

$$\int_{\partial E} \left(|D_\tau \varphi|^2 - |B_{\partial E}|^2 \varphi^2 \right) d\mathcal{H}^{N-1} > 0 \qquad \text{for all } \varphi \in T^\perp(\partial E) \setminus \{0\},$$

where $T^\perp$ is defined in (3.1). Then there exists $\gamma_0 > 0$ with the following property: for all $\gamma \in (0, \gamma_0)$ there exists $E_\gamma \subset \mathbb{T}^N$ such that $|E_\gamma| = |E|$ and E_γ is an isolated α-local minimizer of J_γ (where α is defined in (3.3)). Moreover, for all $\beta \in (0, 1)$ we have $E_\gamma \to E$ in $C^{3,\beta}$ as $\gamma \to 0$.

Proof. We only sketch the proof, since many of the arguments have been already used before. Let X be as in Remark 5.8. We start by observing that the functionals $\hat{J}_\gamma : X \to \mathbb{R} \cup \{+\infty\}$ defined as

$$\hat{J}_\gamma(u) := \begin{cases} J_\gamma(F) & \text{if } u = u_F, \text{ for some } F \text{ with } |F| = |E|, \\ +\infty & \text{otherwise in } X, \end{cases}$$

are equi-coercive and Γ-converge with respect to the distance α to the functional $\hat{J}_0$ defined as

$$\hat{J}_0(u) := \begin{cases} P_{\mathbb{T}^N}(F) & \text{if } u = u_F, \text{ for some } F \text{ with } |F| = |E|, \\ +\infty & \text{otherwise in } X. \end{cases}$$

Recall that by Theorem 4.1-(ii) the set E is an isolated α-minimizer for $\hat{J}_0$. Therefore, we may apply Theorem 5.4 (see also Remark 5.5-(i)) to deduce the existence of a one-parameter family of sets $(E_\gamma)_{\gamma>0}$ such that $|E_\gamma| = |E|$ and $\alpha(E_\gamma, E) \to 0$ as $\gamma \to 0^+$. Moreover, for all γ sufficiently small, say $\gamma < \gamma_0$, E_γ is a local minimizer with respect to the distance α, with minimality radius independent of γ (see Remark 5.5-(ii)); that is, there exists $\delta > 0$ such that

$$E_\gamma \in \operatorname{argmin}\Big\{ J_\gamma(F) : F \subset \mathbb{T}^N,\ |F| = |E|,\ \alpha(E_\gamma, F) \leq \delta \Big\}$$

for all $\gamma \in (0, \gamma_0)$. By Lemma 4.6 and arguing as in Remark 4.7, we infer that there exist positive constants ω and r such that E_γ is strong quasi-minimizer of $P_{\mathbb{T}^N}$ with constant (ω, r) (see Definition 4.8). In turn, by Theorem 4.11 we deduce that $E_\gamma \to E$ in a $C^{1,\beta}$-sense for all $\beta \in (0, 1)$. The $C^{3,\beta}$ convergence follows now from the Euler-Lagrange equations satisfied by E_γ and arguing as in Step 6 of the proof of Theorem 4.3. Finally, set

$$\inf\Big\{ \int_{\partial E} \Big(|D_\tau \varphi|^2 - |B_{\partial E}|^2 \varphi^2 \Big)\, d\mathcal{H}^{N-1} : \varphi \in T^\perp \setminus \{0\},\ \|\varphi\|_{H^1(\partial E)} = 1 \Big\}$$
$$=: m_0 > 0\,.$$

The positivity of m_0 can be shown as in Step 1 of the proof of Theorem 3.2. Using the $C^{3,\beta}$-convergence of E_γ to E and by arguments similar to Step 2 of the proof of Theorem 4.3, one can show that

$$\liminf_{\gamma \to 0^+} \inf \Big\{ \partial^2 J_\gamma(E_\gamma)[\varphi] : \varphi \in T^\perp \setminus \{0\},\ \|\varphi\|_{H^1(\partial E)} = 1 \Big\} \geq m_0\,.$$

Thus, by taking γ_0 smaller if needed, we have that E_γ is strictly stable in the sense of Definition 3.1-(ii) for all $\gamma \in (0, \gamma_0)$. In turn, by Theorem 4.1-(ii), for all such γ's the set E_γ is an isolated α-local minimizer for J_γ. □

Remark 5.14. There are several known examples of subsets of the torus $\mathbb{T}^N$, whose boundary is a strictly stable constant mean curvature hypersuface; among them, we mention of course lamellae, cylinders, balls, but also more complicated structures like the class of triply periodic sets considered in [50] (see Figure 5.2).

5.3 Minimality results in the small mass regime

In this subsection we consider the Periodic Nonlocal Isoperimetric problem for small values of the mass d. We want to show that in this regime

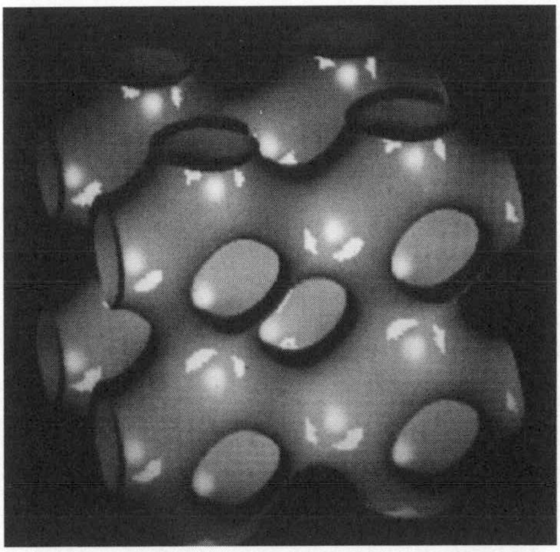

Figure 5.2. An example of triply periodic strictly stable set.

the perimeter term becomes dominant and the global minimizer is given by a single almost round droplet, which resembles a small sphere more and more as $d \to 0^+$. Precisley, we will establish:

Theorem 5.15 (single droplet solutions in the small mass regime). *Fix $\gamma > 0$ and let E_d denote a minimizer of* (5.3). *Then for all d small enough E_d consists of a single (almost) round droplet, and in fact (up to translations)*

$$d^{-1/N} E_d \to B \qquad \text{in the } C^{3,\beta}\text{-sense}$$

for all $\beta \in (0, 1)$, where B is a ball of measure one.

Proof. Throughout the proof, B will denote a fixed ball of unit measure. Moreover, given $s > 0$ and $E \subset s\mathbb{T}^N$ we denote by $v_E^{(s)}$ the potential associated with E as in (2.4), with $\mathbb{T}^N$ replaced by $s\mathbb{T}^N$. Now note that for any $E \subset \mathbb{T}^N$ and $d > 0$ if we set $\widetilde{E} := d^{-1/N} E \subset d^{-1/N}\mathbb{T}^N$, then by scaling we clearly have:

$$v_E(z) = d^{2/N} v_{\widetilde{E}}^{(d^{-1/N})}(d^{-1/N} z) \qquad \text{for all } z \in \mathbb{T}^N. \tag{5.8}$$

In turn, by a change of variables,

$$\int_{\mathbb{T}^N} |\nabla v_E|^2 \, dz = d^{1+2/N} \int_{d^{-1/N}\mathbb{T}^N} \left|\nabla v_{\widetilde{E}}^{(d^{-1/N})}\right|^2 dz. \tag{5.9}$$

Now, let $E \subset \mathbb{T}^N$ be of measure $|E| = d$ and let $\widetilde{E}$ as before, so that $|\widetilde{E}| = 1$. Using (5.9), we have

$$P_{\mathbb{T}^N}(E) + \gamma \int_{\mathbb{T}^N} |\nabla v_E|^2 \, dz$$
$$= d^{\frac{N-1}{N}} \left(P_{d^{-1/N}\mathbb{T}^N}(\widetilde{E}) + \gamma \, d^{3/N} \int_{d^{-1/N}\mathbb{T}^N} \left| \nabla v_{\widetilde{E}}^{(d^{-1/N})} \right|^2 dz \right).$$

We conclude that by scaling (5.3) is equivalent to

$$\min \left\{ P_{d^{-1/N}\mathbb{T}^N}(E) + \gamma \, d^{3/N} \int_{d^{-1/N}\mathbb{T}^N} \left| \nabla v_E^{(d^{-1/N})} \right|^2 dz \right.$$
$$\left. : \; E \subset d^{-1/N}\mathbb{T}^N, \; |E| = 1 \right\}. \tag{5.10}$$

Let now $d_n \searrow 0$ and denote by E_n any minimizer of the following penalized problem:

$$\min \left\{ P_{d_n^{-1/N}\mathbb{T}^N}(E) + \gamma \, d_n^{3/N} \int_{d_n^{-1/N}\mathbb{T}^N} \left| \nabla v_E^{(d_n^{-1/N})} \right|^2 dz \right.$$
$$\left. + \Lambda \big| |E| - 1 \big| \; : \; E \subset d_n^{-1/N}\mathbb{T}^N \right\}, \tag{5.11}$$

where $\Lambda > 0$ is chosen in such a way that any ball B of unit measure is the unique (up to translation) minimizer of

$$\min \left\{ P_{\mathbb{R}^N}(E) + \Lambda \big| |E| - 1 \big| \; : \; E \subset \mathbb{R}^N \right\}. \tag{5.12}$$

Using the Isoperimetric Inequality, it is easy to check that

$$\Lambda := 2 P_{\mathbb{R}^N}(B) \tag{5.13}$$

satisfies the above requirement. Note that by minimality, and using (5.9), we have

$$\begin{aligned} P_{d_n^{-1/N}\mathbb{T}^N}(E_n) &\leq P_{d_n^{-1/N}\mathbb{T}^N}(E_n) + \Lambda \Big| |E_n| - 1 \Big| \\ &\leq P_{d_n^{-1/N}\mathbb{T}^N}(B) + \gamma \, d_n^{3/N} \int_{d_n^{-1/N}\mathbb{T}^N} \left| \nabla v_B^{(d_n^{-1/N})} \right|^2 dz \\ &= P_{\mathbb{R}^N}(B) + \gamma d^{1-1/N} \int_{\mathbb{T}^N} |\nabla v_{d_n^{1/N} B}|^2 \, dz \\ &\leq P_{\mathbb{R}^N}(B) + \gamma d_n^{1-1/N} C |d^{1/N} B| \\ &= P_{\mathbb{R}^N}(B) + \gamma C d_n \, , \end{aligned} \tag{5.14}$$

where $C = C(\mathbb{T}^N)$ is the Lipschitz constant provided by (the periodic analog of) Lemma 4.5. Let now $F_1, F_2 \subset d^{-1/N}\mathbb{T}^N$. Using again (5.9), we have

$$\begin{aligned} &\gamma\, d^{3/N}\left|\int_{d^{-1/N}\mathbb{T}^N}\left|\nabla v_{F_1}^{(d^{-1/N})}\right|^2 dz - \int_{d^{-1/N}\mathbb{T}^N}\left|\nabla v_{F_2}^{(d^{-1/N})}\right|^2 dz\right| \\ &\quad = \gamma\, d^{1-1/N}\left|\int_{\mathbb{T}^N}|\nabla v_{d^{1/N}F_1}|^2\, dz - \int_{\mathbb{T}^N}|\nabla v_{d^{1/N}F_2}|^2\, dz\right| \\ &\quad \le \gamma C d^{1-1/N}|d^{1/N}F_1 \Delta d^{1/N}F_2| = \gamma C d |F_1 \Delta F_2|\,, \end{aligned} \tag{5.15}$$

where $C > 0$ is as before. Arguing as in Remark 4.7, we can conclude that each E_n is a strong quasi-minimizer of the perimeter with constants $(\Lambda + C, R)$ (in fact $(\Lambda + \gamma C d_n, R)$), for any $R > 0$, provided that n is large enough.

We claim now that for n large enough every (measure theoretic) connected component (see for instance [3] for the precise definition) of E_n is bounded uniformly with respect to n, provided that n is large enough. To this end, we use the following property of (ω, R) strong quasi-minimizers: there exists a positive constant θ_0 depending only on N such that for every (ω, R) strong quasi-minimizer E and for every $y \in \partial^* E$, we have

$$P_{B_r(y)}(E) \ge \theta_0\, r^{N-1}\,, \tag{5.16}$$

provided that $r < \min\{R, 1/(2N\omega)\}$ (see [33, Theorem 21.11]. By density the inequality extends to the closure of $\partial^* E$ which coincides with the topological boundary of (a suitable representative of) E. Now fix $r > 0$ such that (5.16) holds for E_n for all n sufficiently large. Let C_n be a connected component of E_n and let $(B^n_k)_k$ be a collection of balls of radius r and centers on ∂C_n, covering ∂C_n. By Vitali's Covering Lemma, we may extract a subfamily of disjoint balls $(B^n_{k_i})_i$ such that the family $(\hat{B}^n_{k_i})_i$ still covers ∂C_n, where $\hat{B}^n_{k_i}$ is the ball having the same center as $B^n_{k_i}$ and radius $5r$. Since by (5.16) we have $P_{B^n_{k_i}}(E_n) \ge \theta_0\, r^{N-1}$ and the perimeters of the E_n's are equibounded by (5.14), we deduce that the number of balls in such subfamily is finite and bounded independently of n. We may clearly suppose that the family $(\hat{B}^n_{k_i})_i$ is minimal, in the sense that if we remove any ball than the reduced family does not cover ∂C_n anymore. Then, necessarily, no ball of $(\hat{B}^n_{k_i})_i$ can be disjoint from the others, since, otherwise, the intersection of C_n with such a ball would intersect a different connected component of C_n, which is impossible. Thus the diameter of $\cup_i \hat{B}^n_{k_i}$ is bounded uniformly with respect to n and in turn C_n itself is uniformly bounded.

Denote now by $\widetilde{E}_n$ the union of all connected components $\widetilde{C}_n$ of the $d_n^{-1/N}\mathbb{T}^N$-periodic extension of E_n, which are compactly contained in $(0, 2d_n^{-1/N})^N$ and intersect $(0, d_n^{-1/N})^N$. By the uniform bound on the diameter of such connected components, one can easily see that for n large enough

$$P_{\mathbb{R}^N}(\widetilde{E}_n) = P_{d_n^{-1/N}\mathbb{T}^N}(E_n) \qquad \text{and} \qquad |E_n| = |\widetilde{E}_n|\,. \tag{5.17}$$

Using (5.17) and recalling that the ball B of unit measure is the unique (up to translations) minimizer of (5.12), we have

$$P_{d_n^{-1/N}\mathbb{T}^N}(E_n) + \Lambda\Big||E_n| - 1\Big| = P_{\mathbb{R}^N}(\widetilde{E}_n) + \Lambda\Big||\widetilde{E}_n| - 1\Big| \geq P_{\mathbb{R}^N}(B)\,.$$

Combining with (5.14), we deduce that

$$P_{\mathbb{R}^N}(\widetilde{E}_n) \to P_{\mathbb{R}^N}(B) \qquad \text{and} \qquad |\widetilde{E}_n| \to 1 \tag{5.18}$$

as $n \to \infty$. It is well known that the above limiting relations imply that, up to translations,

$$\widetilde{E}_n \to B \qquad \text{in } L^1. \tag{5.19}$$

This is a particular case of the quantitative isoperimetric inequality established in [18]. However, (5.19) can be easily proven by more standard arguments based on the Concentration Compactness Principle, which allows one to show that the sequence $\widetilde{E}_n$ is relatively compact and thus, upon translating and extracting a subsequence, it must converge to a set E such that $|E| = 1$ and $P_{\mathbb{R}^N}(E) = P_{\mathbb{R}^N}(B)$. The standard isoperimetric inequality then implies that $E = B$ (up to translations). Recalling that the sets $\widetilde{E}_n$ are strong quasi-minimizers of the perimeter with constants independent of n, we may apply (the periodic analog of) Theorem 4.11 to deduce from (5.19) that in fact $\widetilde{E}_n \to B$ in the $C^{1,\beta}$-sense. In turn, the same holds for the original sequence E_n, upon suitable translations; that is,

$$E_n \to B \qquad \text{in the } C^{1,\beta}\text{-sense} \tag{5.20}$$

for all $\beta \in (0, 1)$. Now observe that by Remark 2.3 and the Sobolev Embedding Theorem we have

$$\begin{aligned}\|\nabla v_{d_n^{1/N}E_n}\|_{C^{0,\beta}(\mathbb{T}^N)} &\leq C\|v_{d_n^{1/N}E_n}\|_{W^{2,N/(1-\beta)}(\mathbb{T}^N)}\\ &\leq C'\|u_{d_n^{1/N}E_n} - m_{d_n^{1/N}E_n}\|_{L^{N/(1-\beta)}} \leq C''d_n^{(1-\beta)/N},\end{aligned}$$

where all the constants involved depend only on $\mathbb{T}^N$. Using (5.8), we can deduce

$$\left\|\nabla v_{E_n}^{(d_n^{-1/N})}\right\|_{C^{0,\beta}(d_n^{-1/N}\mathbb{T}^N)} \leq C''d_n^{-\beta/N}$$

and thus

$$\left\| d_n^{3/N} v_{E_n}^{(d_n^{-1/N})} \right\|_{C^{1,\beta}(d_n^{-1/N}\mathbb{T}^N)} \to 0$$

for all $\beta \in (0,1)$. Having established this, the $C^{3,\beta}$-convergence of E_n to B follows from (5.20) and the Euler-Lagrange equations satisfied by E_n, arguing as in Step 6 of the proof of Theorem 4.3. In order to conclude, it remains to show that $|E_n| = 1$ for n large enough. This can be argued by contradiction. Assume that along a subsequence we have $|E_n| < 1$ and set $\hat{E}_n := \lambda_n(E_n - z_0) + z_0$, where $\lambda_n := (1/|E_n|)^{1/N}$ and z_0 is the center of B. Observe that by n large E_n is convex (this follows from the C^3-convergence to B) and $z_0 \in E_n$, so that $E_n \subset \hat{E}_n$. Denoting by J_n the functional appearing in (5.11), we may estimate

$$\begin{aligned}
J_n(E_n) - J_n(\hat{E}_n) &= (1-\lambda_n^{N-1})P_{\mathbb{R}^N}(E_n) + \Lambda|E_n|(\lambda^N - 1) \\
&+ \gamma d_n^{3/N}\left(\int_{d_n^{-1/N}\mathbb{T}^N} \left|\nabla v_{E_n}^{(d_n^{-1/N})}\right|^2 dz - \int_{d_n^{-1/N}\mathbb{T}^N} \left|\nabla v_{\hat{E}_n}^{(d_n^{-1/N})}\right|^2 dz\right) \\
&\overset{(5.15)}{\geq} (1-\lambda_n^{N-1})P_{\mathbb{R}^N}(E_n) + \Lambda|E_n|(\lambda_n^N - 1) - \gamma C d_n |\hat{E}_n \Delta E_n| \\
&= (1-\lambda_n^{N-1})P_{\mathbb{R}^N}(E_n) + (\Lambda - \gamma C d_n)|E_n|(\lambda_n^N - 1) \\
&\geq (1-\lambda_n^{N-1})\frac{4}{3}P_{\mathbb{R}^N}(B) + \frac{3}{4}\Lambda(\lambda_n^N - 1),
\end{aligned}$$

where the last equality holds for n large enough, thanks to (5.18) and the fact that $d_n \searrow 0$. Recalling (5.13), we easily see that the last expression is positive. Therefore we have shown that $J_n(\hat{E}_n) < J_n(E_n)$ for n large enough, contradicting the minimality of E_n. The same argument shows that we cannot have $|E_n| > 1$ along a subsequence, and thus $|E_n| = 1$ for n large enough. This concludes the proof of the theorem. □

Remark 5.16. (i) Theorem 5.15 is a particular case of the analysis carried out in [13], where also the Neumann case is considered and a precise rate of convergence of the minimizers to the ball together with an asymptotic expansion of the energy are provided.
(ii) If one lets the parameter γ in (5.3) depend on d in a suitable way, it is expected that global minimizers are made up of an increasing number of small round droplets periodically arranged in space. This expectation is supported by experiments and numerical evidence and has received an almost complete analytical validation in two dimensions in the recent works [21, 22], where a careful asymptotic analysis as $d \to 0+$ is performed via Γ-convergence techniques. It is shown, in particular, that the asymptotic location of the droplet centers minimizes a suitable renormalized energy. However, it is still analytically unproven that such

an optimal arrangement is periodic. We remark that the analysis of the aforementioned works borrows ideas and techniques from the Ginzburg-Landau world. We refer also to [7,8] for further related work.

We conclude this subsection by mentioning the following nonlocal isoperimetric problem set in the whole space $\mathbb{R}^N$:

$$\text{minimize} \quad P_{\mathbb{R}^N}(E) + \int_E \int_E \frac{1}{|z_1 - z_2|^\alpha}\, dz_1 dz_2 \tag{5.21}$$

among all sets $E \subset \mathbb{R}^N$ with given volume $|E| = d$. Here α ranges in the interval $(0, N-1)$. The case $N = 3$ and $\alpha = 1$ corresponds to the classical Gamow's liquid drop model of the atomic nuclei. More in general, the case $\alpha = N - 2$ may be thought as the analog of (2.3), with $\Omega = \mathbb{R}^N$. The unboundedness of the ambience space poses some new compactness issues and, in fact, the existence of global minimizers may fail if d is large, as shown in [28,29,32]. This is due to the repulsive effects of the nonlocal term that force minimizing sequences to split into several connected components with diverging mutual distances.

However, in the same spirit of Theorem 5.15, when d is small one expects the perimeter term to dominate and global minimizers to exist and resemble balls. In fact, it has been shown in [26,28,29,41] in some particular cases, and in [5] in the general case that if d is small enough, then global minimizers do exist and are balls. It has been conjectured (see [7]) that in the case $\alpha = N - 2$ either global minimizers are balls or they fail to exist. So far, this is known to be true only for α sufficiently small (see [5]). We conclude by mentioning the recent work [16], where some of the aforementioned results has been extended to the case where the perimeter term in (5.21) is replaced by the so called *fractional* perimeter (see [56]). Moreover, in [16] the parameter α is allowed to range in the interval $(0, N)$.

5.4 Lamellar configurations with a lot of interfaces

In this subsection we consider lamellar configurations with an arbitrary number $k \in \mathbb{N}$ of interfaces in two-dimensional rectangles of the form $(0, a) \times (0, 1)$, $a > 0$. Here the non locality parameter $\gamma > 0$ is any fixed positive constant that we don't assume to be small. Instead, we will play with the thickness parameter a in order to study the local and global minimality. Moreover, we switch back to the Neumann boundary conditions. All the results that we present below, with the exception of Proposition 5.18, are contained in [36].

Throughout the subsection, it is convenient to regard the energy functional J as defined on functions belonging to $BV(\Omega; \{-1, 1\})$, rather then on sets. More precisely, given $\Omega \subset \mathbb{R}^2$ open and bounded, for all $u \in BV(\Omega; \{-1, 1\})$ we write

$$J_\Omega(u) := P_\Omega(\{u = 1\}) + \gamma \int_\Omega |\nabla v|^2 \, dxdy \,, \tag{5.22}$$

where

$$-\Delta v = u - m \text{ in } \Omega, \quad \nabla v \cdot n_\Omega = 0 \text{ on } \partial\Omega, \quad \int_\Omega v = 0, \tag{5.23}$$

with n_Ω denoting, as usual, the outer unit normal to $\partial\Omega$, and $m := \fint_\Omega u$. In (5.22) we have used the symbol J_Ω (instead of J) to stress the dependence of the functional on the domain Ω. We recall that in this framework the volume constraint corresponds to fixing the average $m := \fint_\Omega u$.

By a LAMELLAR CONFIGURATION ON $\Omega_a := (0, a) \times (0, 1)$, $a > 0$, we mean any function $u \in BV(\Omega_a; \{-1, 1\})$ depending only on the y-variable; *i.e.*, such that $u(x, y) := \hat{u}(y)$ for some $\hat{u} \in BV((0, 1); \{-1, 1\})$. Note that, necessarily, the potential v associated with u according to (5.23) is in turn one-dimensional; *i.e.*, $v(x, y) := \hat{v}(y)$, where $\hat{v}$ solves

$$\begin{cases} -\hat{v}'' = \hat{u} - \int_0^1 \hat{u} & \text{in } (0, 1), \\ \hat{v}'(0) = \hat{v}'(1) = 0\,, \\ \int_0^1 \hat{v} = 0\,. \end{cases}$$

The given lamellar configuration can be visualized as a finite collection of horizontal stripes, where the function u takes alternatively the values 1 and -1. Denoting the set of jump points of $\hat{u}$ by $\{y_1, \ldots, y_k\}$, the k-interfaces of such stripes are horizontal straight line segments located at the heights $y_1, \ldots, y_k$ (see Figure 5.3 below).

In the following, in order to simplify the presentation, we assume $m = \fint_\Omega u = 0$. Thus, in particular, the above system reduces to

$$\begin{cases} -\hat{v}'' = \hat{u} & \text{in } (0, 1), \\ \hat{v}'(0) = \hat{v}'(1) = 0\,, \\ \int_0^1 \hat{v} = 0\,. \end{cases} \tag{5.24}$$

We now note that in the case of a lamellar configuration the Euler-Lagrange equation (2.16) reduces to

$$v = \text{const} \qquad \text{on } \partial\{u = 1\} \cap \Omega_a$$

or, equivalently,

$$\hat{v}(y_1) = \hat{v}(y_2) = \cdots = \hat{v}(y_k)\,.$$

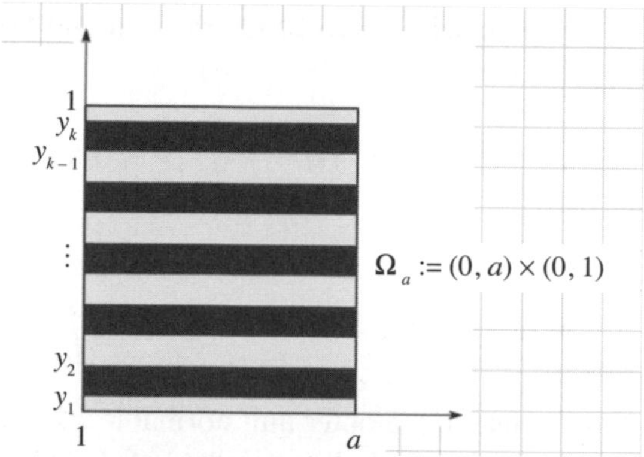

Figure 5.3. An example of *critical* lamellar configuration with k-interfaces. Note that all stripes are equally spaced.

It is straightforward to show that for the solution to (5.24) to satisfy the above condition it is necessary that all the intervals (y_{j-1}, y_j) have equal length. More precisely, one can show that a lamellar configurations with k interfaces (and $m = 0$) is *critical* if and only if

$$y_j = \frac{2j-1}{2k} \qquad j = 1, \ldots, k. \tag{5.25}$$

In the following we will denote by u^k the CRITICAL lamellar configuration with k interfaces, with locations described in (5.25). Note that such a configuration is uniquely determined up to the sign, i.e. up to multiplying u by -1. We make the convention that $u^k = 1$ in the bottom stripe $(0, a) \times (0, y_1)$. We also denote by v^k the associated potential solving (5.24). Using the representation of u^k given by (5.25), one can find the associated potential v^k explicitly and then compute $J_{\Omega_a}(u^k)$. In turn, for any given $\gamma > 0$ one can determine the optimal number $k(\gamma)$ of interfaces which makes the energy J_{Ω_a} minimal among all possible lamellar configurations. Elementary computations show that for every $k \in \mathbb{N}$ one has

$$J_{\Omega_a}(u^k) = a\left(k + \frac{\gamma}{12k^2}\right)$$

and

$$\min_{k \in \mathbb{N}} J_{\Omega_a}(u^k) = J_{\Omega_a}(u^{k(\gamma)}), \tag{5.26}$$

where

$$k(\gamma) \text{ is either } \left[\sqrt[3]{\tfrac{\gamma}{6}}\right] \text{ or } \left[\sqrt[3]{\tfrac{\gamma}{6}}\right] + 1. \tag{5.27}$$

Here $[\cdot]$ stands for the integer part of its argument.

Thus, $u^{k(\gamma)}$ minimizes the energy among one-dimensional configurations. We will see later on that $u^{k(\gamma)}$ is in fact a true volume constrained global minimizer of J_{Ω_a}, provided that Ω_a is sufficiently thin, that is, a is sufficiently small.

Next we address the local minimality of critical lamellar configurations. In view of Theorem 4.1, this amounts to estimating the first eigenvalue of the quadratic form $\partial^2 J_{\Omega_a}(u^k)$. To this aim, let $\Gamma_j := (0,a)\times\{y_j\}$ denote the j-th interface of the lamellar configuration u^k. Set also $\Gamma := \bigcup_{j=1}^k \Gamma_j$ and note that any function $\varphi \in \widetilde{H}(\Gamma)$ (see (2.22)) may be identified with k functions $\varphi_1, \dots, \varphi_k \in H^1(0,a)$, where each φ_j represents the restriction of φ to Γ_j. In turn, the zero average condition reads

$$\sum_{j=1}^k \int_0^a \varphi_j \, dx = 0 .$$

With this identification in mind, we may rewrite $\partial^2 J(u^k)$ in the following way

$$\begin{aligned}\partial^2 J_{\Omega_a}(u^k)[\varphi] = & \sum_{j=1}^k \int_0^a \varphi_j'(x)^2 \, dx \\ & + 8\gamma \int_\Gamma\int_\Gamma G_{\Omega_a}(z_1,z_2)\,\varphi(z_1)\,\varphi(z_2)\, d\mathcal{H}^1(z_1)\, d\mathcal{H}^1(z_2) \\ & + 4\gamma \int_\Gamma \left(\nabla v^k(z)\cdot n_{\{u^k=1\}}(z)\right)\varphi^2(z)\, d\mathcal{H}^1(z)\end{aligned} \tag{5.28}$$

for all $\varphi \in \widetilde{H}(\Gamma)$. We have the following:

Proposition 5.17 (Local minimality of lamellar configurations). *Assume that*

$$a < 2\pi\sqrt{\frac{k}{2\gamma}} . \tag{5.29}$$

Then there exists $\delta > 0$ and $C > 0$ such that

$$J_{\Omega_a}(u) \geq J_{\Omega_a}(u^k) + C\|u - u^k\|^2_{L^1(\Omega_a)}$$

for all $u \in BV(\Omega_a;\{-1,1\})$, with $\fint_{\Omega_a} u = 0$ and $\|u-u^k\|_{L^1(\Omega_a)} \leq \delta$.

Proof. Owing to Theorem 4.1[1], it suffices to show that (5.29) implies the positive definiteness of $\partial^2 J_{\Omega_a}(u^k)$; *i.e.*, $\partial^2 J_{\Omega_a}(u^k)[\varphi] > 0$ for all $\varphi \in$

[1] It should be noticed that here the smoothness assumption on Ω is not satisfied. However, the interfaces of the lamellar configurations only meet the regular part of $\partial\Omega_a$ and this allows one to extend Theorem 4.1 to this case; see [36] for the details.

$\widetilde{H}(\Gamma) \setminus \{0\}$. Here we establish the claim under the additional assumption that

$$\int_{\Gamma_j} \varphi(z)\, d\mathcal{H}^1(z) = \int_0^a \varphi_j(x)\, dx = 0 \qquad \text{for all } j = 1, \ldots, k. \tag{5.30}$$

We refer the reader to [36, Section 4] for the general case. So assume that (5.30) holds. Then we will analyze each term of (5.28) separately. Regarding the first one, we use the one-dimensional version of Poincaré inequality to infer that

$$\int_0^a \varphi_j'(x)^2\, dx \geq \left(\frac{2\pi}{a}\right)^2 \int_0^a \varphi_j(x)^2\, dx\,. \tag{5.31}$$

Next, by the properties of the Green's function G_{Ω_a}, we note that

$$\int_\Gamma \int_\Gamma G_{\Omega_a} \varphi\, \varphi\, d\mathcal{H}^1 d\mathcal{H}^1 = \int_{\Omega_a} |\nabla v_\varphi|^2 dz\,, \tag{5.32}$$

where v_φ denotes the weak solution to

$$-\Delta v_\varphi = \varphi \mathcal{H}^1 \llcorner \Gamma \qquad \text{in } \Omega_a$$

subject to homogeneous Neumann boundary conditions and zero average (see Remark 2.10-(ii)).

It remains to compute the last integral in (5.28). In view of (5.24) and (5.25) we have the alternating pattern

$$(v^k)'(y_1) = -\frac{1}{2k},\ (v^k)'(y_2) = \frac{1}{2k},\ (v^k)'(y_3) = -\frac{1}{2k}, \ldots$$

Recalling that $\nabla v_k \cdot n_{\{u^k=1\}}$ denotes the outer normal derivative with respect to the set $\{u_k = 1\}$, we then have

$$\begin{aligned} \int_\Gamma \left(\nabla v^k \cdot n_{\{u^k=1\}}\right) \varphi^2\, d\mathcal{H}^1 &= \sum_{j=1}^k \int_{\Gamma_j} \left(\nabla v^k \cdot n_{\{u^k=1\}}\right) \varphi^2\, d\mathcal{H}^1 \\ &= (v^k)'(y_1) \int_0^a \varphi_1(x)^2\, dx - (v^k)'(y_2) \int_0^a \varphi_2(x)^2\, dx + \ldots \\ &\quad - (-1)^k (v^k)'(y_k) \int_0^a \varphi_k(x)^2\, dx \\ &= -\frac{1}{2k} \sum_{j=1}^k \int_0^a \varphi_j(x)^2\, dx. \end{aligned} \tag{5.33}$$

Combining (5.31), (5.32) and (5.33) we conclude that

$$\partial^2 J_{\Omega_a}(u^k)[\varphi] \geq \left(\left(\frac{2\pi}{a}\right)^2 - \frac{2\gamma}{k}\right)\sum_{j=1}^{k}\int_0^a \varphi_j(x)^2\,dx + 8\gamma\int_{\Omega_a}|\nabla v_\varphi|^2\,dz.$$

Recalling (5.29), the conclusion follows. □

Note that for a fixed (large) γ one needs a lot of interfaces for (5.29) be satisfied. Viceversa, one can show that for any given $k \in \mathbb{N}$ the critical lamellar configuration with k interfaces u^k becomes unstable if γ is large enough. Thus, increasing k has a stabilizing effect, whereas increasing γ leads to instability. The latter statement is made precise by the following proposition (see [11, Proposition 3.6]).

Proposition 5.18 (Instability for γ large). *Fix $a > 0$ and $k \in \mathbb{N}$. Then, there exists $\gamma_0 > 0$ (depending on k and a) such that u^k is unstable in Ω_a if $\gamma \geq \gamma_0$.*

Proof. The idea of the proof is borrowed from [11]. With the same notation as before, we choose $\varphi \in \widetilde{H}(\Gamma)$ of the form $(\varphi_1, 0, \ldots, 0)$, where $\varphi_1(x) = \sin(\frac{2\pi}{a}x)$ and for $n \in \mathbb{N}$ we set $\varphi_1^n(x) := \sin(\frac{2\pi}{a}nx)$ and take $\varphi^n \in \widetilde{H}(\Gamma)$ of the form $(\varphi_1^n, 0, \ldots, 0)$. Note that

$$\int_0^1 (\varphi_1^n)'^2\,dx = \frac{4\pi^2}{a^2}n^2\int_0^a(\varphi_1^n)^2\,dx = \frac{4\pi^2}{a^2}n^2\int_0^a(\varphi_1)^2\,dx\,.$$

Thus, arguing also as in (5.32) and (5.33), we have

$$\begin{aligned}\partial^2 J_{\Omega_a}(u^k)[\varphi^n] &= \left(\frac{2\pi}{a}\right)^2 n^2\int_0^a(\varphi_1)^2\,dx \\ &\quad - \gamma\left(\frac{2}{k}\int_0^a(\varphi_1)^2\,dx - \int_{\Omega_a}|\nabla v_{\varphi_1^n}|^2\,dz\right),\end{aligned} \tag{5.34}$$

where $v_{\varphi_1^n}$ satisfies (in the weak sense)

$$-\Delta v_{\varphi_1^n} = \varphi_1^n \mathcal{H}^1 \llcorner \Gamma_1 \qquad \text{in } \Omega_a \tag{5.35}$$

subject to homogeneous Neumann boundary conditions and zero mean. Choosing $v_{\varphi_1^n}$ itself in the weak formulation of (5.35), we have

$$\int_{\Omega_a}|\nabla v_{\varphi_1^n}|^2\,dz = \int_{\Gamma_1}\varphi^n\, v_{\varphi_1^n}\,d\mathcal{H}^1\,. \tag{5.36}$$

Thus, using the continuity of the trace on Γ_1 and Poincaré inequality,

$$\int_{\Omega_a} |\nabla v_{\varphi_1^n}|^2\,dz \le \|\varphi^n\|_{L^2(\Gamma_1)}\|v_{\varphi_1^n}\|_{L^2(\Gamma_1)} \le C\|\nabla v_{\varphi_1^n}\|_{L^2(\Omega_a)},$$

from which we infer the boundedness of $(v_{\varphi_1^n})_n$ in $H^1(\Omega_a)$. Therefore, up to extraction of a subsequence, we may assume that $v_{\varphi_1^n} \rightharpoonup w$ weakly in $H^1(\Omega_a)$. In particular, by the compactness of the trace operator, $v_{\varphi_1^n} \to w$ strongly in $L^2(\Gamma_1)$. Using again (5.36) and the fact that $\varphi^n \rightharpoonup 0$ weakly in $L^2(\Gamma_1)$, we conclude that

$$\int_{\Omega_a} |\nabla v_{\varphi_1^n}|^2\,dz \to 0.$$

In particular, we may fix n such that

$$\frac{2}{k}\int_0^a (\varphi_1)^2\,dx - 8\int_{\Omega_a} |\nabla v_{\varphi_1^n}|^2\,dz < 0$$

and find correspondingly $\gamma_0 > 0$ so large that the right-hand side of (5.34) is negative for all $\gamma \ge \gamma_0$. This concludes the proof of the proposition. □

Fix $\gamma > 0$, choose a satisfying

$$0 < a < 2\pi\sqrt{\frac{k(\gamma)}{2\gamma}}, \tag{5.37}$$

with $k(\gamma)$ defined as in (5.26) and (5.27), and consider the sequence of shrinking rectangles $\Omega_{a/n} = (0, a/n) \times (0, 1)$. We want to show that for any given $\gamma > 0$ the optimal lamellar configuration $u^{k(\gamma)}$ is the unique (up to the sign) volume constrained global minimizer of J in $\Omega_{a/n}$, provided n is large enough. More precisely, we will show:

Theorem 5.19 (Global minimality of lamellar configurations). *There exists $\bar{n} \in \mathbb{N}$ such that for all $n \ge \bar{n}$ the optimal lamellar configuration $u^{k(\gamma)}$ is the unique (up to the sign) global minimizer of*

$$\min\left\{ J_{\Omega_{a/n}}(u) \,:\, u \in BV(\Omega_{a/n}; \{-1, 1\}),\, \textstyle\int_{\Omega_{a/n}} u = 0 \right\}. \tag{5.38}$$

The strategy of proof of the theorem consists in two main steps: in the first one we show by a Γ-convergence argument that the global minimizers of $J_{\Omega_{a/n}}$ converge (in a suitable sense) to a lamellar configuration as $n \to \infty$; in the second step we use the stability result established in

Proposition 5.17 to show that global minimizers must eventually coincide with such a lamellar configuration.

We start by addressing the aforementioned Γ-convergence analysis. In order to study the asymptotic behavior of $J_{\Omega_{a/n}}$, it is convenient (as customary in dimension reduction problems) to rescale the functionals over the fixed domain Ω_a. To this aim, recall that $\Omega_{a/n} = (0, a/n) \times (0, 1)$ and notice that, setting $\tilde{u}(x, y) = u(x/n, y)$, $u \in BV(\Omega_{a/n}; \{-1, 1\})$ if and only if $\tilde{u} \in BV(\Omega_a; \{-1, 1\})$. Moreover, given $u \in BV(\Omega_{a/n}; \{-1, 1\})$ and writing $n_{\{\tilde{u}=1\}} = (n_1, n_2)$, it can be checked that

$$\begin{aligned} J_{\Omega_{a/n}} &= \frac{1}{n}\left(\int_{\partial^*\{\tilde{u}=1\}} \sqrt{n^2 n_1^2 + n_2^2}\, d\mathcal{H}^1 + \gamma \int_{\Omega_a} (n^2(\partial_x \tilde{v})^2 + (\partial_y \tilde{v})^2)\, dxdy \right) \\ &=: \frac{1}{n} J_n(\tilde{u}), \end{aligned} \tag{5.39}$$

where we set $\tilde{v}(x, y) = v(x/n, y)$. This follows from a simple change of variable. Note that $\tilde{v}$ solves

$$-(n^2 \partial_{xx} \tilde{v} + \partial_{yy} \tilde{v}) = \tilde{u} \qquad \text{in } \Omega_a,$$

subject to the associated natural Neumann boundary conditions on $\partial\Omega_a$. Equivalently, $\tilde{v}$ is the unique solution to

$$\begin{cases} \displaystyle\int_{\Omega_a} (n^2 \partial_x \tilde{v}\, \partial_x \varphi + \partial_y \tilde{v}\, \partial_y \varphi)\, dxdy \\ \displaystyle= \int_{\Omega_a} \tilde{u}\, \varphi\, dxdy & \text{for all } \varphi \in H^1(\Omega_a), \\ \displaystyle\int_{\Omega_a} \tilde{v} = 0. \end{cases} \tag{5.40}$$

We will study the Γ-convergence of the rescaled functional J_n with respect to the L^1 distance on Ω_a. More precisely, we define $J_n : L^1(\Omega_a) \to [0, +\infty]$ as

$$J_n(u) := \begin{cases} \displaystyle\int_{\partial^*\{u=1\}} \sqrt{n^2 n_1^2 + n_2^2}\, d\mathcal{H}^1 \\ \displaystyle\quad + \gamma \int_{\Omega_a} (n^2(\partial_x v)^2 + (\partial_y v)^2)\, dxdy & \text{if } u \in BV(\Omega_a; \{-1, 1\}) \\ & \text{and } \displaystyle\int_{\Omega_a} u = 0, \\ +\infty & \text{otherwise in } L^1(\Omega_a), \end{cases}$$

where we used the notation $n_{\{u=1\}} = (n_1, n_2)$ and v is the unique solution to (5.40), with $\tilde{u}$ replaced by u. It will be useful to observe that

$$J_n(u) \geq J_1(u) = P_{\Omega_a}(\{u = 1\}) + \gamma \int_{\Omega_a} |\nabla v|^2 \, dxdy \tag{5.41}$$

for all $u \in BV(\Omega_a; \{-1, 1\})$.

We also introduce the limiting functional $J_\infty : L^1(\Omega_a) \to [0, +\infty]$ defined as

$$J_\infty(u) := \begin{cases} J_{\Omega_a}(u) & \text{if } u \text{ is lamellar; i.e, } u(x, y) = \hat{u}(y) \\ & \text{for some } \hat{u} \in BV((0,1); \{-1, 1\}), \\ & \text{with } \int_0^1 \hat{u} = 0, \\ & \\ +\infty & \text{otherwise in } L^1(\Omega_a). \end{cases}$$

Remark 5.20. In words, J_∞ is finite only on one-dimensional configurations, where it coincides with J_{Ω_a}. In particular, the unique (up to the sign) global minimizer of J_∞ is given by the optimal lamellar configuration $u^{k(\gamma)}$, with $k(\gamma)$ described in (5.26) and (5.27).

We are now ready to state and prove the following:

Proposition 5.21. *We have that*

$$J_n \xrightarrow{\Gamma(L^1)} J_\infty$$

as $n \to \infty$.

Proof. We start by establishing the liminf inequality. To this purpose, let $u_n \to u$ in $L^1(\Omega_a)$. We claim that

$$J_\infty(u) \leq \liminf_n J_n(u_n)\,. \tag{5.42}$$

First, notice that without loss of generality we may assume that the right-hand side of (5.42) is finite, otherwise the inequality is trivial. Moreover, up to extracting a (not relabeled) subsequence, we can also assume that

$$\liminf_n J_n(u_n) = \lim_n J_n(u_n) < +\infty\,. \tag{5.43}$$

Note that, in turn we have

$$\begin{aligned} +\infty > \lim_n J_n(u_n) &\geq \liminf_n \int_{\partial^*\{u_n=1\}} \sqrt{n^2 n_1^2 + n_2^2}\, d\mathcal{H}^1 \\ &\geq \liminf_n \int_{\partial^*\{u_n=1\}} n|n_1|\, d\mathcal{H}^1 = \frac{1}{2} \liminf_n n|\partial_x u_n|(\Omega_a)\,, \end{aligned}$$

where $|\partial_x u_n|(\Omega_a)$ denotes the the *total variation* of the distributional x-derivative of u_n, which is a measure. Due to the presence of the factor n, it must hold that $|\partial_x u_n|(\Omega_a) \to 0$ and in turn, by lower semicontinuity of the total variation, $|\partial_x u|(\Omega_a) = 0$. This implies that the limiting function u is lamellar; *i.e.*, $u(x, y) = \hat{u}(y)$ for some $\hat{u} \in BV((0, 1); \{-1, 1\})$, with $\int_0^1 \hat{u} = 0$. This shows that the Γ-limit (if any) of $(J_n)_n$ is finite only on one-dimensional configurations. Now, again by (5.43), we also have

$$+\infty > \liminf_n \int_{\Omega_a} (n^2(\partial_x v_n)^2 + (\partial_y v_n)^2)\, dxdy$$

$$\begin{cases} \geq \liminf_n \int_{\Omega_a} n^2(\partial_x v_n)^2\, dxdy\,, \\ \geq \liminf_n \int_{\Omega_a} |\nabla v_n|^2\, dxdy\,. \end{cases} \tag{5.44}$$

The bottom inequality on the right-hand side of (5.44) implies that $(v_n)_n$ is bounded in $H^1(\Omega_a)$. Thus, up to extracting a further (not relabeled) subsequence if needed, we may assume that

$$v_n \rightharpoonup v \qquad \text{weakly in } H^1(\Omega_a) \tag{5.45}$$

for some v.

On the other hand, using the top inequality on the right-hand side of (5.44) we deduce that

$$\int_{\Omega_a} (\partial_x v_n)^2\, dxdy \to 0$$

as $n \to \infty$, due to the presence of the factor n^2. In turn, by lower semicontinuity and recalling (5.45), we have

$$\int_{\Omega_a} (\partial_x v)^2\, dxdy = 0.$$

Thus, v is one-dimensional, that is,

$$v(x, y) = \hat{v}(y) \qquad \text{for some } \hat{v} \in H^1(0, 1), \text{ with } \int_0^1 \hat{v} = 0. \tag{5.46}$$

We claim that v is the potential associated with u according to (5.23) or, equivalently, that $\hat{v}$ solves (5.24). Indeed, using (5.40) with v_n and u_n in place of $\tilde{v}$ and $\tilde{u}$, respectively, and choosing $\varphi(x, y) = \hat{\varphi}(y)$ for any $\hat{\varphi} \in H^1(0, 1)$, we get

$$\int_{\Omega_a} \partial_y v_n(x, y)\, \hat{\varphi}'(y)\, dxdy = \int_{\Omega_a} u_n(x, y)\, \varphi(y)\, dxdy\,.$$

Taking into account (5.45) and (5.46), we can pass to the limit in the above identity to obtain

$$\int_0^1 \hat{v}' \hat{\varphi}' \, dy = \int_0^1 \hat{u} \, \hat{\varphi} \, dy \qquad \text{for all } \hat{\varphi} \in H^1(0,1),$$

that is, $\hat{v}$ solves (5.24). This establishes that v is the potential associated with u.

Using now the lower semicontinuity of the perimeter and of the Dirichlet energy and recalling (5.41), we can estimate

$$\begin{aligned}\liminf_n J_n(u_n) &\geq \liminf_n \left(P_{\Omega_a}(\{u_n = 1\}) + \gamma \int_{\Omega_a} |\nabla v_n|^2 \, dxdy \right) \\ &\geq P_{\Omega_a}(\{u = 1\}) + \gamma \int_{\Omega_a} |\nabla v|^2 \, dxdy = J_{\Omega_a}(u) = J_\infty(u),\end{aligned}$$

where last equality follows from the very definition of J_∞ and the fact that u is lamellar. This establishes (5.42).

The construction of the recovery sequence is trivial: it suffices to take $u_n \equiv u$ and observe that

$$\lim_n J_n(u) = J_\infty(u) \qquad \text{for all } u \in L^1(\Omega_a). \tag{5.47}$$

The verification of (5.47) is left as an (easy) exercise to the reader. This concludes the proof of the proposition. □

We are now in a position to prove the main result of this subsection.

Proof of Theorem 5.19. Let u_n be a solution to (5.38) and denote by $\tilde{u}_n$ the function defined by $\tilde{u}_n(x, y) := u_n(x/n, y)$.

Then $\tilde{u}_n \in BV(\Omega_a; \{-1, 1\})$ and, recalling (5.39),

$$J_n(\tilde{u}_n) = \min \left\{ J_n(u) : u \in BV(\Omega_a; \{-1,1\}), \textstyle\int_{\Omega_a} u = 0 \right\}.$$

Since

$$\sup_n P_{\Omega_a}(\{\tilde{u}_n = 1\}) \leq \sup_n J_n(\tilde{u}_n) \leq \sup_n J_n(u^{k(\gamma)}) = J_{\Omega_a}(u^{k(\gamma)}) < +\infty,$$

by the Compactness Theorem 2.1 we may extract a (not relabeled) subsequence such that $\tilde{u}_n \to \tilde{u}$ in $L^1(\Omega_a)$. Now, by Propositions 5.3 and 5.21, we infer that $\tilde{u}$ satisfies

$$J_\infty(\tilde{u}) = \min_{L^1(\Omega_a)} J_\infty,$$

and thus, by Remark 5.20, either $\tilde{u} = u^{k(\gamma)}$ or $\tilde{u} = -u^{k(\gamma)}$. By multiplying some of the functions of $(\tilde{u}_n)_n$ by -1 if needed, we may assume without loss of generality that $\tilde{u} = u^{k(\gamma)}$ and that the whole sequence $(\tilde{u}_n)_n$ satisfies

$$\tilde{u}_n \to u^{k(\gamma)} \qquad \text{in } L^1(\Omega_a). \tag{5.48}$$

Now we go back to the original sequence $(u_n)_n$ of functions defined on $\Omega_{a/n}$. Fix $n \in \mathbb{N}$. By an even reflection of u_n with respect to the vertical side $\{a/n\} \times (0,1)$ we get a new function $u_n^{(2)}$ defined on the rectangle $(a/n, 2a/n) \times (0,1)$. We repeat the procedure and reflect $u_n^{(2)}$ with respect to $\{2a/n\} \times (0,1)$ to get a function $u_n^{(3)}$ defined on $(2a/n, 3a/n) \times (0,1)$. By iterating this reflection procedure $(n-1)$ times, we define a function $u_n^{(n)}$ defined on $((n-1)a/n, a) \times (0,1)$ and, in turn, we can consider the function $u_n^r \in BV(\Omega_a; \{-1,1\})$ defined by

$$\begin{cases} u_n^r = u_n & \text{in } \Omega_{a/n}, \\ u_n^r = u_n^{(i)} & \text{in } ((i-1)a/n, ia/n) \times (0,1) \text{ for } i = 2, \ldots, n, \end{cases}$$

see figure 5.4. Let v_n be the potential associated with u_n according to

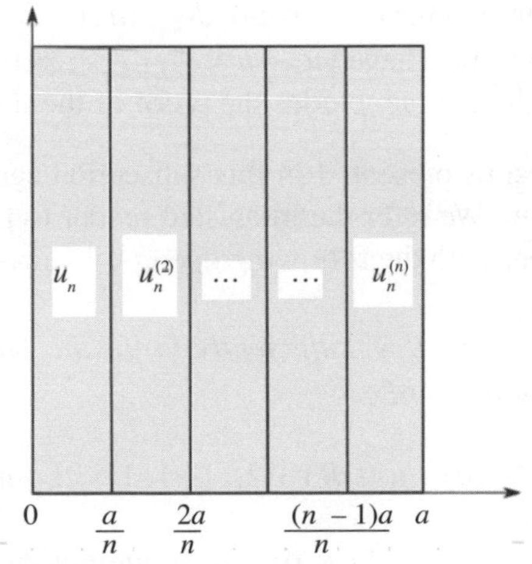

Figure 5.4. The construction of the function u_n^r.

(5.23) and define $v_n^r \in H^1(\Omega_a)$ by the same reflection procedure used for u_n^r. It is immediate to check that v_n^r is the potential associated with u_n^r. Thus, it follows that

$$J_{\Omega_a}(u_n^r) = n J_{\Omega_{a/n}}(u_n)\,. \tag{5.49}$$

Moreover, one can also check that

$$\|u_n^r - u^{k(\gamma)}\|_{L^1(\Omega_a)} = n\|u_n - u^{k(\gamma)}\|_{L^1(\Omega_{a/n})} = \|\tilde{u}_n - u^{k(\gamma)}\|_{L^1(\Omega_a)}\,, \tag{5.50}$$

where the last equality follows from an obvious change of variable. Recalling (5.48), we conclude that

$$u_n^r \to u^{k(\gamma)} \qquad \text{in } L^1(\Omega_a). \tag{5.51}$$

Now Proposition 5.17 and (5.37) yield the existence of $\delta > 0$ and $C > 0$ such that

$$J_{\Omega_a}(u) \geq J_{\Omega_a}(u^{k(\gamma)}) + C\|u - u^k\|^2_{L^1(\Omega_a)}$$

for all $u \in BV(\Omega_a; \{-1, 1\})$, with $\fint_{\Omega_a} u = 0$ and $\|u - u^{k(\gamma)}\|_{L^1(\Omega_a)} \leq \delta$. Therefore, by (5.51) we have that

$$J_{\Omega_a}(u_n^r) \geq J_{\Omega_a}(u^{k(\gamma)}) + C\|u_n^r - u^{k(\gamma)}\|^2_{L^1(\Omega_a)},$$

provided n is large enough. Taking into account (5.49), (5.50), (5.51), and the obvious fact that $J_{\Omega_a}(u^{k(\gamma)}) = nJ_{\Omega_{a/n}}(u^{k(\gamma)})$, the previous inequality is equivalent to

$$J_{\Omega_{a/n}}(u_n) \geq J_{\Omega_{a/n}}(u^{k(\gamma)}) + Cn\|u_n - u^{k(\gamma)}\|^2_{L^1(\Omega_{a/n})}$$

for n large enough. On the other hand, $J_{\Omega_{a/n}}(u_n) \leq J_{\Omega_{a/n}}(u^{k(\gamma)})$ by minimality and thus, we must have $\|u_n - u^{k(\gamma)}\|_{L^1(\Omega_{a/n})} = 0$, that is, $u_n = u^{k(\gamma)}$ for n large enough. This concludes the proof of the theorem. □

Most of the results presented in this subsection can be generalized to higher dimensions. We refer the interested reader to [36, Section 6]. We conclude the paper with the following open problem.

Open Problem: *For all γ sufficiently large the function $u^{k(\gamma)}$ is the unique global minimizer of*

$$\min\left\{J_{\Omega_1}(u) : u \in BV(\Omega_1; \{-1, 1\}), \int_{\Omega_1} u = 0\right\},$$

where, we recall, $\Omega_1 = (0, 1) \times (0, 1)$. An analogous statement holds in dimension $N \geq 3$.

References

[1] E. ACERBI, N. FUSCO and M. MORINI, *Minimality via second variation for a nonlocal isoperimetric problem*, Comm. Math. Phys. **322** (2013), 515–557.

[2] G. ALBERTI, R. CHOKSI and F. OTTO, *Uniform energy distribution for an isoperimetric problem with long-range interactions*, J. Amer. Math. Soc. **22** (2009), 569–605.

[3] L. AMBROSIO, V. CASELLES, S. MASNOU and J. M. MOREL, *Connected components of sets of finite perimeter and applications to image processing*, J. Eur. Math. Soc. (JEMS) **3** (2001), 39–92.

[4] L. AMBROSIO, N. FUSCO and D. PALLARA, "Functions of Bounded Variation and Free Discontinuity Problems", Oxford Mathematical Monographs. The Clarendon Press, Oxford University Press, New York, 2000.

[5] M. BONACINI and R. CRISTOFERI, *Local and global minimality results for a nonlocal isoperimetric problem on* $\mathbb{R}^N$, SIAM J. Math. Anal. **46** (2014), 2310–2349.

[6] A. BRAIDES, "Γ-Convergence for Beginners", Oxford Lecture Series in Mathematics and its Applications, Vol. 22. Oxford University Press, Oxford, 2002.

[7] R. CHOKSI and M. PELETIER, *A. Small volume-fraction limit of the diblock copolymer problem: I. Sharp-interface functional*, SIAM J. Math. Anal. **42** (2010), 1334–1370.

[8] R. CHOKSI and M. PELETIER, *A. Small volume-fraction limit of the diblock copolymer problem: II. Diffuse-interface functional*, SIAM J. Math. Anal. **43** (2011), 739–763.

[9] R. CHOKSI and X. REN, *O*n the derivation of a density functional theory for microphase separation of diblock copolymers, J. Statist. Phys. **113** (2003), 151–176.

[10] R. CHOKSI and P. STERNBERG, *Periodic phase separation: the periodic Cahn-Hilliard and isoperimetric problems*, Interfaces Free Bound. **8** (2006), 371–392.

[11] R. CHOKSI and P. STERNBERG, *On the first and second variations of a nonlocal isoperimetric problem*, J. reine angew. Math. **611** (2007), 75–108.

[12] M. CICALESE and G. LEONARDI, *A selection principle for the sharp quantitative isoperimetric inequality*, Arch. Ration. Mech. Anal. **206** (2012), 617–643.

[13] M. CICALESE and E. SPADARO, *Droplet minimizers of an isoperimetric problem with long-range interactions*, Comm. Pure Appl. Math. **66** (2013), 1298–1333.

[14] G. DAL MASO, "An Introduction to Γ-Convergence", Progress in Nonlinear Differential Equations and their Applications, Vol. 8, Birkhäuser Boston, Inc., Boston, MA, 1993.

[15] L. ESPOSITO and N. FUSCO, *A remark on a free interface problem with volume constraint*, J. Convex Anal. **18** (2011), 417–426.

[16] A. FIGALLI, N. FUSCO, F. MAGGI, V. MILLOT and M. MORINI, *Isoperimetry and stability properties of balls with respect to nonlocal energies*, Comm. Math. Phys. **336** (2015), 441–507.

[17] A. FIGALLI, F. MAGGI and A. PRATELLI, *A mass transportation approach to quantitative isoperimetric inequalities*, Invent. Math. **182** (2010), 167–211.

[18] N. FUSCO, F. MAGGI and A. PRATELLI, *The sharp quantitative isoperimetric inequality*, Ann. of Math. **168** (2008), 941–980.

[19] N. FUSCO and M. MORINI, *Equilibrium configurations of epitaxially strained elastic films: second order minimality conditions and qualitative properties of solutions*, Arch. Rational Mech. Anal. **203** (2012), 247–327.

[20] D. GILBARG and N. S. TRUDINGER, "Elliptic Partial Differential Equations of Second Order", Second Edition, Grundlehren der Mathematischen Wissenschaften, Vol. 224, Springer Verlag, Berlin 1983.

[21] D. GOLDMAN, C. B. MURATOV and S. SERFATY, *The Γ-limit of the two-dimensional Ohta-Kawasaki energy. I. Droplet density*, Arch. Ration. Mech. Anal. **210** (2013), 581–613.

[22] D. GOLDMAN, C. B. MURATOV and S. SERFATY, *The Γ-limit of the two-dimensional Ohta-Kawasaki energy. Droplet arrangement via the renormalized energy*, Arch. Ration. Mech. Anal. **212** (2014), 445–501.

[23] K. GROSSE-BRAUCKMANN, *Stable constant mean curvature surfaces minimize area*, Pacific. J. Math. **175** (1996), 527–534.

[24] H. HADWIGER, *Gitterperiodische Punktmengen und Isoperimetrie*, Monatsh. Math. **76** (1972), 410–418.

[25] H. HOWARDS, M. HUTCHINGS and F. MORGAN, *The isoperimetric problem on surfaces*, Amer. Math. Monthly **106** (1999), 430–439.

[26] V. JULIN, *Isoperimetric problem with a Coulomb repulsive term*, Indiana Univ. Math. J. **63** (2014), 77–89.

[27] V. JULIN and G. PISANTE, *Minimality via second variation for microphase separation of diblock copolymer melts*, J. Reine Angew. Math., to appear.

[28] H. KNÜPFER and C. B. MURATOV, *On an isoperimetric problem with a competing nonlocal term I: The planar case*, Comm. Pure Appl. Math. **66** (2013), 1129–1162.

[29] H. KNÜPFER and C. B. MURATOV, *On an isoperimetric problem with a competing nonlocal term II: The general case*, Comm. Pure Appl. Math. **67** (2014), 1974–1994.

[30] R. V. KOHN and P. STERNBERG, *Local minimisers and singular perturbations*, Proc. Roy. Soc. Edinburgh Sect. A **111** (1989), 69–84.

[31] N. S. LANDKOF, "Foundations of Modern Potential Theory", Springer-Verlag, 1972.

[32] J. LU and F. OTTO, *Nonexistence of a minimizer for Thomas-Fermi-Dirac-von WeizsŁcker model*, Comm. Pure Appl. Math. **67** (2014), 1605–1617.

[33] F. MAGGI, "Sets of Finite Perimeter and Geometric Variational Problems. An Introduction to Geometric Measure Theory", Cambridge Studies in Advanced Mathematics, Vol. 135, Cambridge University Press, Cambridge, 2012.

[34] L. MODICA, *The gradient theory of phase transitions and minimal interface criterion*, Arch. Rational Mech. Anal. **98** (1987), 123–142.

[35] F. MORGAN and A. ROS, *Stable constant-mean-curvature hypersurfaces are area minimizing in small L^1 neighbourhoods*, Interfaces Free Bound. **12** (2010), 151–155.

[36] M. MORINI and P. STERNBERG, *Cascade of minimizers for a nonlocal isoperimetric problem in thin domains*, SIAM J. Math. Anal. **46** (2014), 2033–2051.

[37] S. MÜLLER, *Singular perturbations as a selection criterion for periodic minimizing sequences*, Calc. Var. Partial Differential Equations **1** (1993),169–204.

[38] C. B. MURATOV, *Theory of domain patterns in systems with long-range interactions of Coulomb type*, Phys. Rev. E **66** (2002), 066108.

[39] C. B. MURATOV, *Droplet phases in non-local Ginzburg-Landau models with Coulomb repulsion in two dimensions*, Commun. Math. Phys. **299** (2010), 45–87.

[40] C. B. MURATOV and V. V. OSIPOV, *General theory of instabilities for patterns with sharp interfaces in reaction-diffusion systems*, Phys. Rev. E **53** (1996), 3101–3116.

[41] C. B. MURATOV and A. ZALESKI, *On an isoperimetric problem with a competing non-local term: quantitative results*, Ann. Global Anal. Geom. **47** (2015), 63–80.

[42] T. OHTA and K. KAWASAKI, *Equilibrium morphology of block copolymer melts*, Macromolecules **19** (1986), 2621–2632.

[43] L. SIMON, "Lectures on Geometric Measure Theory", Proceedings of the Centre for Mathematical Analysis, Australian National University, Canberra, 1983.

[44] X. REN and J. WEI, *Concentrically layered energy equilibria of the di-block copolymer problem*, European J. Appl. Math. **13** (2002), 479–496.

[45] X. REN and J. WEI, *On energy minimizers of the diblock copolymer problem*, Interfaces Free Bound. **5** (2003), 193–238.

[46] X. REN and J. WEI, *Stability of spot and ring solutions of the diblock copolymer equation*, J. Math. Phys. **45** (2004), 4106–4133.

[47] X. REN and J. WEI, *Wriggled lamellar solutions and their stability in the diblock copolymer problem*, SIAM J. Math. Anal. **37** (2005), 455–489.

[48] X. REN and J. WEI, *Many droplet pattern in the cylindrical phase of diblock copolymer morphology*, Rev. Math. Phys. **19** (2007), 879–921.

[49] X. REN and J. WEI, *Spherical solutions to a nonlocal free boundary problem from diblock copolymer morphology*, SIAM J. Math. Anal. **39** (2008), 1497–1535.

[50] M. ROSS, *Schwartz' P and D surfaces are stable*, Differential Geom. Appl. **2** (1992), 179–195.

[51] E. N. SPADARO, *Uniform energy and density distribution: diblock copolymers' functional*, Interfaces Free Bound. **11** (2009), 447–474.

[52] P. STERNBERG and I. TOPALOGLU, *On the global minimizers of a nonlocal isoperimetric problem in two dimensions*, Interfaces Free Bound. **13** (2011), 155–169.

[53] I. TAMANINI, *Boundaries of Caccioppoli sets with Hölder-continuous normal vector*, J. Reine Angew. Math. **334** (1982), 27–39.

[54] E. L. THOMAS, D. M. ANDERSON, C. S. HENKEE andD. HOFFMAN, *Periodic area-minimizing surfaces in block copolymers*, Nature **334** (1988), 598–601.

[55] I. TOPALOGLU, *On a nonlocal isoperimetric problem on the two-sphere*, Comm. Pure Appl. Anal. **12**, 597-620, (2013).

[56] E. VALDINOCI, *A fractional framework for perimeters and phase transitions*, Milan J. Math. **81** (2013), 1–23.

[57] B. WHITE, *A strong minimax property of nondegenerate minimal submanifolds*, J. Reine Angew. Math. **457** (1994), 203–218.

CRM Series
Publications by the Ennio De Giorgi Mathematical Research Center Pisa

The Ennio De Giorgi Mathematical Research Center in Pisa, Italy, was established in 2001 and organizes research periods focusing on specific fields of current interest, including pure mathematics as well as applications in the natural and social sciences like physics, biology, finance and economics. The CRM series publishes volumes originating from these research periods, thus advancing particular areas of mathematics and their application to problems in the industrial and technological arena.

Published volumes

1. Matematica, cultura e società 2004 (2005). ISBN 88-7642-158-0
2. Matematica, cultura e società 2005 (2006). ISBN 88-7642-188-2
3. M. GIAQUINTA, D. MUCCI, *Maps into Manifolds and Currents: Area and* $W^{1,2}$-, $W^{1/2}$-, BV*-Energies*, 2006. ISBN 88-7642-200-5
4. U. ZANNIER (editor), *Diophantine Geometry*. Proceedings, 2005 (2007). ISBN 978-88-7642-206-5
5. G. MÉTIVIER, *Para-Differential Calculus and Applications to the Cauchy Problem for Nonlinear Systems*, 2008. ISBN 978-88-7642-329-1
6. F. GUERRA, N. ROBOTTI, *Ettore Majorana. Aspects of his Scientific and Academic Activity*, 2008. ISBN 978-88-7642-331-4
7. Y. CENSOR, M. JIANG, A.K. LOUISR (editors), *Mathematical Methods in Biomedical Imaging and Intensity-Modulated Radiation Therapy (IMRT)*, 2008. ISBN 978-88-7642-314-7
8. M. ERICSSON, S. MONTANGERO (editors), *Quantum Information and Many Body Quantum systems*. Proceedings, 2007 (2008). ISBN 978-88-7642-307-9
9. M. NOVAGA, G. ORLANDI (editors), *Singularities in Nonlinear Evolution Phenomena and Applications*. Proceedings, 2008 (2009). ISBN 978-88-7642-343-7

– Matematica, cultura e società 2006 (2009). ISBN 88-7642-315-4

10. H. HOSNI, F. MONTAGNA (editors), *Probability, Uncertainty and Rationality*, 2010. ISBN 978-88-7642-347-5

11. L. AMBROSIO (editor), *Optimal Transportation, Geometry and Functional Inequalities*, 2010. ISBN 978-88-7642-373-4
12*. O. COSTIN, F. FAUVET, F. MENOUS, D. SAUZIN (editors), *Asymptotics in Dynamics, Geometry and PDEs; Generalized Borel Summation*, vol. I, 2011. ISBN 978-88-7642-374-1, e-ISBN 978-88-7642-379-6
12**. O. COSTIN, F. FAUVET, F. MENOUS, D. SAUZIN (editors), *Asymptotics in Dynamics, Geometry and PDEs; Generalized Borel Summation*, vol. II, 2011. ISBN 978-88-7642-376-5, e-ISBN 978-88-7642-377-2
13. G. MINGIONE (editor), *Topics in Modern Regularity Theory*, 2011. ISBN 978-88-7642-426-7, e-ISBN 978-88-7642-427-4
– Matematica, cultura e società 2007-2008 (2012). ISBN 978-88-7642-382-6
14. A. BJORNER, F. COHEN, C. DE CONCINI, C. PROCESI, M. SALVETTI (editors), *Configuration Spaces*, Geometry, Combinatorics and Topology, 2012. ISBN 978-88-7642-430-4, e-ISBN 978-88-7642-431-1
15 A. CHAMBOLLE, M. NOVAGA E. VALDINOCI (editors), *Geometric Partial Differential Equations*, 2013. ISBN 978-88-7642-343-7, e-ISBN 978-88-7642-473-1
16 J. NEŠETŘIL, M. PELLEGRINI (editors), *The Seventh European Conference on Combinatorics, Graph Theory and Applications*, EuroComb 2013. ISBN 978-88-7642-524-0, e-ISBN 978-88-7642-525-7
17 L. AMBROSIO (editor), *Geometric Measure Theory and Real Analysis*, 2014. ISBN 978-88-7642-522-6 , e-ISBN 978-88-7642-523-3
18 J. MATOUŠEK, J. NEŠETŘIL, M. PELLEGRINI (editors), *Geometry, Structure and Randomness in Combinatorics*, 2015. ISBN 978-88-7642-524-0, e-ISBN 978-88-7642-525-7
19 N. FUSCO, A. PRATELLI (editors), *Free Discontinuity Problems*, 2016. ISBN 978-88-7642-592-9, e-ISBN 978-88-7642-593-6

Volumes published earlier

Dynamical Systems. Proceedings, 2002 (2003)
Part I: *Hamiltonian Systems and Celestial Mechanics*.
ISBN 978-88-7642-259-1
Part II: *Topological, Geometrical and Ergodic Properties of Dynamics*.
ISBN 978-88-7642-260-1

Matematica, cultura e società 2003 (2004). ISBN 88-7642-129-7

Ricordando Franco Conti, 2004. ISBN 88-7642-137-8

N.V. KRYLOV, *Probabilistic Methods of Investigating Interior Smoothness of Harmonic Functions Associated with Degenerate Elliptic Operators*, 2004. ISBN 978-88-7642-261-1

Phase Space Analysis of Partial Differential Equations. Proceedings, vol. I, 2004 (2005). ISBN 978-88-7642-263-1

Phase Space Analysis of Partial Differential Equations. Proceedings, vol. II, 2004 (2005). ISBN 978-88-7642-263-1

Fotocomposizione “CompoMat” Loc. Braccone, 02040 Configni (RI) Italia
Finito di stampare presso le Industrie Grafiche della Pacini Editore S.r.l.
Via A. Gherardesca, 56121 Ospedaletto, Pisa, nel dicembre 2016